Springer
Tokyo
Berlin
Heidelberg
New York
Barcelona
Budapest
Hong Kong
London
Milan
Paris
Santa Clara
Singapore

M. Morio · H. Kikuchi · O. Yuge (Eds.)

Malignant Hyperthermia

Proceedings of the 3rd International Symposium on Malignant Hyperthermia, 1994

With 49 Figures

Springer

MICHIO MORIO, M.D., PH.D.
Professor Emeritus, Hiroshima University
and
Director General, Chugoku Rosai General Hospital
Kure, Hiroshima, 737-01 Japan

HIROSATO KIKUCHI, M.D., PH.D.
Chairman and Professor, The First Department of Anesthesiology,
Toho University School of Medicine
Ohta-ku, Tokyo, 143 Japan

OSAFUMI YUGE, M.D., PH.D.
Professor and Chairman, Department of Anesthesiology and Critical Care Medicine,
Hiroshima University School of Medicine
Hiroshima, 734 Japan

ISBN 978-4-431-68348-3 ISBN 978-4-431-68346-9 (eBook)
DOI 10.1007/978-4-431-68346-9

Printed on acid-free paper

Typesetting: Best-set Typesetter, Ltd., Hong Kong

Preface

It was decided at the Fifth International Workshop on Malignant Hyperthermia (V-IWMH) held in Munich (1990) that the Seventh International Workshop on Malignant Hyperthermia (VII-IWMH) would be held in Hiroshima in July 1994. As eminent scientists would be traveling to Hiroshima for this workshop, the co-organizers, Dr. H. Kikuchi and Dr. O. Yuge, and I, the general organizer, considered it most worthwhile to convene in advance of the VII-IWMH a symposium with designated speakers to review the results of research made to date on malignant hyperthermia and related problems so that the VII-IWMH which followed would be of greater benefit. Thus, on July 16 and 17, 1994, the Third International Symposium on Malignant Hyperthermia (III-ISMH) was held with designated lectures on 25 subject fields.

A number of symposia and workshops on malignant hyperthermia have been held before in various countries. There was thus some debate on whether to designate this symposium the Third International Symposium on Malignant Hyperthermia. The first symposium was held in Toronto, Canada, and the second symposium in Denver, Colorado, USA. Considering ours to be the successor of the second symposium, plans were developed to publish the proceedings with all the presentations in report form as in the previous symposia. Thus, at the outset, requests were made to all the invited symposiasts making a presentation to submit a written review article.

As may be expected, there were some symposiasts who submitted their articles well in advance; however, some were much delayed, causing publication to be later than planned. Nonetheless, we wish to express our gratitude to all the authors for making it possible to publish this volume as a record of the research results obtained to date on malignant hyperthermia.

This book also incorporates the most interesting discourse held during the free-discussion session of the VII-IWMH which were transcribed and then edited. They are appended to the proceedings together with the titles of the presentations made at the VII-IWMH. In addition, authors of interesting papers selected from the presentations made at the VII-IWMH were requested to submit their paper in report form. These are also included in this book.

As the contents of the free-discussion session would be tape-recorded and transcribed and then become a part of the proceedings, the Chairperson requested at the outset that each speaker kindly give his name and affiliation. Unfortunately, not all gave their names as requested and therefore the organizers made editorial changes and identified the speakers other than the chairpersons as A, B, C, etc. and were not able to check the statements with each of the speakers.

As for research on malignant hyperthermia, it was expected with the commencement of studies on chromosomes that the pathogenesis of malignant hyperthermia would be delineated at an early date. However, it was found even at the present ISMH and IWMH that delineation of the pathogenesis involved is still very difficult and that many outstanding areas of research require further exploration.

In particular, with regard to the clinical diagnosis of malignant hyperthermia, a comprehensive, easy-to-understand clinical definition is required. Though the clinical grading scale of Dr. Larach et al. is of much significance as a scale, complete agreement has yet to be reached worldwide on the details of its contents. This is because the body (core) temperature, blood gas, CO_2 output, and oxygen consumption have not been determined in all malignant hyperthermia cases so far, and because it is difficult to employ the scale at this time. As for monitoring of anesthetic equipment, development of more advanced monitoring is desired as advancements are made in computer technology.

As for muscle testing, there is a method of using muscle bundles and also one that uses skinned fiber. The method employed by the European Malignant Hyperthermia Group differs from that used by the North American Malignant Hyperthermia Group and the Japanese Group. For future research it is necessary to standardize the method, but an internationally agreed-upon method has yet to be developed. For the accurate diagnosis of malignant hyperthermia, clinical definition and muscle testing in conjunction with chromosome analysis are needed.

Though malignant hyperthermia-related syndrome is considered to be unconnected with malignant hyperthermia, it is possible that the involvement of the central nervous system, the pyrexia mechanism, and the rhabdomyolysis mechanism may prove useful in the study of malignant hyperthermia.

To ascertain the incidence of malignant hyperthermia in the world, we called upon various countries to present reports on the epidemiology of malignant hyperthermia. We had presentations from countries that had not reported previously, indicating that there is increased international understanding of the risk of malignant hyperthermia.

In closing, I wish to express my profound appreciation to the authors for their contributions to this proceedings and pray that this publication will prove useful in future research of malignant hyperthermia.

Michio Morio
General Organizer
III-ISMH & VII-IWMH

Contents

List of Contributors

ACCORSI, A. (p 33)
Italian MH Registry, USL 25, 40010 Bentivoglio, Italy

ADNET, P.J. (p 197)
Malignant Hyperthermia Investigation Unit, Centre Hospitalier Universitaire, Lille Hôpital B, 59037 Lille, France

ALLEN, GREGORY C. (p 39)
North American Malignant Hyperthermia Registry, Department of Anesthesiology, Penn State University College of Medicine, Hershey, PA 17033, USA

ARAKI, MAKOTO (p 67)
Department of Neurology, Toranomon Hospital and Okinaka Memorial Institute for Medical Research, Tokyo, 105 Japan

BELANI, KUMAR G. (p 221)
University of Minnesota Hospital and Clinic, Anesthesiology, Minneapolis, MN 55455, USA

BELLO, N. (p 197)
Malignant Hyperthermia Investigation Unit, Centre Hospitalier Universitaire, Lille Hôpital B, 59037 Lille, France

BENDIXEN, DIANA (p 21)
The Danish Malignant Hyperthermia Register, Herlev University Hospital, 2730 Herlev, Denmark

BRANCADORO, V. (p 33)
MH Laboratory, II Policlinico, 80121 Naples, Italy

BRITT, BEVERLEY A. (pp 3, 147, 221)
Malignant Hyperthermia Investigation Unit, Department of Anaesthesia, University of Toronto, Toronto General Hospital, Toronto, Ontario M5G 2C4, Canada

CHEAH, KHAY S. (pp 137, 163)
Department of Agriculture, The University of Melbourne, Parkville, Victoria, 3052 Australia

DENBOROUGH, MICHAEL A. (pp 43, 221)
Division of Biochemistry and Molecular Biology, The John Curtin School of Medical Research, Australian National University, Canberra, ACT 2601, Australia

DEUFEL, THOMAS (p 221)
Kinderklinik der Universität Münster, 48129 Münster, Germany

DI MARTINO, D. (p 33)
MH Laboratory, USL 40, 80131 Naples, Italy

DONNER, EVA (p 25)
Vienna Malignant Hyperthermia Investigation Unit, Department of Anesthesia and Critical Care Medicine B, School of Medicine, University of Vienna, Vienna, 1090 Austria

ELLIS, F. RICHARD (p 63)
Academic Unit of Anaesthesia, St. James's University Hospital, University of Leeds, Leeds LS9 7TF, UK

ENDO, MAKOTO (pp 75, 105, 221)
Department of Pharmacology, Faculty of Medicine, University of Tokyo, Tokyo, 113 Japan

FLETCHER, JEFFREY E. (pp 87, 119, 221)
Department of Anesthesiology, Hahnemann University, Philadelphia, PA 19102, USA

FOSTER, PAUL S. (p 107)
Division of Biochemistry and Molecular Biology, The John Curtin School of Medical Research, Australian National University, Canberra, ACT 2601, Australia

FRODIS, WANDA (p 147)
Malignant Hyperthermia Investigation Unit, Department of Anaesthesia, University of Toronto, Toronto General Hospital, Toronto, Ontario M5G 2C4, Canada

FUHRMANN, LINDA J. (p 39)
North American Malignant Hyperthermia Registry, Department of Anesthesiology, Penn State University College of Medicine, Hershey, PA 17033, USA

GEORGE, ALFRED L. (p 87)
Vanderbilt University Medical Center, Nashville, TN 37232-2372, USA

GILBERT, JOHN (p 95)
Departments of Medicine (Neurology) and Neurobiology, Duke University Medical Center, Durham, NC 27710, USA

GRONERT, GERALD A. (pp 159, 221)
Department of Anesthesia, University of California, Davis, CA 95616, USA

HACKL, WERNER (p 25)
Ludwig-Bolzmann Institute of Experimental Anesthesia and Research in Intensive Care, Wien-Linz, Vienna Division, Vienna, 1090 Austria

HAYASHI, TERUO (p 187)
Department of Psychiatry and Neurosciences, Hiroshima University School of Medicine, Hiroshima, 734 Japan

HEILINGER, DORRIT (p 25)
Vienna Malignant Hyperthermia Investigation Unit, Department of Anesthesia and Critical Care Medicine B, School of Medicine, University of Vienna, Vienna, 1090 Austria

HEIMAN-PATTERSON, TERRY (p 87)
Thomas Jefferson University, Philadelphia, PA 19102, USA

HOGAN, SIMON P. (p 107)
Division of Biochemistry and Molecular Biology, The John Curtin School of Medical Research, Australian National University, Canberra, ACT 2601, Australia

IDA, MASAYOSHI (p 67)
Department of Neurology, Toranomon Hospital and Okinaka Memorial Institute for Medical Research, Tokyo, 105 Japan

JEDLICKA, ANNE E. (p 87)
Department of Anesthesiology and Critical Care Medicine, The Johns Hopkins Medical Institutions, Baltimore, MD 21287-7834, USA

JOHN, J. EDWARD, III (p 95)
Departments of Internal Medicine (Cardiovascular Division), Molecular Physiology, and Biological Physics, University of Virginia, Charlottsville, VA 22908, USA

KAWAMOTO, MASASHI (p 49)
Department of Anesthesiology and Critical Care Medicine, Hiroshima University School of Medicine, Hiroshima, 734 Japan

KIKUCHI, HIROSATO (pp 49, 221)
First Department of Anesthesiology, Toho University School of Medicine, Tokyo, 143 Japan

KOJIMA, SUSUMU (p 67)
Department of Neurology, Toranomon Hospital and Okinaka Memorial Institute for Medical Research, Tokyo, 105 Japan

KOZAK-RIBBENS, GENEVIÈVE (pp 129, 221)
Centre de Résonance Magnétique et Médicale, Centre National de la Recherche Scientifique Unité de Recherche Associée 1186, Laboratoire Associé à l'Assistance Publique à Marseille, Faculté de Médecine, 13005 Marseille, France

KRIVOSIC-HORBER, RENEE (pp 197, 221)
Malignant Hyperthermia Investigation Unit, Centre Hospitalier Universitaire, Lille Hôpital B, 59037 Lille, France

LARACH, MARILYN GREEN (pp 39, 221)
North American Malignant Hyperthermia Registry, Department of Anesthesiology, Penn State University College of Medicine, Hershey, PA 17033, USA

LEHMANN-HORN, FRANK (p 203)
Department of Applied Physiology, University of Ulm, 89069 Ulm, Germany

LEVITT, ROY C. (p 87)
Department of Anesthesiology and Critical Care Medicine, The Johns Hopkins Medical Institutions, Baltimore, MD 21287-7834, USA

MACLENNAN, DAVID H. (p 79)
Banting and Best Department of Medical Research, University of Toronto, Charles H. Best Institute, Toronto, Ontario M5G 1L6, Canada

MAEHARA, YASUHIRO (p 49)
Department of Anesthesiology and Critical Care Medicine, Hiroshima University School of Medicine, Hiroshima, 734 Japan

MAURITZ, WALTER (p 25)
Ludwig-Bolzmann Institute of Experimental Anesthesia and Research in Intensive Care, Wien-Linz, Vienna Division, Vienna, 1090 Austria

MCCARTHY, TOMMIE V. (p 221)
Department of Biochemistry, University College Cork, Lee Maltings, Ireland

MICKELSON, JAMES R. (p 221)
Department of Veterinary PathoBiology, University of Minnesota, St. Paul, MN 55108, USA

MILDON, C. ANNE (p 147)
Malignant Hyperthermia Investigation Unit, Department of Anaesthesia, University of Toronto, Toronto General Hospital, Toronto, Ontario M5G 2C4, Canada

MOORMAN, J. RANDALL (p 95)
Departments of Internal Medicine (Cardiovascular Division), Molecular Physiology, and Biological Physics, University of Virginia, Charlottsville, VA 22908, USA

MORI, KENJIRO (pp 183, 221)
Department of Anesthesia, Kyoto University Hospital, Kyoto, 606 Japan

MORIO, MICHIO (pp 49, 221)
Chugoku Rosai General Hospital, Kure, Hiroshima, 737-01 Japan

MORONI, I. (p 33)
Neurology Institute, Ospedale Maggiore, 20100 Milan, Italy

MOUNSEY, J. PAUL (p 95)
Departments of Internal Medicine (Cardiovascular Division), Molecular Physiology, and Biological Physics, University of Virginia, Charlottsville, VA 22908, USA

MUKAIDA, KEIKO (p 49)
Department of Anesthesiology and Critical Care Medicine, Hiroshima University School of Medicine, Hiroshima, 734 Japan

NAKAO, MASAKAZU (p 49)
Department of Anesthesiology and Critical Care Medicine, Hiroshima University School of Medicine, Hiroshima, 734 Japan

NELSON, THOMAS E. (p 173)
Department of Anesthesia, Bowman Gray School of Medicine, Wake Forest University, Winston-Salem, NC 25157-1009, USA

ØRDING, HELE (p 21)
The Danish Malignant Hyperthermia Register, Herlev University Hospital, 2730 Herlev, Denmark

OLCKERS, ANTONEL (p 87)
Department of Anesthesiology and Critical Care Medicine, The Johns Hopkins Medical Institutions, Baltimore, MD 21287-7834, USA
Department of Human Genetics and Developmental Biology, University of Pretoria, Pretoria 0001, South Africa

PEDUTO, V. (p 33)
Department of Anaesthesia, University of Cagliari, Cagliari, Italy

REYFORD, H. (p 197)
Malignant Hyperthermia Investigation Unit, Centre Hospitalier Universitaire, Lille Hôpital B, 59037 Lille, France

ROSENBERG, HENRY (pp 59, 87, 221)
Department of Anesthesiology, Hahnemann University, Philadelphia, PA 19102, USA

ROSES, ALLEN D. (p 95)
Departments of Medicine (Neurology) and Neurobiology, Duke University Medical Center, Durham, NC 27710, USA

SAUBERER, ANDREA (p 25)
Vienna Malignant Hyperthermia Investigation Unit, Department of Anesthesia and Critical Care Medicine B, School of Medicine, University of Vienna, Vienna, 1090 Austria

SCOTT, ELIZABETH (p 147)
Malignant Hyperthermia Investigation Unit, Department of Anaesthesia, University of Toronto, Toronto General Hospital, Toronto, Ontario M5G 2C4, Canada

SCHULTE-SASSE, UWE (p 213)
German Hotline for Malignant Hyperthermia Emergencies, Department of Anaesthesia and Critical Care Medicine, Heilbronn Community Hospital of the University of Heidelberg, 74078 Heilbronn, Germany

SHINGU, KOH (p 183)
Department of Anesthesia, Kansai Medical University, Osaka, 570 Japan

TAKAGI, AKIO (p 67)
Department of Neurology, Toranomon Hospital and Okinaka Memorial Institute for Medical Research, Tokyo, 105 Japan

TEGAZZIN, VINCENZO (pp 33, 221)
MH Laboratory, S. Antonio Hospital, 35127 Padua, Italy

VITA, GARY M. (p 87)
Department of Anesthesiology and Critical Care Medicine, The Johns Hopkins Medical Institutions, Baltimore, MD 21287-7834, USA

WATANABE, TOMOJI (p 67)
Department of Neurology, Toranomon Hospital and Okinaka Memorial Institute for Medical Research, Tokyo, 105 Japan

WIELAND, STEVEN J. (p 119)
Department of Anatomy, Hahnemann University, Philadelphia, PA 19102, USA

XU, PUTING (p 95)
Departments of Medicine (Neurology) and Neurobiology, Duke University Medical Center, Durham, NC 27710, USA

YAMAWAKI, SHIGETO (p 187)
Department of Psychiatry and Neurosciences, Hiroshima University School of Medicine, Hiroshima, 734 Japan

YUGE, OSAFUMI (pp 49, 221)
Department of Anesthesiology and Critical Care Medicine, Hiroshima University School of Medicine, Hiroshima, 734 Japan

YUO, SANG HO (p 221)
Summit Medical Center, Oakland, CA, USA

Part 1. History of Malignant Hyperthermia

History of Malignant Hyperthermia

Beverley A. Britt

Key words. Malignant hyperthermia—History

Early Clinical Cases

Unexpected fever during anesthesia has been described since the beginning of the twentieth century. The very first cases of malignant hyperthermia (MH) were reported by several workers on a single page in the *Journal of the American Medical Association* in 1900 [1–5].

In 1992, Harrison and Isaacs [6] published, in the journal *Anaesthesia*, several documents written in 1919 by attending anesthetists concerning two deaths during anesthesia. The first, a report by the anesthetist E. Penny, noted that the anesthetics for a male patient with a ruptured kidney were chloroform and ether. When the surgeon was about to begin the operation, a curious muscular rigidity was noted in the arms, legs, and abdominal muscles. Just as the kidney had been exposed, a violent and persistent spasm of all the respiratory muscles of the thorax occurred and movements of the chest ceased suddenly and absolutely. Artificial respiration was at once begun but no reestablishment of natural breathing could be obtained and the patient died.

This report was sent to Dr. G.A. Jones, who had anesthetized this patient's mother 4 years previously in 1915. Dr. Jones then wrote a letter to her husband in which he stated that the operation was a uterine suspension and that the anesthetic was chloroform and ether. Near the end of the second stage of the operation a marked tendency to spasm of the muscles of the arms and the jaw developed, and apparently also those of respiration, because breathing stopped fairly quickly. The pupils were not dilated as in a chloroform overdose. Stimulants and a tracheotomy were tried but without avail. At the time breathing stopped, the pulse was rather rapid. A sister later told Dr. Jones that the patient had had attacks of muscle spasms.

In 1925, a third member of the family died while anesthetized for an appendectomy. Some years later a first cousin of the patient, anesthetized for a ruptured kidney, also died while anesthetized. A granddaughter of this third patient died more recently during anesthesia with halothane 45 min after the onset of a fulminant episode of MH characterized by tachycardia, muscle rigidity, cyanosis, and a hyperthermia of 42°C.

Malignant Hyperthermia Investigation Unit, Department of Anaesthesia, University of Toronto, Toronto General Hospital, Toronto, Ontario M5G 2C4, Canada

Subsequently three relatives had caffeine halothane contracture tests (CHCTs) positive for MH and three more distant relatives have recently had CHCTs negative for MH. Although the first four of these cases occurred many years ago, the cases were not described until 1992 so they played no role in developing our understanding of the genetic nature of MH.

Other early reactions were reported by Hewer [7], Guedel [8], Bigler [9], Mangiardi [10], Cullen [11], Lee [12], and Brown [13]. Hewer [7] discussed a series of 22 patients afflicted with "late ether convulsions" that were associated with intense muscle activity and which were followed by high temperatures. In Guedel's cases [8], patient temperature rose within a few hours to between 108°F and 110°F and all patients died. Necroscopy, when done, showed only cerebral edema. Cullen [11] described "ether convulsions" and high fevers in adolescents and robust young adults. He listed as contributory factors high endogenous or exogenous temperatures, retention of carbon dioxide, and hypoxia. Lee [12] observed that "ether convulsions" associated with excessive rises in temperature had been reported with increasing frequency since 1926 and that many of the "convulsions" occurred in children and young adults.

The first breakthrough in realizing that MH was a genetic disorder was made by Denborough [14]. In 1962, Denborough et al. described 11 MH reactions within one family in the *British Journal of Anaesthesia* [14]. In this family the proband was a young male who, during halothane anesthesia for a fractured leg, developed tachycardia, cyanosis, and a hot, sweaty skin. He remained comatose for 30 min following anesthesia, and survived after a stormy postoperative course. He had been very apprehensive before induction because 10 relatives had died during anesthesia. For all 10, the anesthetic agents used were ethyl chloride and ether. In the 3 best-documented cases the deaths were preceded by convulsions and fulminant fevers. The pattern of inheritance was compatible with an autosomal dominant gene.

A second, much larger family was discovered in northern Wisconsin by Locher and was reported in 1969 by Britt et al. [15]. In this family 22 reactions had occurred, extending back to 1922, and 7 of these reactions had been fatal. It is of interest, however, that one grandmother in this family, who was an obligatory carrier because 3 descendants and 19 collateral relatives had had MH reactions, herself had 19 uneventful anesthetizations, some of very long duration and some with agents such as ether, chloroform, ethylene, cyclopropane, and halothane. This patient demonstrated that possession of the MH trait does not guarantee an MH reaction every time a triggering agent is given.

From this time onward similar large families began to be reported. For instance, Britt described a family, first detected by Y. Juteau in 1971, of 138 individuals at risk of whom 21 had had MH reactions during anesthesia [16]. Eight of these persons had died during or shortly after completion of surgery. Most had experienced rigidity, fever, tachycardia, and cyanosis. The total number of patients in this kindred has now been extended to several thousand. Of more than 100 who have been biopsied in Toronto, about half have been found to be positive for MH.

Malignant Hyperthermia Symposia

The very first symposium on MH was organized by Dr. R.A. Gordon, held in Toronto in 1964, and published in 1966 in the *Canadian Anesthetists' Society Journal* [17]. This first symposium consisted exclusively of case reports presented by Canadian anesthetists. It was realized at this meeting that common factors were muscle rigidity, fever, tachycardia, cyanosis, hyperventilation, hyperkalemia, and respiratory and metabolic acidosis.

The first International Symposium on Malignant Hyperthermia open to all MH investigators was held in Toronto in 1971. More than 200 people attended although only 20 had been expected. A great ferment of ideas took place at this symposium. Researchers from Australia, New Zealand, South Africa, Japan, the United Kingdom, Europe, and North America who had been working in isolation got together for the first time and exchanged views in a relatively unstructured environment. The second International MH Symposium was held in Denver, Colorado, in 1977. The proceedings of both these symposia were subsequently published in book form [18,19]. Since then international symposia have continued to be held at 3- to 4-year intervals until the current symposium here in Japan.

Triggers of Malignant Hyperthermia

The early triggering agents of MH were ether, chloroform, ethylene, and cyclopropane. The first case of a reaction caused by halothane was that described by Denborough et al. [14]. The initial case of MH triggered by succinylcholine alone was reported by Patterson in 1962 [20]. Eventually it was noticed that MH reactions were more likely to occur when halothane administration preceded succinylcholine infusion and less apt to occur when succinylcholine was injected before halothane inhalation had been started [21].

MH reactions began to be reported after newer halogenated hydrocarbons in addition to older agents such as halothane and the even older nonhalogenated ether, chloroform, and cyclopropane. For example, in 1975 Pan et al. [22] reported a case of MH triggered by enflurane. In 1981, Shulman et al. [23] found that sevoflurane triggered MH reactions in malignant hyperthermia-susceptible (MHS) pigs. Joseph et al. [24] in 1982 reported MH reactions triggered by isoflurane. Wedel et al. [25] found that desflurane triggered MH reactions in susceptible pigs but was not as potent a trigger as halothane.

Halogenated chemicals other than those used during general anesthesia were shown by Denborough et al. to be capable of inducing stress-induced MH reactions when they reported a case of nonanesthetic MH triggered in a man who was working in a factory making fire extinguishers [26].

Agents Safe for MHS Patients

Fortunately, a considerable number of safe anesthetic agents as well as dangerous agents have been found for patients with MH; examples are local anesthetics, nitrous oxide, barbiturates, narcotics, tranquilizers [27], propofol [28], and ketamine [29].

Clinical Signs

By the mid 1960s, the clinical signs and laboratory abnormalities (muscle rigidity, tachycardia, ventricular arrhythmias, hyperventilation, cyanosis, fever, hyperkalemia, and respiratory and metabolic acidosis) of MH were fairly well known. Myoglobinuria and creatine kinase (CK) elevations after MH reactions in human patients were then reported by Carpenter et al. [30]. This information was analyzed in a retrospective statistical review by Britt and Kalow in 1970 [31]. Wade [32] described, at the MH Symposium in 1971, the first case of MH in which myoglobinuria had become so severe that renal failure had ensued, necessitating renal dialysis. Daniels et al. [33] published the first case of MH associated with disseminated intravascular coagulopathy (DIC) that seemed to have been induced mainly by a sudden disappearance of platelets and red blood cells.

In 1978, Donlan et al. [34] noted the phenomenon of isolated masseter muscle rigidity (MMR) after succinylcholine as being possibly related to MH. The relationship of MMR to MH has subsequently been disputed by Littleford and Patel et al. [35] and by Vanderspek and Wilton [36]. However, it is necessary for anesthetists to clearly distinguish between the normal initial increase in muscle tone produced by succinylcholine and the severe muscle rigidity that makes intubation difficult or even impossible. It is also important to remember that some cases of MMR clearly do advance to generalized muscle rigidity and all the other signs of a full-blown MH reaction. Because one cannot predict which cases will progress and which will not, safety precautions mandate stopping triggering drugs the moment MMR begins. If absolutely essential, anesthesia can be continued with nontriggering agents.

Nonanesthetic Clinical Abnormalities Associated with MH

In 1970, Isaacs and Barlow [37] observed that, quite apart from anesthesia, there was a greater than expected incidence of CK elevations in relatives of a patient who had died of MH. In a study of a large number of individuals who had undergone the caffeine halothane contracture test (CHCT), Britt et al. [38] showed that CK measurement was moderately useful for screening but was not valid for diagnosis of MH because there were so many false-positive and false-negative results.

The first publication of a preexisting myopathy in an MH patient was described in 1964 by Saidman et al. [39]. The patient had generalized muscle weakness before receiving a halothane anesthetic. The anesthetic episode was characterized by seizures, a temperature of 42.4°C, and fixed and dilated pupils. King et al. [40] reported MH in patients with short stature, lumbar lordosis, ptosis, cryptorchidism, winged scapulae, small chins, crowded lower teeth, lowset ears, and antimongoloid eye slants. This constellation of signs became known as the King syndrome. An association between MH and central-core disease, another myopathy, was first described in 1973 by Denborough et al. [41]. Microscopic abnormalities in MH muscle supporting a concomitant myopathy were first detailed by Harriman et al. [42]. Oka et al. first reported an association between MH and Duchenne muscular dystrophy in 1982 [43], and Brownell et al. [44] then published a report summarizing an MH reaction in a child who was subsequently found to have Duchenne muscular dystrophy. Heiman-Patterson et al. [45] described a family of three siblings, two of whom had both MH and myotonia congenita. Otsuka et al. [46] published a case of MH occurring in a child with central-core disease (CCD) during anesthesia with sevoflurane.

Until the Second International Symposium it had been believed that stress-induced hyperthermic reactions did not occur in humans. However, at this symposium

Wingard [47] described a family suffering not only anesthetic-induced reactions but also stress-induced reactions. Next, Denborough [48] noted an association between heatstroke and MH. Other reports of stress-induced reactions in humans then began to appear. For instance, Ikegaki et al. [49] reported the sudden death during sports of a boy belonging to a known MH family, and Britt [50] described fatal heatstroke-like reactions in two siblings who had previously each had anesthetic-induced reactions. Both parents of these siblings appeared to be affected with MH [50].

Occurrence of MH in Species Other Than Humans

Hall and Woolf et al. [51] discovered that MH occurs in pigs as well as humans. They reported MH reactions in three young Landrace siblings during halothane-succinylcholine anesthesia being given for a pharamacological experiment in the laboratory of Mme. Zaimis in Cambridge [51]. Each piglet developed rigidity, tachycardia, cyanosis, and a very high fever, and died. Investigators then began to detect MH in other varieties of pigs. For example, Ahern et al. [52] described halothane-induced MH reactions in Pietrain pigs. At the first MH symposium, Jones et al. [53] described the first case of an anesthetic-induced reaction in a Poland China pig, and the identity of pork stress syndrome and human MH was postulated by T.E. Nelson [54].

In addition to humans and pigs, MH reactions started to be recorded in other species; for instance, in dogs by Short and Paddleford [55], in cats by De Jong et al. [56], by Klein in racehorses [57], and in a giraffe by Citinos et al. [58]. Most current scientific research on MH utilizes pigs or dogs.

Etiology of MH

MH was shown to be a muscle disorder by Satnick [59], who recounted a case in which all parts of the body, except those distal to a tourniquet, became rigid. The part of the limb distal to the tourniquet remained flaccid [59].

In 1970, Kalow and Britt et al. [60] described in *Lancet* the caffeine halothane contracture test (CHCT). They observed that the dose of caffeine required to induce a 1.0-g contracture in MHS human muscle either in the absence or in the presence of 1% halothane was smaller than normal [60]. They postulated that this hypercontractility of MHS muscle resulted from a rise in myoplasmic calcium, and in this same article they also identified an error in calcium transport in the muscle as the MH defect [60]. They recorded two human cases in which calcium uptake into the sarcoplasmic reticulum (SR) was reduced in the presence of halothane [60]. It is of course now known that the error is an increased release of calcium from the SR rather than increased uptake into the SR. However, this article was the first to direct our attention to the SR [60]; that calcium release was mistaken for calcium uptake was probably the result of the impurity of the SR homogenate and the crude methods of measurement used at that time.

Work that helped establish increased calcium release from the SR as the trigger of MH reactions included that of Takagi et al. [61] who observed in MHS human single chemically skinned skeletal muscle fibers that halothane released calcium from the SR at doses lower than normal. Endo et al., who had first discovered the phenomenon of calcium-induced calcium release (CICR) in 1970 [62], in 1983 [63] discovered that the CICR of MHS human muscle fibers was more sensitive than normal to calcium.

Greater CICR was shown by Ohta et al. [64] and by Nelson et al. [65] to be also present in MHS pig skinned skeletal muscle fibers.

Lopez et al. [66] directly measured elevated cytoplasmic calcium in MHS pig muscle by means of electrodes. Using the same techniques, Lopez et al. also reported that dantrolene reduced myoplasmic calcium in MHS humans [67].

The permeability of heavy MHS pig SR was discovered by Ohnishi to be greater than normal [68]. Halothane and caffeine further increased, while tetracaine and ruthenium red decreased, this permeability [68]. O'Brien [69], using a microtechnique, found that caffeine induced twofold more calcium release, three times as fast, from MHS than from normal terminal cisternae. Lowering the temperature to 25°C was found by Nelson [70] to reduce MH pig halothane contractures, while increasing the temperature to 38°C increased CICR. Fill et al. [71] noted that the calcium release channels of MH SR were excessively sensitive to caffeine. Halothane was shown by Nelson [72] to cause calcium release channels in planar lipid bilayers to open, and it also increased their conductance.

Other defects in skeletal muscle were also detected. For instance, Foster et al. [73] reported that pig SR was deficient in inositol 1,4,5-triphosphate phosphatase ($InsP_3$-5ase) with, therefore, increased $InsP_3$-5ase opening of SR calcium channels. As a result, MHS pig muscle contains more inositol 1,4,5-triphosphate and cyclic adenosine monophosphate (cAMP) than normal [74]. Ervasti et al. [75] reported that binding of dihydropyridine (DHP) receptor antagonists to DHP was lowered. He suggested, therefore, that the DHP receptor content was reduced in MHS muscle. Adnet et al. [76] showed that Bay K 8644, which is a dihydropyridine receptor agonist, enhanced halothane contractures in MH muscle.

Membranes within MH muscle cells other than the SR were found to be defective. For example, Mickelson et al. [77] demonstrated that calcium uptake into MHS pig sarcolemma was reduced.

MH Defects Detected in Tissues Other Than Skeletal Muscle Cells

The question has been raised as to whether the DIC first described by Daniels [33] was secondary to the various metabolic and biochemical abnormalities arising in the skeletal muscle or was caused by a primary defect in the blood cells. This conundrum first began to be answered by Harrison [78], who observed increased fragility in red blood cells isolated from MHS pigs several weeks after their exposure to halothane, and was later confirmed by O'Brien et al. [79], who used MH dog red blood cells. Other abnormalities in MHS red blood cells have also been detected. Halothane thus was observed, using paramagnetic resonance spectroscopy, to cause a greater than normal decrease in the rotational correlation time of spin-labeled MH human red blood cells [80].

Defects in other types of blood cells were also detected; for instance, halothane was observed by Klip et al. [81] to cause excessive release of calcium from intracellular stores to the cytoplasm of pig lymphocytes. Furthermore, the rate of extrusion of calcium from MHS pig lymphocytes was noted to be twofold greater than normal by O'Brien et al. [82]. Fink et al. [83,84] reported that halothane caused a rise greater than normal in cytosolic calcium in human and pig MHS platelets. Defects were observed in still other tissues; for example, E. Kochs et al. [85,86] described abnormalities in electrical activity, blood flow, and electroencephalograms (EEGs) of MH pig brains. Thus, although the most severe expression of MH is in the skeletal muscle, other cell types do not appear to be entirely normal.

Treatment of MH

In the early days of MH, treatment of acute reactions consisted of stopping the anesthetic agents and starting various symptomatic and sometimes not very effective measures, such as changing to a clean machine, hyperventilating with oxygen, applying external and internal cooling, and administering sodium bicarbonate, insulin, and diuretics.

Beldavs was the first to use successfully a drug that had a specific action at the site of the primary defect, namely procainamide in the treatment of an MH reaction [87]. The patient had developed profound rigidity and a temperature of 112°F but with the aid of procainamide survived. Because procainamide was known to act by preventing release of calcium from the SR, its success helped to confirm that MH was caused by a defect in the SR. The use of procainamide, however, did not become universal because the large doses required to produce a beneficial effect in skeletal muscle caused profound hypotension from actions on smooth and heart muscle.

The efficacy of dantrolene in the therapy of MH reactions was discovered by Harrison [88]. He first tried treating MHS pigs suffering halothane-induced reactions with a homemade formulation of dantrolene in 1975. The first pig responded favorably immediately to the dantrolene but then began to regress. By the time Harrison had made more dantrolene (a time-consuming process), the pig had expired. For the next pig he manufactured much more and the animal survived as did all further pigs. Anderson and Jones [89] demonstrated that in vitro dantrolene acted by inhibiting halothane-induced contractures in MHS pig muscle. Britt et al. [90] demonstrated that dantrolene inhibited caffeine-induced contractures in MHS muscle. Dantrolene rapidly replaced procainamide as the treatment of choice in MH reactions because it was far more efficacious and did not cause hypotension.

Kolb et al. [91] led the first multihospital study of the efficacy of dantrolene in treatment of human MH reactions. Sixty-five hospitals took part in this study, and in emergencies dantrolene was transported to other nearby hospitals. This study clearly showed that survival from MH reactions was markedly enhanced by the use of dantrolene. It was this work that persuaded the Health Protection Branch in Canada (June 1979) and the Food and Drug Administration (FDA) in the United States (September 1979) to approve dantrolene for general use. When dantrolene first came on the market it was generally believed that prophylactic dantrolene was required before and during elective anesthesia of MHS patients. However, we had anesthetized many MHS patients without any serious difficulties before the advent of dantrolene [92]. It is now generally recognized that prophylactic dantrolene is unnecessary.

Dershwitz and Streter [93] have pointed out that azumolene, a water-soluble analog of dantrolene, is superior because it can be prepared in a much more concentrated formulation without the need for last-minute reconstitution. It also does not induce thrombophlebitis.

Another group of drugs that, like procainamide, did not prove useful in the therapy of MH were the calcium channel blockers. Various clinical reports failed to show that they improved survival. Furthermore, in vivo verapamil was found by Gallant et al.

[94] to cause profound bradycardia and hypotension during halothane-induced MH reactions in pigs, and in vitro high doses of verapamil caused contractures in MHS but not in normal pig muscle.

Improved monitoring, which arrived in the operating room in the middle 1980s, made early detection of MH much easier and so significantly reduced mortality from MH. In particular, Baudendistel et al. [95] showed that the earliest diagnosis of MH was best made by a rising end-tidal carbon dioxide.

The question first raised by Gronert et al. [96] of whether sympathomimetics are safe for MHS patients remains unresolved. Gronert considered that these agents were perfectly safe. However, Britt reported in a retrospective study that the use of sympathomimetics during MH reactions in humans increased mortality [27]. Moreover, Urwyler et al. [97] observed that in vitro ephedrine exacerbated halothane-induced contractures of MHS human muscle.

New Diagnostic and Etiological Techniques

The highly sophisticated technique of ^{31}P nuclear magnetic resonance (NMR) showed that 3% halothane and 2 m*M* caffeine decreased phosphocreatine (PCr) and increased phosphate (Pi) of MHS but not of normal pig muscle [98]. Foster et al. [99] made the same observations. Additionally, Foster et al. demonstrated that at 10 m*M* caffeine or during anoxia there was also a later decrease in adenosine triphosphate (ATP) [99]. During a halothane-induced MH pig reaction, NMR exhibited a rapid and large rise in Pi and a rapid and large fall in pH [100]. In pig muscle, PCr and pH were decreased and Pi was increased. Caffeine or A23187 increased these effects and PCr actually decreased to undetectable levels [101]. A large fall in PCr and a rise in Pi occurred during halothane MH reactions in piglets [102].

By the same technique, Kozak-Reiss et al. [103] observed, in heatstroke patients who had CHCT tests positive for MH, that ischemic exercise produced a decrease in muscle pH and an increase in PCr. Olgin et al. [104] found that MHS human muscle manifested on NMR a decreased PCr/Pi and increased ATP/PCr ratios at rest and a decreased rate of recovery of the Pcr/Pi ratio after exercise. Murakawa et al. [105] used technetium-99 m pyrophosphate scintigraphy 2 days after an MH reaction to reveal abnormal uptake into thing, arm, and chest muscles but not into heart muscle.

The age of MH molecular genetics began with two back-to-back articles by MacLennan et al. [106] and by McCarthy et al. [107]. Using restriction fragment length polymorphism (RFLP) studies, they proposed the human ryanodine gene at 19q13.1 to be the site of the MH defect [106,107]. Healy et al. [108] confirmed by means of RFLP studies that the Q12–Q13.2 region of chromosome 19 segregated with human MH. MacKenzie [109] found that a genetic linkage between the RYR and MH genes could only be observed if the reference points for the CHCT were lowered. Moreover, Levitt et al. [110] examined three human families by RFLP sudies, and in only one family did the MH defect segregate with the ryanodine gene. Levitt et al. in another report also found four families linked to 19q12-q13.2, five families linked to 17q11.2-q24 in the region of the sodium channel gene, and two families not linked to either of these chromosomal regions [111]. Sudbrak et al. [112], using linkage studies, excluded the DHP receptor gene on chromosome 1 and the sodium channel gene on chromosome 17 from being candidates for the human MH gene. These studies confirmed the early report of Ellis [113] at the First MH Symposium in 1971 that human MH was heterogenous.

In pigs, Harbitz et al. [114] assigned the MH gene to the syntetic ryanodine gene on chromosome 6(6P→q21). Sequencing studies in pigs showed that the nucleotide cytosine was replaced by thymine at position 1843 on the ryanodine gene. This caused the argenine amino acid at position 615 of the ryanodine gene to be replaced by cysteine [115]. Fletcher et al. [116] reported that pigs possessing the MH mutation on chromosome 6 did not always develop a reaction when challenged with halothane. Therefore, Fletcher postulated that a modulator gene might also be necessary. It is also possible that this failure to develop a reaction in pigs known to possess the ryanodine mutation might be caused by an absence of as yet undefined environmental triggers, as for example emotional stress or muscle exercise or injury.

Purified MHS pig ryanodine receptor was found by Mickelson to have a greater than normal affinity for ryanodine [117,118]. Valdivia et al. [119] reported that MHS human ryanodine receptors had an affinity higher than normal for ryanodine at Ca $0.3\,\mu M$ but not when Ca was greater than $30\,\mu M$. The number of high-affinity ryanodine binding sites was noted by Hawkes et al. [120] to be increased in MHS pig muscle. These findings supported the hypothesis that the MH defect lies in the ryanodine receptor.

Standardization of the Caffeine Halothane Contracture Test

The European MH group was established in 1985 [121]. Their objective was to standardize the CHCT throughout Europe [121]. Next, the North American Malignant Hyperthermia Registry (NAMHR) was formed and its first report was produced [122]. As well as standarization of the CHCT, this organization had as its objective determination of which methods produced the best diagnostic discrimination. Fletcher et al. [123] found that the European CHCT protocol yielded a higher incidence of false positives and false negatives than the North American protocol. On behalf of the NAMHR, Larach [124,125] reported on the standardization of the CHCT in North America. Severity of human MH reactions by means of a Delphi scoring was developed by NAMHR to aid in determining sensitivity [126], a parameter previously impossible to measure in human patients.

Conclusions

In conclusion, a tremendous amount has been learned about MH since its existence was first clearly demonstrated by M.A. Denborough in 1962 [14]. The mortality rate has been reduced from more than 80% in the first half of this century to near the vanishing point today. Much remains to be done to better define the nature of the ryanodine receptor defect, to locate the genetic defect for all human patients, to develop a noninvasive diagnostic test, to get an improved version of dantrolene such as azumolene approved for clinical use, and to better educate patients, medical personnel, and the public about malignant hyperthermia.

References

1. Gibson CL (1900) Heat-stroke as a post operative complication. JAMA 35:1685
2. Johnson AB (1900) Heat-stroke as a post operative complication. JAMA 35:1685
3. Brewer GE (1900) Heat-stroke as a post operative complication. JAMA 35:1685
4. Tuttle JP (1900) Heat-stroke as a post operative complication. JAMA 35:1685
5. Moschcowitz AV (1900) Heat-stroke as a post operative complication. JAMA 35:1685
6. Harrison GG, Isaacs H (1992) Malignant hyperthermia. An historical vignette. Anaesthesia 47:54–56
7. Hewer CL (1926) Recent advances in anaesthesia and analgesia (including oxygen therapy), 6th edn. Churchill, London
8. Guedel A (1937) Inhalation anaesthesia. A fundamental guide. MacMillan, New York, p 133
9. Bigler JA (1951) Body temperature during anesthesia in infants and children. JAMA 146:551
10. Mangiardi JL (1951) Experience with postoperative temperatures above 108°F: use of hypertonic glucoses. Am J Surg 81:189–192
11. Cullen SC (1951) Anesthesia in general practice, 3rd edn. Year Book, Chicago, p 88
12. Lee JA (1953) A synopsis of anaesthesia, 3rd edn. Wright, Bristol
13. Brown RC (1954) Hyperpyrexia and anaesthesia. Br Med J 4:1526–1527
14. Denborough MA, Forster JFA, Lovell RRH, Maplestone PA, Villiers JD (1962) Anaesthetic deaths in a family. Br J Anaesth 34:395
15. Britt BA, Locher WG, Kalow W (1968) Hereditary aspects of malignant hyperthermia. Can Anaesth Soc J 16:89–98
16. Britt BA (1982) Malignant hyperthermia—a review. In: Milton AS (ed) Handbook of experimental pharmacology, pyretics and antipyretics, vol 60. Springer, Berlin, pp 547–615
17. Gordon RA (1966) Malignant hyperpyrexia during general anaesthesia. Can Anaesth Soc J 13:415–46
18. Gordon RA, Britt BA, Kalow W (eds) (1973) International symposium on malignant hyperthermia. Thomas, Springfield, pp 30–58
19. Aldrete JA, Britt BA (eds) (1978) Second international symposium on malignant hyperthermia. Grune and Stratton, New York
20. Patterson IS (1962) Generalized myotonia following suxamethonium. Br J Anaesth 34:340
21. Lerman J (1988) Controversies in paediatric anaesthesia. Can J Anaesth 35:S18–S22
22. Pan TH, Wollock AR, Demarco JA (1975) Malignant hyperthermia associated with enflurane anesthesia: a case report. Anesth Analg 54:47
23. Shulman M, Braverman B, Ivankovich AD, Gronert GA (1981) Sevoflurane triggers malignant hyperthermia in swine. Anesthesiology 54:259–260
24. Joseph MM, Shah K, Viljoen JF (1982) Malignant hyperthermia associated with isoflurane anaesthesia (letter). Anesth Analg 61:711–712
25. Wedel DJ, Iaizzo PA, Milde JH (1990) Desflurane is not a potent trigger of malignant hyperthermia in susceptible swine (abstract). J Neurol Sci 98(suppl):527.
26. Denborough MA, Hopkinson KC, Banney DG (1988) Firefighting and malignant hyperthermia. Br Med J 296:1442–1443
27. Britt BA (1991) Malignant hyperthermia—a review. In: Schonbaum E, Lomax P (eds) International encyclopedia of pharmacology and therapeutics. Thermoregulation: pathology, pharmacology and therapy. Pergamon, New York, pp 179–292
28. Denborough MA, Hopkinson KC (1988) Propofol and malignant hyperpyrexia (letter). Lancet i:191
29. Dershwitz M, Sreter FA, Ryan JF (1989) Ketamine does not trigger malignant hyperthermia in susceptible swine. Anesth Analg 61:501–503
30. Carpenter GG, Auerbach VH, DiGeorge AM, Mayer BW (1966) Rhabdomyolysis after routine administration of succinylcholine in children (abstract). Society for Pediatric Research, 36th annual meeting, Atlantic City, NJ, April 29–30, p 175

31. Britt BA, Kalow W (1970) Malignant hyperthermia: a statistical review. Can Anaesth Soc J 17:293–315
32. Wade JG (1973) Late treatment of malignant hyperthermia. In: Gordon RA, Britt BA, Kalow W (eds) International symposium on malignant hyperthermia. Thomas, Springfield, pp 441–450
33. Daniels HC, Polayes IM, Villar RV, et al (1969) Malignant hyperthermia with disseminated intravascular coagulation during general anaesthesia: a case report. Anesth Analg 48:877
34. Donlon JV, Newfield p, Sreter F, Ryan JF (1978) Implications of masseter spasm after succinylcholine. Anesthesiology 49:298–301
35. Littleford JA, Patel LR, Bose D, Cameron CB, McKillop C (1991) Masseter muscle spasm in children: implications of continuing the triggering anesthetic. Anesth Analg 72:151–160
36. Vanderspek AFL, Wilton N (1985) Suxamethonium spasm. Br J Anaesth 57:353–357
37. Isaacs H, Barlow MB (1970) The genetic background to malignant hyperpyrexia revealed by serum creatine phosphokinase estimations in asymptomatic relatives. Br J Anaesth 42:1078–1084
38. Britt BA, Endrenyi L, Peters PL, et al (1976) Screening of malignant hyperthermic susceptible families by CPK measurement and other clinical investigations. Can Anaesth Soc J 23:263–284
39. Saidman LJ, Harvard ES, Eger EI II (1964) Hyperthermia during anesthesia. JAMA 190:1029
40. King JO, Denborough MA, Zapf PW (1972) Inheritance of malignant hyperpyrexia. Lancet i:365–370
41. Denborough MA, Dennett X, Anderson RM (1973) Central-core disease and malignant hyperpyrexia. Br Med J i:272–273
42. Harriman DGF, Sumner DW, Ellis FR (1973) Malignant hyperpyrexia myopathy. Q J Med 168:639–664
43. Oka S, Igarashi Y, Takagi A, et al (1982) Malignant hyperpyrexia and Duchenne muscular dystrophy: a case report. Can Anaesth Soc J 29:627–629
44. Brownell AKW, Päasuke RT, Elash A, Fowlow SB, Seagram CGR, Diewold RJ, Friesen C (1983) Malignant hyperthermia in Duchenne muscular dystrophy. Anesthesiology 58:180–182
45. Heiman-Patterson T, Martino C, Rosenberg H, Fletcher J, Tahmoush A (1988) Malignant hyperthermia in myotonia congenita. Neurology 38:810–812
46. Otsuka H, Komura Y, Mayumi T (1991) Malignant hyperthermia during sevoflurane anesthesia in a child with central core disease. Anesthesiology 75:699–701
47. Wingard D (1978) Some observations on stress susceptible patients. In: Aldrete JA, Britt BA (eds) Second international symposium on malignant hyperthermia. Grune and Stratton, New York, pp 363–372
48. Denborough MA (1982) Heat stroke and malignant hyperthermia. Med J Aust i:204–205
49. Ikegaki S, Kiyaozaki K, Hachiya H, Kuze S, Ito Y (1983) Death of a boy due to malignant hyperthermic syndrome following strenuous sport. Hiroshima J Anesth 19:6
50. Britt BA (1988) Combined anaesthetic and stress induced malignant hyperthermia in two offspring of malignant hyperthermic susceptible parents. Anesth Analg 67:393–399
51. Hall LW, Woolf N, Bradley JWP, Jolly DW (1966) Unusual reaction to suxamethonium chloride. Br Med J ii:1305
52. Ahern CP, Somers CJ, Wilson P, McLoughlin JV (1977) The prevention of acute malignant hyperthermia in halothane sensitive Pietrain pigs by low doses of neuroleptic drugs. In: Proceedings of the international conference on products and diseases of farm animals. Center for Agricultural Publications and Documentation, Wageningen, Netherlands, pp 161–171

53. Jones EW, Kerr DD, Nelson TE (1973) Malignant hyperthermia—observations in Poland China pigs. In: Gordon RA, Britt BA, and Kalow W (eds) International symposium on malignant hyperthermia. Thomas, Springfield, pp 198–207
54. Nelson TE (1973) Procine stress syndromes. In: Gordon RA, Britt BA, Kalow W (eds) International symposium on malignant hyperthermia. Thomas, Springfield, pp 191–197
55. Short CE, Paddleford RR (1972) Malignant hyperthermia in the dog. Anesthesiology 39:462
56. De Jong RH, Heavner JE, Amory DW (1974) Malignant hyperpyrexia in the cat. Anesthesiology 41:608–609
57. Klein LV (1975) Case report: a hot horse. Vet Anesth 2:41
58. Citinos SB, Bush M, Phillips LG (1984) Dystocia and fatal hyperthermic episode in a giraffe. J Am Vet Med Assoc 185:1440–1442
59. Satnick JH (1969) Hyperthermia under anaesthesia with regional muscle flaccidity. Anesthesiology 30:472
60. Kalow W, Britt BA, Terreau ME, et al (1970) Metabolic error of muscle metabolism after recovery from malignant hyperthermia. Lancet ii:895–898
61. Takagi A, Sugita H, Toyokura Y, Endo M (1976) Malignant hyperthermia: effect of halothane on single-skinned muscle fibers. Proc Jpn Acad 52:603–606
62. Endo M, Tanaka M, Ogawa Y (1970) Calcium-induced release of calcium from the sarcoplasmic reticulum of skinned skeletal muscle fibres. Nature 228:34
63. Endo M, Yagi S, Ishizuka T (1983) Changes in the Ca-induced Ca release mechanism in the sarcoplasmic reticulum of the muscle from a patient with malignant hyperthermia. Biomed Res 4:83–92
64. Ohta T, Endo M, Nakano T (1989) Ca-induced Ca release in malignant hyperthermia-susceptible pig skeletal muscle. Am J Physiol 256:C358–367
65. Nelson TE, Lin M, Volpe P (1991) Evidence for intraluminal Ca^{++} regulatory site defect in sarcoplasmic reticulum from malignant hyperthermia pig muscle. J Pharmacol Exp Ther 256:645–649
66. Lopez JR, Jones DE, Allen PD (1986) (Ca^{2+}) 1 in muscles of malignant hyperthermia susceptible pigs determined in vivo with Ca^{2+} selective microelectrodes. Muscle & Nerve 9:85–6
67. Lopez JR, Medina P, Alamo L (1987) Dantrolene sodium is able to reduce the resting ionic (Ca^{2+}) 1 in muscle from humans with malignant hyperthermia. Muscle & Nerve 10:77–79
68. Ohnishi ST (1987) Effects of halothane, caffeine, dantrolene and tetracaine on the calcium permeability of skeletal sarcoplasmic reticulum of malignant hyperthermic pigs. Biochim Biophys Acta 897:261–268
69. O'Brien PJ (1990) Malignant hyperthermia: microassays for muscle Ca-channel hypersensitivity and halothane-inhibition of Ca-sequestration (abstract). J Neurol Sci 98(suppl):412
70. Nelson TE (1990) Porcine malignant hyperthermia: critical temperatures for in vivo and in vitro responses. Anesthesiology 73:449–454
71. Fill M, Stefani E, Nelson TE (1991) Abnormal human sarcoplasmic reticulum Ca^{2+} release channels in malignant hyperthermic skeletal muscle. Biophys J 59:1085–1090
72. Nelson TE (1992) Halothane effects on human malignant hyperthermia skeletal muscle single calcium-release channels in plantar lipid bilayers. Anesthesiolgy 76:588–595
73. Foster PS, Claudianos C, Geseni E, Hopkinson KC, Denborough MA (1989) Inisotol 1,4,5-trisphosphate phosphatase deficiency in malignant hyperpyrexia in swine. Lancet ii:124–126
74. Scholz J, Steinfath M, Roewer N, Patten M, Troll U, Schmitz W, Scholz H, Schulte Am Esch J (1993) Biochemical changes in malignant hyperthermia susceptible swine: cyclic AMP, inositol phosphates, A1, B1- and B2-adrenoceptors in skeletal and cardiac muscle. Acta Anesthesiol Scand 37:575–583

75. Ervasti JM, Claessens MT, Mickelson JR, Louis CF (1989) Altered transverse tubule dihydropyridine receptor binding in malignant hyperthermia. J Biol Chem 264:2711-2717
76. Adnet PJ, Krivosic-Horber RM, Adamantidis MM, Reyford H, Cordonnier C, Haudecoeur G (1991) Effects of calcium-free solution, calcium antagonists, and the calcium agonist Bay K 8644 on mechanical responses of skeletal muscle from patients susceptible to malignant hyperthermia. Anesthesiology 75:413–419
77. Mickelson JR, Ross JA, Hyslop RJ, Gallant EM, Louis CF (1987) Skeletal muscle sarcolemma in malignant hyperthermia: evidence for a defect in calcium regulation. Biochim Biophys Acta 897:364–374
78. Harrison GG (1973) Erythrocyte osmotic fragility in hyperthermia susceptible swine. Br J Anaesth 45:131–133
79. O'Brien PJ, Forsyth GW, Olexson DW, et al (1984) Canine malignant hyperthermia susceptibility: erythrocyte defects—osmotic fragility, glucose-6-phosphate dehydrogenase deficiency and abnormal Ca^{2+} homeostasis. Can J Comp Med 48:381–389
80. Ohnishi ST, Katagi H, Ohnishi T, et al (1988) Detection of malignant hyperthermia susceptibility using a spin label technique on red blood cells. Br J Anaesth 61:565–568
81. Klip A, Ramlal T, Walker D, et al (1987) Cytoplasmic calcium increase by anesthetic in pig lymphocytes. Towa rds a diagnosis of malignant hyperthermia. Anesth Analg 66:381–385
82. O'Brien PJ, Kalow BI, Ali N, Lassaline LA, Lumsden JH (1990) Compensatory increase in calcium extrusion activity of untreated lymphocytes from swine susceptible to malignant hyperthermia. Am J Vet Res 51:1038–1043
83. Fink HS, Hofmann JG, Hentschel H. et al (1992) Abnormalities in the regulation of blood platelet free cytosolic calcium in malignant hyperthermia. I. Human platelets. Cell Calcium 13:149–155
84. Fink HS, Maak S, Von Lengerken G, et al (1992) Abnormalities in the regulation of blood platelet free cytosolic calcium in malignant hyperthermia. II. Pig platelets. Cell Calcium 13:157–162
85. Kochs E, Hoffman WE, Roewer N, et al (1990) Alterations in brain electrical activity may indicate the onset of malignant hyperthermia in swine. Anesthesiology 73:1236-1242
86. Kochs E, Nollen H, Schulte am Esch J (1990) Changes in cerebral blood flow and EEG in experimental malignant hyperthermia. Anasth Intensivther Notfallmed 25:347-353
87. Beldavs J, Small V, Cooper DA, Britt BA (1971) Postoperative malignant hyperthermia: a case report. Can Anaesth Soc J 18:202
88. Harrison GG (1975) Control of the malignant hyperpyrexic syndrome in MHS swine by dantrolene sodium. Br J Anaesth 47:62
89. Anderson IL, Jones EW (1976) Porcine malignant hyperthermia: effect of dantrolene sodium on in vitro halothane-induced contraction of susceptible muscle. Anesthesiology 44:57
90. Britt BA, Scott E, Frodis W, et al (1984) Dantrolene. In vitro studies in malignant hyperthermia susceptible (MHS) and normal muscle. Can Anaesth Soc J 31:130–154
91. Kolb MA, Horne ML, Martz R (1982) Dantrolene in human malignant hyperthermia: a multicenter study. Anesthesiology 56:254–262
92. Cunliffe M, Lerman J, Britt BA (1986) Is prophylactic dantrolene indicated for MHS patients undergoing elective surgery? Can Anaesth Soc J 33:S113
93. Dershwitz M, Sreter FA (1990) Azumolene reverses episodes of malignant hyperthermia in susceptible swine. Anesth Analg 70:253–255
94. Gallant EM, Foldes F, Rempel WM (1985) Verapamil is not a therapeutic adjunct to dantrolene in porcine malignant hyperthermia. Anesth Analg 64:601–606
95. Baudendistel L, Goudsouzian N, Cote C, et al (1984) End-tidal CO_2 monitoring. Anaesthesia 39:1000–1003
96. Gronert GA, Milde JH, Theye RA (1977) Role of sympathetic activity in porcine malignant hyperthermia. Anesthesiology 47:411–415

97. Urwyler A, Censier K, Seeberger MD, Rothenbuhler JM, Kaufmann MA, Drewe J (1993) In vitro effect of ephedrine, adrenaline, noradrenaline and isoprenaline on halothane-induced contractures in skeletal muscle from patients potentially susceptible to malignant hyperthermia. Br J Anaesth 70:76–79
98. Galloway GJ, Denborough MA (1984) Phosphorus-31 nuclear magnetic resonance studies of muscle metabolism in malignant hyperthermia. Br J Anaesth 56:663–664
99. Foster PS, Hopkinson K, Denborough MA (1989) ^{31}P-NMR spectroscopy: the metabolic profile of malignant hyperpyrexic porcine skeletal muscle. Muscle & Nerve 12:390–396
100. Kozak-Reiss G, Bendahan D, Fontanarava E, Confort-Gouny S, Cozzone P, Monin G, Miril A, and Rodet A (1990) A spontaneous non-rigid MH crisis in an anaesthetized pig studied by P-31 NMR spectroscopy (abstract). J Neurol Sci 98(suppl):527
101. Kozak-Reiss G, Desmoulin F, Martin CF, Monin G, Renou JP, Canioni P, Cozzone PJ (1991) In vitro correlation between force and energy metabolism in porcine malignant hyperthermic muscle studied by ^{31}P NMR. Arch Biochem Biophys 287:312–319
102. Gareau PJ, Janzen EG, Towner RA, Stewart WA (1993) In vivo ^{31}P-NMR spectroscopy studies of halothane-induced procine stress syndrome. No effect of c-phenyl *n*-tert-butyl nitrone (PBN). Free Radical Res Commun 19:43–50
103. Kozak-Reiss G, Gascard, Redouane-Benichou K (1986) Malignant hyperthermia: diagnostic means and the possibilities opened up by NMR spectroscopy. Ann Fr Anesth Reanim 5:584–589
104. Olgin J, Argov Z, Rosenberg H (1988) Nonivasive evaluation of malignant hyperthermia susceptibility with phosphorus nuclear magnetic resonance spectroscopy. Anesthesiology 68:507–513
105. Murakawa M, Hatano Y, Mori K, et al (1989) Technetium-99m pyrophosphate scintography in a patient with malignant hyperthermia (letter). J Nucl Med 30:718–719
106. MacLennan D, Duff C, Zorzato F, et al (1990) Ryanodine receptor gene is a candidate for predisposition to malignant hyperthermia. Nature 343:559–561
107. McCarthy TV, Healy JMS, Heffron JJA, Lehane M, Deufel T, Lehmann-Horn F, Farrall M, Johnson K (1990) Localization of the malignant hyperthermia susceptibility locus to human chromosome 19q12–13.2. Nature 343:562–564
108. Healy JMS, Lehane M, Heffron JJA (1990) Localization of the malignant hyperthermia susceptibility locus to human chromosome 19q12–113.2 Biochem Soc Trans 18: 326
109. MacKenzie AE (1991) A comparison of the caffeine halothane muscle contracture test with the molecular genetic diagnosis of malignant hyperthermia. Anesthesiology 75:4–8
110. Levitt RC, Nouri N, Jedlicka AE (1991) Evidence for genetic heterogeneity in malignant hyperthermia susceptibility. Genomics 11:543–547
111. Levitt RC, Olckers A, Meyers S (1992) Evidence for the localization of a malignant hypertermia susceptibility locus (MHS2) to human chromosome 17q. Genomics 14:562–566
112. Sudbrak R, Golla A, Hogan K, Powers P, Gregg R, Chesne ID, Lehmann-Horn, F, Deufel T (1993) Exclusion of malignant hyperthermia susceptibility (MHS) from a putative MHS2 locus on chromosome 17q and of the A1, B1 and U subunits of the dehydropyridine receptor calcium channel as candidates for the molecular defect. Hum Mol Genet 2:857–862
113. Ellis FR (1978) Multifactorial inheritance of malignant hyperthermia susceptibility. In: Aldrete JA, Britt BA (eds) Second international symposium on malignant hypethermia. Grune and Stratton, New York, pp 329–338
114. Harbitz I, Chowdhary B, Thomsen PD, et al (1990) Assignment of the porcine calcium release channel gene, a candidate for the malignant hyperthermia locus, to the 6P11-Q21 segment of chromosome 6. Genomics 8:243–248
115. Fujii J, Otsu K, Zorzato F (1991) Identification of a mutation in porcine ryanodine receptor associated with malignant hyperthermia. Science 253:448–451

116. Fletcher JE, Calvo PA, Rosenberg H (1993) Phenotypes associated with malignant hypethermia susceptibility in swine genotyped as homozygous or heterozygous for the ryanodine receptor mutation. Br J Anaesth 71:410–417
117. Mickelson JR, Gallant EM, Litterer LA (1988) Abnormal sarcoplasmic reticulum ryanodine receptor in malignant hyperthermia. J Biol Chem 263:9310–9315
118. Mickelson JR, Litterer LA, Louis CF (1990) Isolation and reconstitution of the ryanodine receptor (RYR) from malignant hyperthermia susceptible (MHS) and normal pigs (abstract). J Neurol Sci 98:508
119. Valdivia HH, Hogan K, Coronado R (1991) Altered binding site for Ca^{2+} in the ryanodine receptor of human malignant hyperthermia. Am J Physiol 261:C237–C245
120. Hawkes MJ, Nelson TE, Hamilton SL (1992) [^{3}H] Ryanodine as a probe of changes in the functional state of the Ca^{2+}-release channel in malignant hyperthermia. J Biol Chem 267:6702–6709
121. European M.H. Group (1985) Laboratory diagnosis of malignant hyperpyrexia susceptibility (MHS) (letter). Br J Anaesth 57:1038–1046
122. Larach MA, Rosenberg H, Larach DR, Broennle AM (1987) Prediction of malignant hyperthermia susceptibility by clinical signs. Anesthesiology 66:547–550
123. Fletcher JE, Conti PA, Rosenberg H (1991) Comparison of North American and European Malignant Hyperthermia Group halothane contracture testing protocols in swine. Acta Anaesth Scand 35:483–487
124. Larach MG (1989) Standardization of the caffeine halothane muscle contracture test. Anesth Analg 69:511–515
125. Larach MG, Landis JR, Bunn JS, Diaz M, North American MH Registry (1992) Prediction of malignant hyperthermia susceptibility in low risk subjects: an epidemiological investigation of caffeine halothane contracture responses. Anesthesiology 76:16–27
126. Larach MG, Localio AR, Allen GC, Denborough MA, Ellis FR, Gronert GA, Kaplan RF, Muldoon SM, Nelson TE, Ørding H, Rosenberg H, Waud B (1994) A clinical grading scale to predict malignant hyperthermia susceptibility. Anesthesiology 80:771–779

Part 2. Clinical Classification and Incidence of Fulminant and Abortive Malignant Hyperthermia

The Incidence of Malignant Hyperthermia in Denmark

HELLE ØRDING and DIANA BENDIXEN

The Danish Malignant Hyperthermia Register

The Danish Malignant Hyperthermia Register (DMHR) was created in 1977 [1] with the following aims:

To collect information about cases of malignant hyperthermia (MH)
To investigate these patients and their families
To spread information about MH to patients, attending doctors, and dentists
To educate future anesthesiologists in various aspects of MH.

The clinical presentation of MH in humans is highly variable [2–4]. The reason for this is not well understood, but factors such as age of the patient, preoperative anxiety, combination and sequencing of anesthetic agents, and genetic factors are thought to have influence. In the DMHR, we distinguish between *fulminant MH*, in which the clinical history makes it unlikely that the signs and symptoms could result from anything but MH, and *abortive forms of MH*, in which symptoms and signs are less severe and may be caused by MH but also by other factors. In both cases, masseter muscle rigidity may or may not be present. This terminology is not very precise, but it is practical and has been used consistently for many years.

To put things into perspective when discussing the incidence of MH, Denmark is a small country with 5 million inhabitants and a fairly uniform genetic background. Each year approximately 400 000 anesthetics are given. In a previous investigation [5], we found that the incidence of MH depended on the type of anesthesia and the clinical classification of MH. Fulminant MH in a mixed surgical population occurred in 1 in 250 000 anesthetics when all types of anesthesia were included, and in 1 in 60 000 when a combination of potent inhalational agents and suxamethonium was used. MH was only observed with these agents. Abortive forms of MH were observed in 1 in 17 000 and in 1 in 4500 anesthetics, respectively, with these agents. Masseter muscle rigidity was observed in 1 in 12 000 administrations of suxamethonium. Our data did not permit calculation of any age-specific incidence of MH.

Number of Referrals and Types of Suspected MH

The total number of anesthetics given in Denmark is roughly 400 000 per year, and this number has not changed during the past 10 years. The number of referrals to the

The Danish Malignant Hyperthermia Register, Herlev University Hospital, DK-2730 Herlev, Denmark

DMHR in three 5-year periods is shown in Table 1. A 40% reduction in the number of cases referred during the past 5 years is apparent. This reduction is mainly seen in the number of cases with masseter muscle rigidity, but the number of fulminant cases is also smaller in the last 5-year period compared with the two previous periods. There was one death from MH in each of the first two 5-year periods (in 1979 and 1985, respectively). This was a reduction compared to the early 1970s when one to two deaths occurred each year in Denmark. Since 1985 there has been no death from MH in Denmark.

The reduction of the number of cases with fulminant MH mainly results from better monitoring; capnometry especially is of great value in this respect. However, a change in the drugs used for anesthesia is also thought to affect the reduction in frequency of both fulminant and abortive MH. In Table 2, available data on the number and types of anesthetics administered in Denmark in the years 1983 and 1991 are compared. A significant reduction in the use of general anesthesia has occurred in later years, with a concomitant increase in the use of regional anesthesia. The use of potent inhalational agents has been reduced by 22% from 1983 to 1991.

Isoflurane, which is thought to be a weaker trigger of MH than halothane, was not used in Denmark in 1983 but is now widely used. The extent of administration of suxamethonium cannot be documented because there are no statistical data on this point. However, from various sources it is known that the use of suxamethonium was considerably reduced after the introduction of atracurium and vecuronium. The 40% reduction in the number of referrals is mainly seen in the groups with masseter

TABLE 1. Number of patients referred to the Danish Malignant Hyperthermia Register and clinical presentation of malignant hyperthermia (MH) in three 5-year periods [7].

Type of MH	1978–1982	1983–1987	1988–1992	n
Fulminant MH	8	7	2	17
MMR + mild symptoms	35	31	21	87
MMR without other symptoms	34	36	13	83
Mild symptoms without MMR	38	43	33	114
Total	115	117	69	301

Fulminant MH and masseter muscle rigidity (MMR) have decreased in recent years.

TABLE 2. Total number of anesthetics and distribution of different types used in Denmark.

Year	1983	1991
Number of anesthetics	386 250	384 440
General anesthesia	88.3%	79.0%*
Regional anesthesia	11.7%	20.9%*
Use of potent inhalatonal agents	33.7%	26.7%*
Halothane	>90%	46.1%*
Enflurane	<5%	21.1%*
Isoflurane	<5%	32.9%*

Use of potent inhalational agents is given as percentage of the total number of anesthetics; use of each potent agent is given as percentage of the total number of inhalational anesthetics.

* Difference from similar data for 1983 is statistically significant ($P < 0.01$).

muscle rigidity (Table 1), and I postulate that reduction in the use of suxamethonium is the main reason for this general lower incidence of abortive MH, although reduction in the use of potent inhalational agents adds to the overall reduction of MH. Fulminant MH rarely occurs now.

In our previous report [5], we found that 75% of our referrals had received a combination of potent inhalational agents and suxamethonium, and this has not changed during subsequent years. Because of the lack of exact information about the use of suxamethonium, it is not possible to calculate any new specific incidence of clinical MH without contacting all departments of anesthesia in our country, as we did for our previous study [5]. Interestingly, however, preliminary new data on anesthesia in Denmark seem to indicate that the use of inhalational anesthesia is again increasing (in 1993 an inhalational agent was used in 34% of all anesthetics and in 44% of cases with general anesthesia), while the use of suxamethonium is further decreasing. Perhaps, therefore, the differential influence of these agents on the incidence of clinical MH will become more clear in the future.

Age and Sex of MH Probands

Health care statistics in Denmark [6] not previously available have allowed reanalysis of some data concerning the influence of sex and age on the occurrence of MH [7]. Comparing age and sex of the 48 MH-susceptible probands described by Ørding et al. [1] with data on surgery in Denmark (Table 3) shows that boys aged 0–14 years are *not* experiencing MH more frequently than girls of the same age (chi square = 0.0073; $P > .05$); that is, the greater number of boys with clinical MH appears to result from the greater number of operations in boys at this age. Looking solely at the three types of surgery in which MH was most frequently encountered (abdominal, 38%; otolaryngological, 25%; and orthopedic surgery, including surgery for trauma, 21%), there is still no sex difference in the frequency of MH (chi square = 0.0004; $P > .05$). In the age group 15–44 years, men do develop clinical MH more frequently than women, taking into account the number of operations performed (chi square = 7.38; $P < .01$) (Table 3). However, if not all types of surgery are included for this calculation but only the three types encountered in most of our probands, the sex difference disappears (chi square = 0.09; $P > .05$). Thus, the male preponderance in this age

TABLE 3. Number of operations performed per year in Denmark in different age groups, and age and sex of the 48 MH-susceptible probands described in [1].

Age group (yr)	Number of operations		MH (male/female)	Significance (male/female)
	Male	Female		
0–14				
All	23 841	13 839	17/11*	NS
Subgroup	14 254	9 858		NS
15–44				
All	65 797	188 308	11/9	$P < .01$
Subgroup	27 305	28 270		NS

"All" refers to all opertions; "subgroup" refers to three types of surgery (abdominal, otolaryngological, and orthopedic surgery, including surgery for trauma) performed in 84% of the probands.

* Significant difference in incidence of MH between the two age groups; NS, not significant.

group may be more technical than real in that it probably results from the very large number of short surgical procedures performed with intravenous agents in women of this age group [7]. These results are different from those of Kikuchi et al. [8].

Age, however, seems to be important for the development of clinical MH (Table 3). In the age group 0–14 years, MH was significantly more frequent than in the older age group, taking into account both the total number of operations performed and the three types of surgery mostly associated with MH (chi square = 18.0; $P < .001$). Even though anesthesia with trigger agents may be relatively more frequent in the younger age group (no data available for any analysis), the age difference in the incidence of MH seems so large that it is highly unlikely that it is not a real difference.

In conclusion:

The incidence of clinically suspected MH seems to have been reduced during recent years because of better monitoring and a change in anesthetic techniques employed.

With equal exposure to MH triggering agents, the incidence of MH is similar in males and females.

The incidence of clinical MH is greater in children than in older age groups.

Through better education, monitoring, and treatment, it is possible to decrease mortality from MH to nearly zero.

References

1. Ørding H, Ranklev E, Fletcher R (1984) Investigation of malignant hyperthermia in Denmark and Sweden. Br J Anaesth 56:1183–1190
2. Hackl W, Mauritz W, Schemper M, Winkler M, Sporn P, Steinbereithner K (1990) Prediction of malignant hyperthermia susceptibility: statistical evaluation of clinical signs. Br J Anaesth 64:425–429
3. Larach MG, Rosenberg H, Larach DR, Broennle AM (1987) Prediction of malignant hyperthermia susceptibility by clinical signs. Anesthesiology 66:547–550
4. Rosenberg H (1988) Clinical presentation of malignant hyperthermia. Br J Anaesth 60:268–273
5. Ørding H (1985) Incidence of malignant hyperthermia in Denmark. Anesth Analg 64:700–704
6. Sundhedsstyrelsen (the National Board of Health) (1990) Sundhedsstatistik. National Board of Health, Denmark
7. Ørding H (1995) Investigation of malignant hyperthermia susceptibility in Denmark. Thesis, University of Copenhagen, Denmark
8. Kikuchi H, Morio M, Shinozaki M, Ishihara S (1978) Statistical considerations of malignant hyperthermia in Japan. In: Aldrete JA, Britt BA (eds) Second international symposium on malignant hyperthermia. Grune and Stratton, New York, pp 483–498

Malignant Hyperthermia in Austria 1984–1993

Dorrit Heilinger[1], Eva Donner[1], Andrea Sauberer[1], Werner Hackl[2], and Walter Mauritz[2]

Introduction

The Vienna Malignant Hyperthermia (MH) Investigation Unit (which functions as the Austrian National MH Registry) was founded in 1976 by Karl Steinbereithner and Paul Sporn. The unit employs three anesthesiologists, one nurse, two technicians, and one secretary (all part-time) and is funded by the Austrian Government, Ministry of Health, the Ludwig-Boltzmann Society, and the University of Vienna. Diagnostic testing of MH probands and family members has been performed since 1976. Initially, halothane and halothane-caffeine tests were done, but since the founding of the European MH Group in 1983, the European Protocol [1] has been used. In 1985 a program to provide safe anesthesia for MH patients was established [2]. A hot-line for MH-related problems has operated 24 hours a day since 1987. This paper summarizes the results of our studies on epidemiology and clinical course of MH in Austria. Only patients evaluated according to the European Protocol are included in this report.

Patients and Methods

During the last 10 years (January 1 1984–December 31 1993) 160 suspected MH reactions were reported to our unit. Following initial contact (usually a phone call) an MH report form was sent to the institution where the suspected reaction had been observed. This MH report form was designed to give us information on the proband (e.g., age, sex, previous anesthetic experience, surgery, family history), the clinical course of the reaction (e.g., heart rate, ventilation, temperature, blood gases and laboratory values, duration of anesthesia, treatment, outcome), and statistical data on the date, time, and location of the reaction. Upon return of the completed form, the family was invited to undergo diagnostic testing. In 152 cases the proband or his/her parents agreed to be tested, while in 8 cases no tests were done. These latter reactions were excluded from further evaluation (Table 1). According to the clinical information received the reactions were classified as:

[1] Vienna Malignant Hyperthermia Investigation Unit, Department of Anesthesia and Critical Care Medicine B, School of Medicine, University of Vienna, A-1090 Austria

[2] Ludwig-Boltzmann Institute of Experimental Anesthesia and Research in Intensive Care, Wien-Linz, Vienna Division, Vienna, A-1090 Austria

TABLE 1. Malignant hyperthermia-like reactions in Austria 1984–1993.

Reactions	Number
Reported	160
Investigated	152 (100%)
Classified as "MH"	79 (52%)
Classified as "no MH"	73 (48%)

MH, Malignant hyperthermia.

"fulminant": full-blown, life-threatening reaction with at least three of the following: cardiac symptoms, acidosis, hypercarbia, fever, rigidity
"masseter spasm" (MS) as the only symptom
"abortive": reaction with only minor symptoms (e.g., MS plus other symptoms, post-operative fever).

A total of 593 people were tested according to the European Protocol. Muscle biopsies were done under local or regional anesthesia in probands, if they were older than 14 years, and/or in their parents. A muscle biopsy was always done in both parents of probands if possible. In addition, family screening was done by investigating relatives of parents or probands who had given a positive test result. According to the European Protocol the test results were classified as:

MHS (susceptible): halothane and caffeine tests positive
MHEh (equivocal): halothane test positive, caffeine test negative
MHEc (equivocal): halothane test negative, caffeine test positive
MHN (negative): halothane and caffeine tests negative.

A reaction was classified as "MH" if the proband or at least one parent of the proband tested either MHS or MHE. A reaction was classified "no MH" if the proband or both parents tested MHN. The clinical symptoms of reactions classified as "MH" or "no MH," respectively, were compared by means of a logistic regression analysis [3]. The information gained from questionnaires from subjects with proven MH reactions was used to estimate the incidence of MH in Austria and to describe the epidemiology of MH. Analysis of variance, chi-square test, and Mann-Whitney-U-test were used where appropriate; a P less than 0.05 was considered significant. This report is a summary of a number of studies; a detailed description of methods and results has been given in a recent book by Hackl [4].

Results

The classification of reactions reported between 1984 and 1993 is given in Table 1. About half of the reactions reported were classified as MH; all fulminant reactions, however, were classified MH. The 79 proven MH reactions form the basis for this report.

The incidence of typical clinical signs in both "MH" and "no MH" reactions are given in Fig. 1. Masseter spasm, rigidity, tachycardia, arrhythmia, cyanosis and myoglobinuria were the most common features in all reactions. The incidences of MS and tachycardia were not significantly different between the groups. Rigidity, cyanosis, and myoglobinuria were observed significantly more often in the MH group.

No symptom or group of symptoms, however, established the diagnosis of MH without diagnostic testing [3].

The number of MH reactions reported annually varied between 5 and 12 per year (Fig. 2) with an average of 8 per year, but there was no trend toward either an increase or a decrease in the incidence of MH. The MH reactions, however, were not distributed evenly throughout Austria. If the percentage of reactions reported by each county is compared to the percentage of the Austrian population living in these counties (Fig. 3), regional differences in the incidence of MH become obvious. For example, Vorarlberg, with only 4% of the Austrian population, reported 13% of all MH reactions. Vienna, the capital, and Burgenland reported the expected number of reactions, while all other counties reported less than the expected number (Fig. 3).

Seventeen MH reactions (22%) were classified as fulminant, and three of the patients died. This gives an overall mortality of 4% and a mortality of 18% for fulminant cases. Masseter spasm as the only symptom was found in 17 reactions (22%) but was the most common symptom in reactions later classified as "no MH." The vast majority of cases (45 out of 79; 57%) were classified as abortive (Table 2).

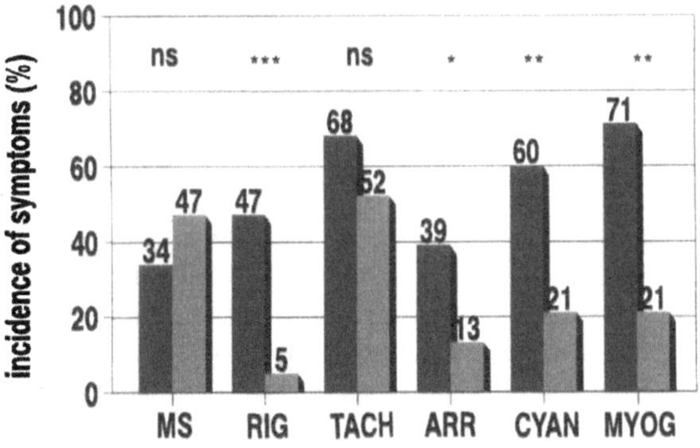

FIG. 1. The clinical course of malignant hyperthermia (MH, *dark bars*), and no MH (*light bars*) reactions. The most common features in all reactions are shown as a percentage of total ($n = 79$) reactions, and include masseter spasm (*MS*), rigidity (*RIG*), tachycardia (*TACH*), arrhythmia (*ARR*), cyanosis (*CYAN*), and myoglobinuria (*MYOG*) *ns*, not significant; *, $P < 0.02$; **, $P < 0.003$; ***, $P < 0.001$

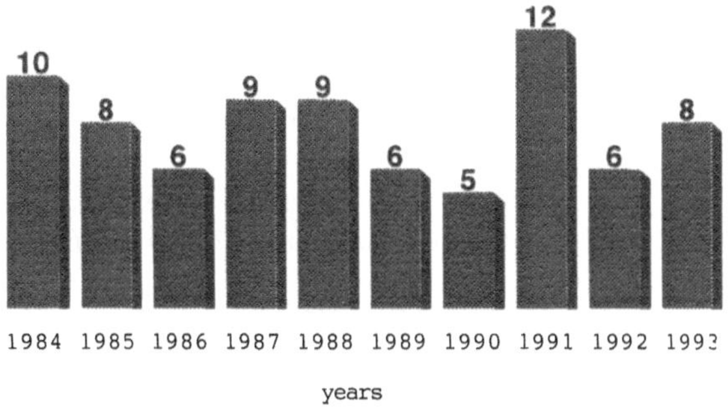

FIG. 2. Number of malignant hyperthermia (MH) reactions reported annually between 1984 and 1993 in total Austrian population. Only reactions classified as "MH" are included

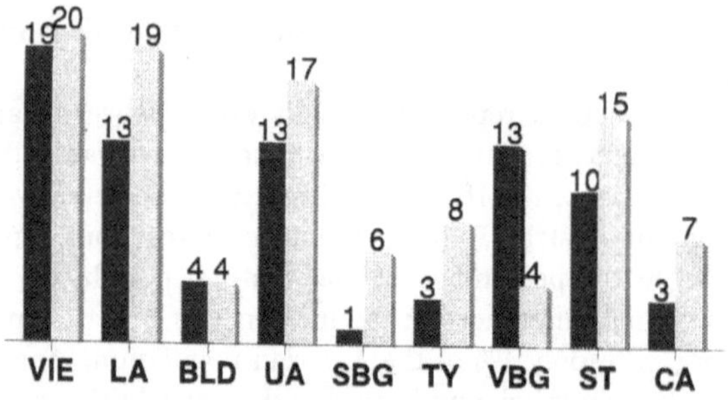

FIG. 3. Regional differences in the incidence of malignant hyperthermia (MH) reactions, expressed as percentage of total MH reactions (*dark bars*) versus percentage of total population (light bars). *VIE*, Vienna; *LA*, Lower Austria; *BLD*, Burgenland; *UA*, Upper Austria; *SBG*, Salzburg; *TY*, Tyrol; *VBG*, Vorarlberg; *ST*, Styria; *CA*, Carinthia

TABLE 2. Incidence and mortality of MH in Austria 1984–1993.

Classification	Number of cases	Incidence[a]	No. of deaths
Abortive	45 (57%)	1:67000	0
Masseter spasm	17 (22%)	1:175000	0
Fulminant	17 (22%)	1:175000	3
Overall	79 (100%)	1:37500	3

[a] An estimated 300000 anesthetic procedures are performed per year.

The overall incidence of MH (i.e., the incidence of all reactions classified as MH) was 1:37500 (Table 2), but there were significant regional differences as previously mentioned (Fig. 3). The incidence of MH in the Vienna General Hospital (the authors' institution) was 1:33400 anesthetics; it was as low as 1:2600 in the City Hospital of Bludenz, Vorarlberg.

The combination of halothane and succinyl choline was the most common trigger of MH (Table 3). Halothane or succinyl choline alone were less common; all other triggers played almost no role.

Almost half of all MH reactions occurred during ear, nose, and throat (ENT) surgery (Table 4); MH was also frequently seen during appendectomy, orthopedic surgery, and correction of strabismus.

Of the patients who experienced an MH crisis, 60% were children aged 10 or younger (Fig. 4); the youngest patient was 8 months old. The number of crises decreased with increasing age; only three reactions occurred in patients over the age of 40. Of the 79 patients, 59 (75%) were male ($P < .01$); the male/female ratio (3:1 for all age groups) tended to increase with age.

Test results in probands and parents of probands (Fig. 5) also suggest linkage between male gender and MH. Of the 24 probands (21 male), 46% tested MHS and 54% tested MHE. Fathers of probands ($n = 57$) gave strikingly similar results: they tested MHS in 46% and MHE in 34%; MHN (21%) was the least common finding. In the mothers of probands ($n = 47$), however, the most common diagnosis was MHN (53%) followed by MHS and MHE. The test results were not influenced by the clinical classification of MH.

Linkage between male gender and MH becomes even more probable if the results from all diagnostic tests done in MH families are considered (Table 5). Males are more

FIG. 4. Age and sex distribution of malignant hyperthermia (MH) reactions, showing the numbers of males (*light bars*) and females (*dark bars*) with MH in different age groups

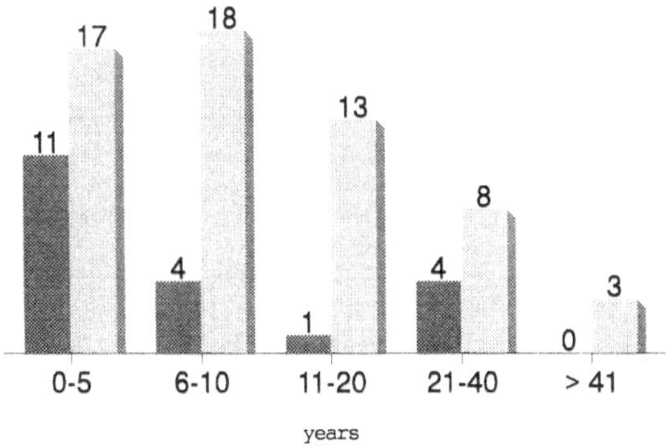

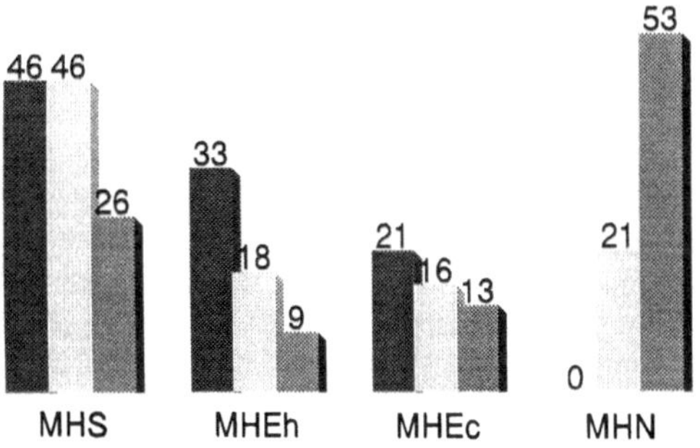

FIG. 5. Results of diagnostic testing for malignant hyperthermia (MH) in probands and both their parents. Results were classified as MH susceptible (*MHS*), MH equivocal for halothane (*MHEh*), MH equivocal for caffeine (*MHEc*), or MH negative (*MHN*), and expressed as a percentage of total probands (n = 24, filled bars), fathers (n = 57, lightly shaded bars), and mothers (n = 47, darkly shaded bars)

TABLE 3. Agents that trigger malignant hyperthermia.

Triggering agents	Number of cases
Halothane + succinyl choline	53 (67%)
Halothane alone	10 (12%)
Succinyl choline alone	8 (10%)
Enflurane + succinyl choline	5 (6%)
Isoflurane + succinyl choline	1 (1%)
Exercise	2 (2%)

TABLE 4. Type of surgery resulting in malignant hyperthermia.

Surgical procedure	Number of cases
ENT (tonsillectomy, adenotomy)	37 (47%)
Appendectomy	12 (15%)
Orthopedic (including trauma)	11 (14%)
Correction of strabismus	7 (9%)
Other abdominal surgery	4 (5%)
Other surgery (vascular, dermatological)	6 (8%)
None (exercise induced)	2 (2%)

ENT, ear, nose, and throat.

TABLE 5. Diagnostic testing of families with malignant hyperthermia.

Test results	Female	Male
MHS	67 (21%)	96 (35%)*
MHE	66 (21%)	78 (28%)
MHN	184 (58%)	102 (37%)*
Total	317	276

MH, malignant hyperthermia; MHS, MH susceptible; MHE, MH equivocal; MHN, MH negative.
* $P < 0.05$.

	Maternal			
Paternal	MHS	MHEh	MHEc	MHN
MHS	**2**	**0**	**2**	15
MHEh	**0**	**2**	**0**	5
MHEc	**0**	**1**	**2**	5
MHN	9	1	2	0

FIG. 6. Results of diagnostic testing in both parents of probands. In 20% of cases, both parents showed abnormal test results (indicated in *bold* type). *MH*, malignant hyperthermia; *MHS*, MH susceptible; *MHEh*, MH equivocal for halothane; *MHEc*, MH equivocal for caffeine; *MHN*, MH negative

likely to test MHS and less likely to test MHN than females, and these differences are statistically significant.

In 46 out of 79 MH crises (58%), both parents of probands underwent in vitro testing. The results from these family tests are shown in Fig. 6. In 9 out of 46 cases both parents gave abnormal test results. Even if all MHEc results are excluded, there remain two couples where both parents are MHS and two couples where both the MHEh. The clinical course in the children of these parents was not uncommon; six reactions were classified as "abortive" and three were "MS only."

Discussion

Malignant hyperthermia will most likely remain a major concern for anesthesiologists well into the next century. Our results show that fulminant reactions, which made up more than 50% of all reactions in earlier reports [5,6], have been decreasing over recent years but there is no clear trend toward a general decrease of MH reactions. The incidence of MH at the Vienna General Hospital was 1:23 600 during the period 1976–1985 [6] and was 1:33 400 from 1984–1993. The total number of reported MH cases, however, was higher during the last 10 years than during the 10 years before 1984 [6]. This puzzling finding may reflect an actual decrease in MH reactions and a concomitant increase in reports to the Vienna MH Investigation Unit; to the contrary, it could also indicate an actual increase in cases of abortive MH. There are insufficient data on the number of anesthetics during the last years to make a sound prediction whether the number of MH cases will increase or decrease during the next years.

Regional differences in the incidence of MH in Austria have been reported [6]; during the last 10 years, we observed a trend toward a decrease in these differences.

Vienna and Vorarlberg, which had reported 36% and 20%, respectively, of all crises during the period 1976–1985 [6], are down to 19% and 13%, respectively (Fig. 3). This is probably due to an increasing number of reports from other counties during the 10 years. However, there are some differences that can only be explained by inbreeding; other factors, like different anesthesia techniques or failure to report suspected reactions, seem to be quite unlikely now. Furthermore, four of the nine couples where both parents of probands gave positive test results (2, MHS + MHS; 1, MHEh + MHEh; 1, MHEh + MHEc) came from secluded mountainous areas in Vorarlberg or Tyrol. These areas are inhabited by less than 5% of the Austrian population, and inbreeding is a major problem.

The statistics on triggering agents look almost identical to a list published earlier [6]. Halothane and succinyl choline are widely used for induction of anesthesia in children but became less popular for anesthesia in adults, and this may explain the high number of pediatric MH patients in our group. Isofluorane is now used for more than 50% of all inhalation anesthetics in Austria; only one case of MH, however, was triggered by isofluorane. This may reflect the fact that compared to halothane, isofluorane is a less potent trigger of MH [5,7]. The two cases of exercise-induced MH have previously been reported in detail [8].

ENT surgery, appendectomy, orthopedic surgery, and strabismus lead the list of surgical procedures where MH crises were observed. While the first three diagnoses probably just reflect the relative frequencies of these operations in a mainly pediatric population, this may not be true for strabismus. Previous studies suggest that strabismus may be a frequent problem in MH families [5].

The incidence of MH is significantly higher in children than in adults, and this finding is confirmed by a number of other studies [9–12]. Possible explanations include the preferred use of agents that are MH triggers in children, the increasing number of surgical procedures in children, and the decreasing number of adult anesthetics where trigger agents are used. Children under 10 years made up 45% of all cases of MH reactions in an earlier study [6]; during the last 10 years, reactions in these children increased their share to 60% of all reactions. With the exception of Morio [11], all authors found a marked decrease of MH reactions in patients older than 20 years similar to that observed in our study.

Male patients are significantly more likely to experience MH. The male:female ratio in our study is 3:1; in the studies mentioned above, the rates range from 1.4:1 in Scandinavia [12] to 4:1 in Japan [11]. There are no obvious explanations for the observed linkage between male gender and MH, but the following unconfirmed hypotheses have been proposed:

1. A higher number of anesthetics in male patients. Unfortunately, no data for Austria are available, but in a sample of 1000 anesthetics the male:female ratio was 531:469, which is quite different from the 3:1 ratio observed for MH reactions.
2. Higher total muscle mass, which might faciliate triggering of the reaction.
3. Females might be hormonally protected [5]—even those below age 10?!?
4. Males might be more susceptible due to sex hormones.

It has been verified, however, that MH is not linked to the Y-chromosome, but that males are much more susceptible. This is also clear from the results of our diagnostic test series (Fig. 5 and Table 5). The genetics of MH remain a mystery. Two large Austrian MH families underwent genetic testing. In one of these, the MH trait was clearly linked to a defect on the ryanodine receptor gene on chromosome 19, which was not unexpected. In the other family, however, any defect on chromosome 19 was ruled out. The single gene hypothesis might not explain all cases of MH but it has been very effective in some families. The high number of families where both parents tested MHS or MHE, as well as the findings in families without any defects on the ryanodine receptor gene, would be easier to explain if a multiple gene hypothesis were adopted.

Conclusion

Epidemiological data from Austria show a slight decrease in the incidence of MH with an uneven geographic distribution; typical distribution patterns for age, sex, triggering agent, and type of surgery are found. For family screening, both parents of probands should be tested whenever possible. At present, in vitro contracture tests with halothane and caffeine remain the only reliable method for diagnosis of MH.

References

1. The European Malignant Hyperpyrexia Group. (1984) A protocol for the investigation of malignant hyperpyrexia (MH) susceptibility. Br J Anaesth 56:1267
2. Hackl W, Mauritz W, Winkler M, Sporn P, Steinbereithner K (1990) Anesthesia in malignant hyperthermia susceptible patients without dantrolene prophylaxis—a report of 30 cases. Acta Anaesthesiol Scand 34:534
3. Hackl W, Mauritz W, Schemper M, Winkler M, Sporn P, Steinbereithner K (1990) Prediction of malignant hyperthermia susceptibility: statistical evaluation of clinical signs. Br J Anaesth 64:425
4. Hackl W (1990) Malignant Hyperthermia. Results of diagnostic, clinical, experimental and epidemiological studies. Beitr Anaesth Intens Notfallmed 34. Maudrich Vienna-Munich-Berne
5. Britt BA (1982) Malignant hyperthermia: A review. In: Milton AS (ed) Pyretics and antipyretics. Handbook of experimental pharmacology, vol 60. Springer, Berlin Heidelberg New York, p 547
6. Mauritz W, Sporn P, Steinbereithner K (1986) Epidemiological and clinical aspects of 65 cases of Malignant hyperthermia (MH) in Austria. Anaesthesist 35:639
7. Mauritz W, Gilly H, Sporn P, Steinbereithner K, Weindlmayr-Goettel M (1986) Isofluorane and malignant hyperthermia. In: Lawin P, Van Aken H, Puchstein C (eds) Isofluorane. Anaesthesie und Intensivemedizin, vol 182. Springer, Berlin Heidelberg New York, p 304
8. Hackl W, Winkler M, Mauritz W, Sporn P, Steinbereithner K (1991) Muscle biopsy for diagnosis of malignant hyperthermia susceptibility in two patients with severe exercise induced myolysis. Br J Anaesth 66:138
9. Britt BA, Kalow W (1970) Malignant hyperthermia: a statistical review. Can Anaesth Soc J 17:293
10. Gjengstö H (1971) Die maligne Hyperpyrexie: eine emsthafte Narkosekomplikation. Anaesthesist 20:299
11. Morio M (1980) Malignant hyperthermia (in Japanese with English abstract). Masui (Jpn J Anesthesiol) 29:1
12. Ørding H, Ranklev E, Fletcher R (1984) Investigation of malignant hyperthermia in Denmark and Sweden. Br J Anaesth 56:1183

Clinical Classification and Incidence of Fulminant and Abortive Malignant Hyperthermia Reactions in Italy

V. Tegazzin[1], A. Accorsi[2], I. Moroni[3], D. Di Martino[4], V. Brancadoro[5], and V. Peduto[6]

Summary. A study made throughout Italy during the period 1990–1993 by the Malignant Hyperthermia (MH) Laboratory and the Italian MH Registry showed that the incidence of MH reactions in Italy was 1 in 51 000. These data are not greatly different from the results reported by anesthesiologists in a retrospective investigation in four regions: Piemonte, Emilia Romagna, Veneto, and Tuscany, where the incidence was found to be about 1 in 70 000–75 000 and only the "very suspected MH reactions (fulminant and abortive)" were considered. The referred signs of MH adverse reactions—fulminant and abortive—were, first of all, tachycardia, arrhythmia, fever, rigidity, high values of creatine kinase (CK), myoglobinuria, acidosis, hypoxia, dark venous bleeding, and, when monitored, high end-tidal (ET) CO_2. Following these clinical signs and considering their association with 110 MH reactions, 28 of the probands referred to the MH Italian Registry and to the four MH laboratories in the past 4 years (1990–1993) were classified as fulminant forms and 69 as abortive forms, with an incidence of 1 in 200 000 and 1 in 81 000, respectively.

Introduction

Malignant hyperthermia (MH) was described as a pharmacogenetic disease in 1960 by Denborough and Lovell [1] in an Australian family in which ten members died of anesthesia. The first description in Italy with a clear sequence of clinical signs, in 1968 [2], reported a 16-year-old boy who died in 1964 during a trigger anesthesia for orthopedic surgery for a bomb injury to the hand. This case is important because the anesthetist reported a hypercatabolic reaction as occurring in fulminant MH reactions: first, the exhausting of soda lime with increase of CO_2 production, which is associated with tachycardia and arrhythmia, raising of body temperature from the accelerated metabolism in the muscle tissue, rigidity as a result of depletion of chemical energy (adenosine triphosphate) and increased calcium levels in the myoplasmic space, and unstable blood pressure caused by the contrasting effects of catecho-

[1] MH Laboratory, S. Antonio Hospital, Via Facclolati 71, 35127 Padua, Italy
[2] Italian MH Registry, USL 25, Via Marconl 35, 40010 Bentivoglio (Bo), Italy
[3] Neurology Institute, Ospedale Maggiore, Via F Sforza 35, 20100 Milan, Italy
[4] MH Laboratory, USL 40, Via S. Giacomo del Capri 66, 80131 Naples, Italy
[5] MH Laboratory, II Policlinico, Via S. Pansini 5, 80121 Naples, Italy
[6] Department of Anaesthesia, University of Cagliari, Cagliari, Italy

lamines and catabolytes. Myoglobinuria and high CK levels were not reported because the patient died 130 min after anesthetic induction. Briefly, after 25 min the soda lime became blue; after 90 min, signs such as increased tachycardia, arrhythmia, high temperature, skin mottling, and rigidity appeared. More succinylcholine was administered to resolve the rigidity and the halothane was stopped. Despite resuscitation, death occured 40 min after the appearance of major signs (Table 1).

Subsequently, 52 more cases were reported in the Italian literature from 1965 to 1990 with signs that could be classified as MH and with a mortality of 31.7%. In 1988, at the first meeting organized in Castello di Bentivoglio (Bologna Italy), an Italian MH group of anesthetists, neurologists, and neuropsychiatrists was constituted, and in 1990 during another meeting in Bologna a questionnaire was prepared and sent to 800 departments and anesthesia services to obtain detailed information about MH reactions during trigger anesthesia. The questionnaire followed the sequence of clinical signs reported in the first case described as MH reactions [2] with some more questions, as in Methods.

Methods

The summarizing data requested in the questionaire are listed here:

1. How many MH reactions did you see in the last 10 years?
2. Years in which the MH reaction occurred: 1981, 1982, 1983, 1984, 1985, 1986, 1987, 1988, 1989, 1990?
3. Age of the patient?
4. Type of surgical operation?
5. Drugs administered?
6. Patient survived or did not survive?
7. Sequence of clinical signs reported in the MH reaction such as tachycardia, arrhythmia, acidosis, rigidity, fever, unstable blood pressure, masseter spasm, increased end-tidal CO_2, myoglobinuria, high values of CK?
8. Therapy used in the referred case of MH reaction (Dantrolene)?

TABLE 1. Sequence of signs and drug administration in 16-year-old boy who died of a MH fulminant reaction during trigger anesthesia.

TPS	+								
Succinylcholine	+	+	+				+		
NO_2	+								
O_2	+				31/min	++	+++	+++	
Halothane, 0.2%–1%	+	+	+	+	Stop				
Tachycardia			+	+	+	+			
Arrhythmia					+	+	+	+	+
Variation in BP				130	160	90		–	–
Soda lime	W	W	B	W	W	W/B	B	B	B
Mask	+	+	+	+	+	+	+	+	+
Breathing				+	+	+	+	+	+
Heartrate				120	130	140	160		
Skin mottling					+	+	+	+	+
Temperature, °C						+	+	+	41.5
Rigidity						+	+	+	+

TPS, Thiopentone sodium; BP, blood pressure; W, white; B, blue. Plus (+) indicates use.

9. Recovery days in intensive care unit?
10. Legal problem if the patient died of MH reactions?

At the end of 1993, a second questionnaire prepared by the MH Laboratory of Padua was sent to the other MH laboratories in Milan and Naples (1 and 2) and to the Italian MH Registry to collect data about probands referred to each laboratory from January 1990 to December 1993. In this questionnaire the data requested were:

1. How many fulminant forms?
2. How many abortive forms?
3. How many atypical forms?
4. In vitro contracture test (IVCT) results following European Malignant Hyperthermia Group (EMHG) [3] protocol when performed in each group of patients?

The classification of MH forms had to follow these criteria: (a) *fulminant forms*: presence of major signs such as increase of $ETCO_2$, temperature >39.5°C, rigidity, CK > 10000 IU, myoglobinuria, and minor signs such as tachycardia, arrhythmia, acidosis, unstable blood pressure, masseter spasm, and profuse sweating to have an association of at least five signs including three major signs; (b) *abortive forms*: not more than two major signs such as CK > 10000 and myoglobinuria but never a temperature above 39.5°C, rigidity, or increased $ETCO_2$, plus minor signs, to have not more than five signs referred by the anesthetists; or (c) *atypical forms*: the other probands with minor signs only. The masseter forms when associated with other signs could be included in fulminant or abortive forms. Table 2 reports the presence of major or minor signs in abortive and fulminant (combined) forms in patients referred to our MH laboratory.

Results

We did not obtain complete data from the first questionnaire. For example, in the Piemonte region in a 4-year period, in 213444 anesthesias (22.7% of the total anesthesia) the incidence of MH reactions was reported first as 1 in 129000, and on review of the data, as 1 in 71000. In the Emilia Romagna region in 1982–1986, the incidence was 1 in 75000, and in the Veneto region from 1982 to 1989 (57% of the questionnaires), the incidence was 1 in 80000. In the latter region, the percentage of clinical signs referring to seven patients having a MH reaction were tachycardia (86%), arrhythmia (72%), hypercapnia, acidosis, and high fever (57%), high CK,

TABLE 2. Percentage of major and minor clinical signs of abortive and fulminant MH observed in the probands (42 subjects) investigated at the MH Laboratory of Padua.

Major signs	Percent	Minor signs	Percent
CK > 10.000	65	Tachycardia	75
Myoglobinuria	65	Arrhythmias	15
Rigidity	50	Masseter spasm	15
Fever (temperature > 39.5°C)	20	Acidosis	14.3
Increased $ETCO_2$	9.5	Unstable blood pressure	15
		Profuse sweating	10

CK, Creatine kinase; $ETCO_2$, end-tidal carbon dioxide.

TABLE 3. Number of fulminant and abortive forms of MH classified in Italy in 1990–1993 following the clinical criteria suggested in this chapter and incidence in the population.

Forms	*n*	Incidence
Fulminant	28	1:200 000
Abortive	69	1:81 000

rigidity, and myoglobinuria (43%), and masseter spasm, tachypnea, and cardiac arrest (14%). In the Tuscany region (52 hospitals of 54) in 1986–1990, 11 MH reactions were referred with a total incidence of MH reaction of 1 in 70 000.

From the second questionnaire, 110 probands were referred to the MH Laboratories and the Italian MH Registry; these included 28 fulminant, 69 abortive, and 13 atypical forms with a total incidence of 1 in 51 000 over all Italy, assuming an average of 1 400 000 general anesthesias per year. The incidence of the fulminant and abortive forms was 1 in 200 000 and 1 in 81 000, respectively (Table 3).

Discussion and Conclusion

The difficulty of collecting data from anesthesia departments renders any retrospective epidemiological study regarding MH reactions to trigger anaesthesia incomplete. With the first questionnaire, we obtained reports from four regions only. Well-described cases were reported, particularly from Veneto, with clinical signs quite similar to those reported in the first case reported in the Italian literature. In the second questionnaire we proposed criteria of major and minor signs, but of course the clinical classification is far from giving us the key to diagnose malignant hyperthermia. In fact this is possible only by the in vitro contracture test (IVCT) [3].

The results of the second questionnaire are in agreement with the data from the research of Ørding in Denmark [4] and Hackl and Mauritz et al. in Austria [5] with respect to incidence of the fulminant forms. We also are agreed with the latter authors in considering rigidity as one of the major signs in MH reaction (50% of our probands) [5], but we also agreed with the Michio Morio scale [6] in considering the temperature >40°C (in our classification, we kept temperature >39.5°C) as one of the major clinical signs of hypercatabolic reactions. All our patients diagnosed as having a fulminant form according to the clinical criteria were found susceptible to IVCT while only 25% of the patients with abortive forms were positive.

References

1. Denborough MA, Lovell RRH (1960) Anaesthetic deaths in a family. Lancet 2:45
2. Pollazzon G, Salvadego L (1968) Considerazioni sulla ipertonia muscolare—ipertermia maligna in corso di anestesia generale. Acta Anaesthesiol (Padua) 19:1–7
3. European MH Group (1984) A protocol for the investigation of malignant hyperpirexia (MH) susceptibility. Br J Anaesth 56:1267–1271
4. Ørding H (1985) Incidence of malignant hyperthermia in Denmark. Anesth Analg 64:700–704

5. Hackl W, Mauritz W, Schemper M, Winkler M, Sporn P, Steinbereithner K (1990) Prediction of malignant hyperthermia susceptibility: statistical evaluation of clinical signs. Br J Anaesth 64:425–429
6. Nakao M, Morio M, Mukaida K, Yuge O (1990) Incidence of fulminant malignant hyperthermia in Japan. MH Workshop, Munich

The Epidemiology of Malignant Hyperthermia Events in North America

Marilyn Green Larach[1,2], Linda J. Fuhrmann[1,2], and Gregory C. Allen[1,2]

Introduction

This chapter presents preliminary data drawn from the North American Malignant Hyperthermia Registry concerning the epidemiology of malignant hyperthermia (MH) events in North America.

The North American Malignant Hyperthermia Registry (the Registry) is located in Hershey, PA (USA) at the Pennsylvania State University College of Medicine. The Registry receives standardized report forms from (a) health care providers in the field who are caring for patients at the time of their anesthetic event and (b) the MH biopsy center directors at the time a patient is undergoing biopsy. The standardized report forms are entered into our computerized database to support research studies such as the epidemiology research presented in this chapter. The Registry gathers information on a population of more than 275 000 000 individuals. The data presented are drawn from more than 6 years of Registry data collection.

The Registry currently has enrolled more than 1500 individuals. In 476 cases, 2 or more individuals are enrolled from the same family. The Registry has data from each of the 50 U.S. states and most of the Canadian provinces. Because more than one report has been received from some individuals, nearly 2500 individual report forms are currently entered into the Registry database.

Methods

To perform this study, we selected only those data forms that reported an adverse metabolic or musculoskeletal reaction to anesthesia. There were 811 such reports. We next applied the MH Clinical Grading Scale to rank the likelihood that the anesthetic event was an MH event. This clinical grading scale was developed in a consensual manner by an international panel of 11 experts. [1]. We used a modified clinical grading scale such that points were assigned for abnormalities in blood gas data only if they occurred in the presence of adequate systemic circulation. We ranked anesthetic events with a rank from "1" or "almost never" MH to a rank "5" or "very likely" MH to a rank "6" or "almost certain" MH. We studied only those anesthetic reactions with a rank of 5 or greater, 178 MH events.

[1] North American Malignant Hyperthermia Registry, Hershey, PA 17033, USA
[2] Penn State University College of Medicine, Hershey, PA 17033, USA

For all patients experiencing an anesthetic event with a MH likelihood of 5 or greater, we analyzed the demographics, presence or absence of congenital musculoskeletal abnormalities, prior anesthetic history, type of surgical procedure, case urgency, anesthetic agents used, and whether the patient survived the anesthetic event. For all patients undergoing caffeine halothane contracture testing (CHCT), we analyzed the biopsy outcome as determined by either the MH diagnostic biopsy center or by the proposed Registry protocol for modified contracture thresholds. With this proposed North American protocol, a 3% halothane contracture of 0.5 g or more or a 2-m*M* caffeine contracture of 0.3 g or greater would be considered a positive CHCT outcome. These proposed thresholds would yield a diagnostic test with 100% sensitivity (one-sided confidence interval of 87.3%) and 79% specificity (one-sided confidence interval of 71%) [2].

Results

In our results, 22% of our patients ($n = 178$) had an MH rank of 5 or higher, and 42% were aged 15 years or younger. This percentage is very different from the 5% of patients of this age undergoing an inpatient surgical procedure in the United States [3]. Also, 72% of our patients were male, again different from the 39% of patients that are males undergoing inpatient surgical procedures in the United States [3]. That 84% of our patients were Caucasian does not differ from the 84% Caucasians found in our U.S. population [4]. There were 22% of patients with a musculoskeletal abnormality (including scoliosis, hernia, strabismus, ptosis, pectus, clubbed foot, and other muscle tone abnormalities), and 43% of patients had experienced at least one prior unremarkable general anesthetic.

Twenty-five percent of patients were undergoing ENT (ears, nose, and throat) surgery, which again differs from the 2% of patients undergoing such surgery in the United States on an inpatient basis [3]. Twenty-five percent of patients were scheduled to undergo orthopedic surgery, differing from 7.3% of the surgical population [3]. Twenty-four percent were undergoing emergency surgery; although we have no data with which to compare, we believe that this percentage of patients undergoing emergency surgery is also different from that found in the general U.S. surgical population.

The anesthetics used before that MH event are contained in Table 1. Of note are the 2% of patients in whom no volatile anesthetic or succinylcholine use was reported.

We found a 1% death rate among patients who did not have a CHCT. Forty-seven percent of rank "5" and "6" patients did undergo CHCT; 67% of rank "6" or "almost

TABLE 1. Use of anesthetic agents before the development of an MH event.

Percent of patients	Agent used (+) or not (−)	
64%	+ Volatile anesthetic	+ Succinylcholine
21%	+ Volatile anesthetic	− Succinylcholine
6%	− Volatile anesthetic	+ Succinylcholine
2%	− Volatile anesthetic	− Succinylcholine
5%	Unreported	

certain" MH patients had their MH susceptibility confirmed by the individual MH diagnostic center; and 76% of rank "6" patients would have had their MH susceptibility confirmed if the modified North American CHCT protocol with new contracture thresholds had been applied. It remains to be determined whether the clinical grading scale or the CHCT is misclassifying the MH susceptibility of these patients. We do know that no clinical test, including the North American and the European contracture tests, can perfectly diagnose a patient population.

Conclusions

We found that those patients who experienced a "very likely" or "almost certain" MH event were disproportionately young and male when compared to the U.S. overall inpatient surgical procedure population. The MH event occurred frequently in association with ENT and orthopedic surgical procedures. Our study confirms the initial observations of Britt and Kalow [5] in 1970, although we add the frequent association of MH events with ENT surgery. The reported mortality of MH events in North America is low, at 1% of rank 5 and rank 6 MH events. To calculate incidence figures for MH events in North America and to determine whether or not our MH population is different from the U.S. surgical population, the United States needs to collect data on the number and demographic characteristics of patients having anesthesia. Also, we need to know the anesthetic techniques and agents used to anesthetize U.S. patients.

This study demonstrates that the expression of malignant hyperthermia susceptibility requires more than a genetic susceptibility, which is thought to be inherited in an autosomal dominant fashion and should be present for the lifetime of the individual, and more than a simple exposure to triggering anesthetic agents. There are additional unknown triggering factors. There factors pose a challenge for our colleagues who are investigating the biochemical and molecular genetic basis of MH.

Acknowledgment. The authors thank the health care providers and MH diagnostic center directors who submitted the data that made this study possible.

References

1. Larach MG, Localio AR, Allen GC, Denborough MA, Ellis FR, Gronert GA, Kaplan RF, Muldoon SM, Nelson TE, Ørding H, Rosenberg H, Waud BE, Wedel DJ (1994) A clinical grading scale to predict malignant hyperthermia susceptibility. Anesthesiology 80:771–779
2. Larach MG, Landis JR, Shirk SJ, Diaz M (1992) Prediction of malignant hyperthermia susceptibility in man: improving sensitivity of the caffeine halothane contracture test. Anesthesiology 77:A1052
3. Graves EJ (1991) Detailed diagnoses and procedures. National Hospital Discharge Survey. The National Center for Health Statistics. Vital and Health Statistics 13(115)
4. U.S. Bureau of the Census (1993) Statistical abstract of the U.S., 113th edn. U.S. Bureau of the census, Washington, DC
5. Britt BA, Kalow W (1970) Malignant hyperthermia: a statistical review. Can Anesth Soc J 17:293–315

Clinical Classification and Incidence of Malignant Hyperthermia in Australia

Michael A. Denborough

Classification

We classify malignant hyperthermia (MH) into the three myopathies that predispose to the anesthetic complication. These are as follow.

1. The Evans myopathy, which is called after the family in which MH was first described [1], is by far the most common myopathy to predispose to MH and is inherited as a Mendelian dominant characteristic. The myopathy is usually subclinical, although in later life some affected individuals may show mild muscle wasting, particularly in the lower parts of the thighs [2]. The serum creatine kinase (CK) may be elevated in affected individuals but is often normal. The muscle is usually normal on histochemistry and electron microscopy. Minor, variable, and nonspecific myostructural changes may sometimes be found [3].

2. Central-core disease [4] is an uncommon myopathy that is almost invariably associated with susceptibility to MH and is usually inherited as a Mendelian dominant characteristic. Histochemical and electron microscopic examinations of muscle in affected individuals show striking nonstaining lesions extending along type 1 fibers, and there is often type 1 atrophy. All members of families in which a diagnosis of central-core disease has been made should be regarded as being susceptible to MH until proved otherwise. It seems that the histological changes are a late manife station of central-core disease, because younger individuals in these families who are shown by pharmacological tests to be susceptible to MH may have normal muscle histology.

3. The King-Denborough syndrome [5] is much less common than the Evans myopathy and is found in children, usually boys. In addition to the myopathy, the boys are small for their age and have undescended testes, lumbar lordosis, thoracic kyphosis, and pectus carinatum. The children also have an unusual facial appearance with a small chin, low-set ears, ptosis, and antimongoloid obliquity of the palpebral fissures. This myopathy is probably inherited as a recessive trait.

Division of Biochemistry and Molecular Biology, The John Curtin School of Medical Research, Australian National University, Canberra, ACT 2601, Australia

Incidence

MH is said to occur in 1 in 10 000 to 1 in 50 000 anesthesias [6], but the incidence of the muscle disorders that predispose to MH is probably higher than this. There are two reasons for this. The first is that triggering agents do not always induce MH in susceptible individuals. One patient with clear-cut pharmacological evidence of susceptibility to MH developed the clinical syndrome only on her twelfth exposure to succinylcholine and halothane. The second reason is that MH may be triggered not only by anesthetics, but also by a variety of other mechanisms.

Exercise in Hot Conditions

Vigorous exercise in hot conditions may precipitate MH in susceptible individuals.

Case 1

A 19-year-old male army officer trainee developed heatstroke on 14 April 1980 during a route march [7]. He was unconscious and unresponsive to painful stimuli, with a rectal temperature of 42°C. Investigations showed a metabolic acidosis, myoglobinuria, and disseminated intravascular coagulation. He was given intravenous dantrolene and improved markedly within an hour. His condition gradually returned to normal during the next few days. His serum CK reached a peak of 20 000 IU/l on the second day (normal level, <150 IU/l). Two months later, when he had recovered completely and his serum CK was 36 IU/l, an in vitro muscle test showed that he was susceptible to MH; the 3% halothane contracture was 0.9 g and the 2 m*M* caffeine contracture was 0.5 g. It was realized later, when a detailed family tree was drawn up, that in another branch of his family several members had been shown to be susceptible to MH.

Case 2

A 36-year-old man developed troublesome muscle pains after playing squash and drinking moderate amounts of alcohol [8]. His serum CK was consistently elevated between 500 IU/l and 1600 IU/l. In vitro muscle testing on 21 June 1982 showed that he was susceptible to MH; the 3% halothane contracture was 0.5 g and the 2 m*M* caffeine contracture was 0.6 g. The father and one sister of the patient were also found to be susceptible to MH.

Case 3

During 3 years in the Army, a 22-year-old male soldier experienced about 20 attacks of headaches, lightheadedness, nausea, chest pain, and fatigue after exertion. He also had aching in the lower limbs below the knee that were thought to be shin splints. The symptoms were associated with exercise in hot conditions, such as 3- to 4-km runs or aerobic exercise, and he would feel unwell for about 2–3 days. He was put on desk duty until he recovered. After one of these episodes his serum CK was found to be 18 600 IU/l on the fourth day after exercise. Cardiological testing was normal. In vitro muscle testing on 31 July 1990 showed that he was susceptible to MH; the 3% halothane contracture was 0.5 g, and the 2 m*M* caffeine contracture was 0.5 g. His father was also shown to be susceptible to MH.

Case 4

A 19-year-old male soldier developed heatstroke in Canberra during a combat exercise in May 1992. A punch muscle biopsy was normal but his serum CK was persistently elevated to about 10 000 IU/l. In vitro muscle testing on 13 October 1992 showed that he was susceptible to MH; the 3% halothane contracture was 3.2 g, and the 2 m*M* caffeine contracture was 1.5 g. The muscle started contracting at 0.5 m*M* caffeine.

Case 5

A 23-year-old male professional sprinter developed rhabomyolysis (serum CK, 1500 IU/l) and mild renal failure after training for the 200-m sprint. He had had a similar episode of mild renal failure and rhabdomyolysis 2 years previously after hard physical exercise as a laborer. In vitro muscle testing on 29 March 1994 showed that he was susceptible to MH; the 3% halothane contracture was 6.1 g, and the 2 m*M* caffeine contracture was 0.3 g.

Infections

Case 1

A 32-year-old man developed a "flu-like" illness in 1989 with headache, fever, vomiting, and muscle aches. He developed severe rhabdomyolysis (serum CK, 243 000 IU/l) and died in the intensive care ward in hospital 1 week after his illness began [9]. The only significant finding in the septic screens that were carried out was a fourfold rise in the antibody titer to influenza B virus. His father had died in Pietermaritzburg from a similar illness at the age of 33. The son of the patient had a similar illness at the same time and also had rhabdomyolysis (serum CK, 3550 IU/l), but he recovered completely in a few days. In vitro muscle testing in the son on 2 November 1993 showed that he was susceptible to MH; the 3% halothane contracture was 1.15 g, the muscle contracted at 1 m*M* caffeine, and the 2 m*M* caffeine contracture was 1 g.

It seems likely that the son experienced the attack of viral rhabdomylosis because he is susceptible to MH. It seems very likely also that the fatal rhabdomyolysis in his father, and perhaps in his grandfather, occurred because they were also susceptible to MH.

The Neuroleptic Malignant Syndrome (NMS)

Neuroleptic drugs can also induce MH in individuals predisposed to MH.

Case 1

A 31-year-old man who had been treated with fluphenazine by mouth and intramuscularly became nauseated and drowsy, had a temperature of 38.5°C, and had a grand mal fit on 29 July 1983. His serum CK was 53 000 IU/l, and his urine contained myoglobin. Two months later, when he was fully recovered from this episode and his serum CK was normal (87 IU/l), in vitro muscle testing showed that he was susceptible to MH. The 3% halothane contracture was 0.7 g, and the 2 m*M* caffeine contracture was 0.4 g [8]. His brother was also susceptible to MH. Neuroleptic drugs increase the release of Ca^{2+} from the sarcoplasmic reticulum (SR), diminish its uptake, and inhibit ATPase, which regulates the uptake of Ca^{2+} into the SR[10].

The Sudden Infant Death Syndrome (SIDS)

The MH myopathy also predisposes to SIDS. Six of 16 parents of SIDS children in Canberra were susceptible to MH [11], and a significant excess of anesthetic deaths was found in SIDS families in the United States [12]. We have recently extended these observations and have shown that the link between MH and SIDS occurs through overheating [13]. Four MH-susceptible piglets (diagnosed pharmacologically and by DNA testing) and four MH-negative piglets were heated in a neonatal incubator in sleeping bags. The temperature of the incubator was gradually increased from room temperature to 39°–40°C and then maintained at this temperature for another 30 min. Two of the MH-susceptible piglets died, one after 1 h and 12 min and the other after 1 h and 27 min. The other two MH-susceptible piglets appeared stressed after the procedure and took about 30 min to recover completely. All four control pigs survived and recovered quickly, within 10–15 min.

Miscellaneous

Case 1

A 65-year-old man had pain in his lower limbs on exertion. The pulses in his legs were normal. A microtubular myopathy was suspected, and a muscle biopsy was requested. Pharmacological testing of the muscle at the time of the biopsy on 1 June 1993 showed that he was susceptible to MH. The 3% halothane contracture was 1.5 g and the 2 m*M* caffeine contracture was 0.5 g. The muscle started contracting at 1 m*M* caffeine. A 21-year-old son who had the Sturge–Weber syndrome was also susceptible to MH.

Case 2

A 31-year-old man developed severe pain in his thighs while resting at home on a Sunday afternoon. No predisposing factors were identified. On arrival at hospital his serum CK was 8000 IU/l. He gave a past history of episodes of pain in his thighs after exertion, about two to three times per year, and had had a severe episode when he cycled to Queensland. On that occasion the pain was so severe that he had to put his bicycle on the train and return to Canberra by this means himself. In vitro muscle testing on 23 February 1993 showed that he was susceptible to MH; the 3% halothane contracture was 2.5 g and the 2 m*M* caffeine contracture was 0.85 g. The muscle started contracting at 1 m*M* caffeine.

Environmental and Occupational Factors: Case 3

In June 1985 the 45-year-old father of a 12-year-old girl who had survived an episode of MH was found to have a serum CK of 650 IU/l [14]. In vitro muscle testing on 2 July showed that he too was susceptible to MH; the 3% halothane contracture was 9.1 g, and the 2 m*M* contracture was 3.35 g. The muscle started contracting at 0.25 m*M* caffeine. In 1987 he was working in a factory that made fire extinguishers where one of his tasks was to discharge bromochlorodifluoromethane (BCF) from the extinguishers before filling them. This work was done in the open air but it was hard to avoid inhaling some of the gas. He complained of malaise and of stiffness and weakness in his forearms and hands during the 18 months that he had been in the job. The symptoms worsened during the week, being worst on Fridays, and improved at weekends. Serum CK activity was 1056 IU/l on one Saturday in March 1987 and 544 IU/l on the following Monday. Physical examination showed no abnormality. Because of the structural similarity between BCF and halothane, the effect of BCF was studied on muscle contracture in vitro and was found to be identical to halothane. The patient was advised to change his job, which he did, and his symptoms immediately improved. It seems as though this MH-susceptible patient had recurrent episodes of rhabdomyolysis from exposure to BCF.

Summary

MH is classified by us into the three myopathies that predispose to the anesthetic complication. These are Evans myopathy, central-core disease, and the King-Denborough syndrome.

MH is said to occur in 1 in 10000 to 1 in 50000 anesthesias, but the incidence of the muscle disorders that predispose to MH is probably much higher than this. Triggering agents do not always induce MH in susceptible individuals, and MH may be triggered not only by anesthetics but also by a variety of other mechanisms. These include vigorous exercise in hot conditions, alcohol, neuroleptic drugs, overheating in infants, and occupational drugs with a structure similar to halothane. MH may also occasionally occur spontaneously.

References

1. Denborough MA, Lovell RRH (1960) Anaesthetic deaths in a family. Lancet 2:45
2. Denborough MA, Ebeling P, King JO, Zapf P (1970) Myopathy and malignant hyperpyrexia. Lancet 1:1138–1140
3. Harriman DGF, Sumner DW, Ellis FR (1973) Malignant hyperthermia myopathy. Q J Med 42:639–664
4. Denborough MA, Dennett X, Anderson RMcD (1973) Central-core disease and malignant hyperpyrexia. Br Med J 1:272–273
5. King JO, Denborough MA (1973) Anesthetic-induced malignant hyperpyrexia in children. J Pediatr 83:37–40
6. Rosenberg H (1982) The clinical syndrome of malignant hyperpyrexia. In: Proceedings of the Third International Workshop on Malignant Hyperpyrexia. Banff, Alberta, Canada, p 3
7. Denborough MA (1982) Heat stroke and malignant hyperpyrexia. Med J Aust 1:204–205
8. Denborough MA, Collins SP, Hopkinson KC (1984) Rhabdomyolysis and malignant hyperpyrexia. Br Med J 288:1878
9. Denborough MA, McLean A, Morgan G, Hopkinson KC (1994) Fatal inherited rhabdomyolysis and malignant hyperthermia. Lancet 343:236–237
10. Collins SP, White MD, Denborough MA (1988) Calmodulin antagonist drugs and malignant hyperpyrexia. Clin Exp Pharmacol Physiol 15:473–477
11. Denborough MA, Galloway GJ, Hopkinson KC (1982) Malignant hyperpyrexia and sudden infant death. Lancet 2:1068–1069
12. Peterson DR, Davis N (1986) Sudden infant death syndrome and malignant hyperthermia diathesis. Aust Paediatr J 22(Suppl):33–35
13. Denborough MA, Hopkinson KC (1994) Death caused by overheating in piglets susceptible to malignant hyperthermia. Med J Aust 160:731–732
14. Denborough MA, Hopkinson KC (1988) Firefighting and malignant hyperthermia. Br Med J 296:1442–1443

Clinical Classification and Incidence of Malignant Hyperthermia in Japan

OSAFUMI YUGE[1], MICHIO MORIO[2], HIROSATO KIKUCHI[3], KEIKO MUKAIDA[1], YASUHIRO MAEHARA[1], MASAKAZU NAKAO[1], and MASASHI KAWAMOTO[1]

Introduction

Denborough and Lovell [1] first described malignant hyperthermia (MH) as an inherited syndrome in 1960. At present, it is generally accepted that MH is triggered by many anesthetics. Succinylcholine chloride (SCC) and volatile anesthetics have been especially implicated as important triggering drugs [2,3]. With these triggering drugs, induced hypermetabolism produces tachycardia, increased O_2 consumption and CO_2 production, premature ventricular contraction, hypotension and hypertension, cyanosis, tachypnea, muscle rigidity, and hyperthermia as the signs of MH. Also seen as complications of MH are electrolyte imbalances, myoglobinuria, hyperkalemia, creatine phosphokinase (CPK) elevation, impaired coagulation, renal failure, and severe metabolic and respiratory acidosis.

The incidence of MH has been estimated to be approximately 1 in 14 000 to 250 000 general anesthesia events [4–6]. The purpose of this study is to investigate the reporting patterns of MH cases in Japan, to classify these published cases to fulminant MH (f-MH) and abortive MH (a-MH), following the MH criteria of Morio et al. [7], to identify episode-triggering drugs and surgical procedures, and to review the clinical classification and incidence of MH in Japan.

Methods: 1

The retrospective analysis of published MH cases in Japan covers 1961 to 1994. These reported MH cases have been classified as f-MH and a-MH following the criteria of Morio et al. [7] (Table 1).

The following data were collected for each case: age, sex, survival, types of surgery, clinical symptoms and signs of MH, anesthetics and other drugs used, and laboratory findings (blood gas analysis, end-tidal CO_2 [ET-CO_2], serum potassium, and serum myoglobin).

The comparative analysis has utilized laboratory findings and clinical signs between dead and surviving patients in f-MH and between f-MH and a-MH. In these

[1] Department of Anesthesiology and Critical Care Medicine, Hiroshima University School of Medicine, Hiroshima, 734 Japan
[2] Chugoku Rosai General Hospital, Kure, Hiroshima, 737-01 Japan
[3] First Department of Anesthesiology, Toho University School of Medicine, Tokyo, 143 Japan

TABLE 1. Criteria of Morio for malignant hyperthermia (MH) in Japan[a].

A. Temperature increase
1. Maximum body temperature ≥40°C
2. 40°C > Maximum body temperature ≥38°C and increasing rate of body temperature ≥0.5°C/15 min.
B. Clinical presentations of MH except for temperature, tachycardia, arrhythmia, respiratory or metabolic acidosis, muscle rigidity, myoglobinuria, etc.

[a]Fulminant MH (f-MH), criteria A and B; abortive MH (a-MH), criteria B only.

comparative studies, Mann-Whitney's U-test or the chi square test were used at a significant difference of $P < .05$.

Methods: 2

In this study, the Ca-induced Ca release (CICR) rate was measured in 31 f-MH and 33 a-MH patients by the skinned fiber method [8].

Results

The number of f-MH cases in Japan during the past 34 years (1961–1994) is 300. The overall mortality rate of f-MH cases was 32.7%, but in the 5 years from 1991 to 1994 it decreased to approximately 12.5% (Fig. 1).

There were 73 cases of a-MH in Japan from 1987 to 1994. All cases of a-MH survived.

In f-MH cases, the male-to-female ratio was 3.3:1. The mortality rate in male patients was slightly higher than in female (male, 35.0%; female, 24.6%) (Fig. 2). In f-MH cases, the most frequently used anesthetic was halothane (196 cases, 61.4%) and the second most frequent was enflurane (67 cases, 21%) (Fig. 3). Several cases have been reported to occur under the effects of isoflurane (21 cases, 6.6%) or sevoflurane (11 cases, 3.5%). SCC was the neuromuscular relaxant most commonly used (235 cases, 78.2%) (Fig. 3).

Most cases of f-MH occurred during musculoskeletal (86 cases, 28.6%) and abdominal surgery (70 cases, 23.3%), followed by otolaryngeal (ENT) surgery (41 cases, 13.7%), nervous system surgery (28 cases, 9.3%), cardiovascular surgery (14 cases, 4.7%), and eye surgery (13 cases, 4.3%) (Fig. 4).

The Clinical sign that was found in most f-MH cases was unexplained tachycardia (96%), followed by muscle rigidity (65.1%), arrhythmia (73.2%), myoglobinuria (67.8%), and masseter spasm (49.8%) (Fig. 5).

There were significant differences between dead and surviving patients in 300 f-MH cases in maximum body temperature, increasing rate of body temperature, level of pH, base excess, serum potassium, and serum myoglobin (Table 2).

There were also significant differences in maximum body temperature, increasing rate of body temperature, level of $PaCO_2$, pH, base excess (BE), serum potassium, serum myoglobin, and incidence of muscle rigidity between a-MH and f-MH cases (Table 3).

In 31 f-MH patients, Ca-induced calcium release (CICR) rates were measured (Table 4). The rate of CICR was accelerated in 24 patients; however, the rate of CICR

TABLE 2. Clinical findings in dead and surviving f-MH patients.

Value:		
Age	ns	
Maximum body temperature	$P < .001$	Died > survived
Increasing rate of body temperature	$P < .01$	Died > survived
$Paco_2$	ns	
pH	$P < .05$	Died < survived
Base excess	$P < .001$	Died < survived
Serum potassium	$P < .001$	Died > survived
Serum CPK	ns	
Serum myoglobin	$P < .05$	Died > survived
Urine myoglobin	ns	
Incidence:		
Muscle rigidity	ns	
Masseter spasm	ns	
Cola-colored urine	ns	

ns, not significant; CK, creatine kinase.

TABLE 3. Clinical findings in f-MH and a-MH.

Value:		
Age	$P < .01$	f-MH > a-MH
Maximum body temperature	$P < .001$	f-MH > a-MH
Increasing rate of body temperature	$P < .001$	f-MH > a-MH
$Paco_2$	$P < .001$	f-MH > a-MH
pH	$P < .001$	f-MH < a-MH
Base excess	$P < .001$	f-MH < a-MH
Serum potassium	$P < .01$	f-MH > a-MH
Serum CPK	ns	
Serum myoglobin	$P < .05$	f-MH > a-MH
Urine myoglobin	ns	
Incidence:		
Muscle rigidity	$P < 0.001$	f-MH > a-MH
Masseter spasm	ns	
Cola-colored urine	ns	

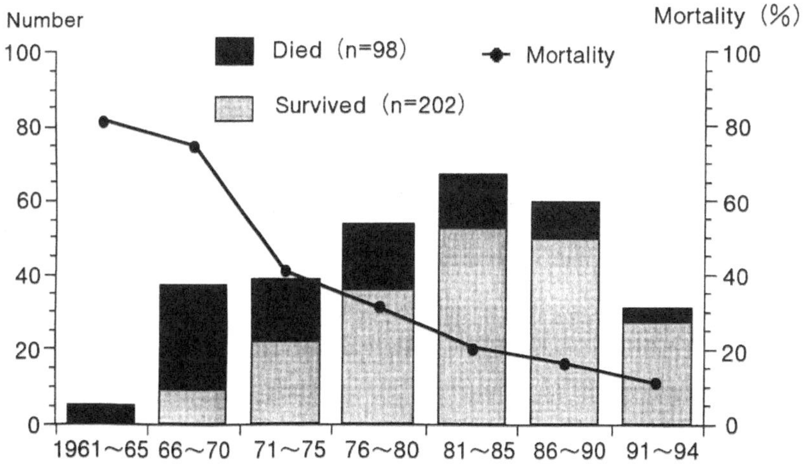

FIG. 1. Distribution of fulminant malignant hyperthermia (f-MH) patients during years from

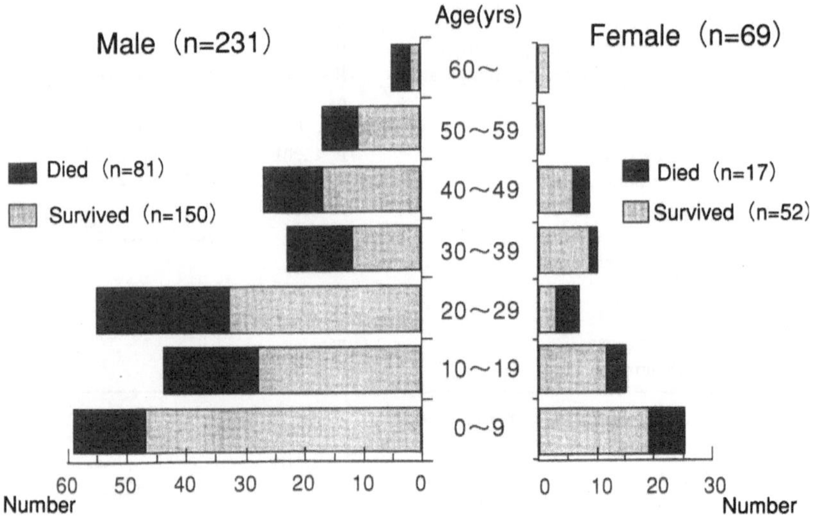

FIG. 2. Sex and age distribution in f-MH patients during years from 1961 to 1994 in Japan

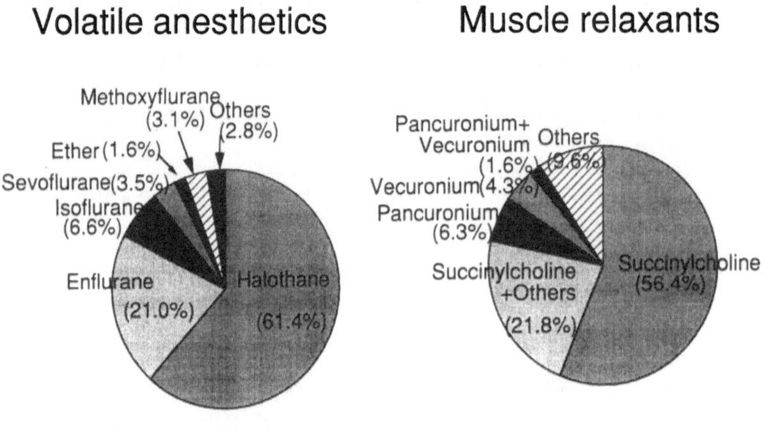

FIG. 3. Agents used for anesthesia in 300 f-MH cases in Japan

TABLE 4. Clinical classification relative to rate of Ca-induced Ca release (CICR).

	Rate of CICR		Total
	Accelerated	Unaccelerated	
f-MH	24	7	31
a-MH	3	30	33

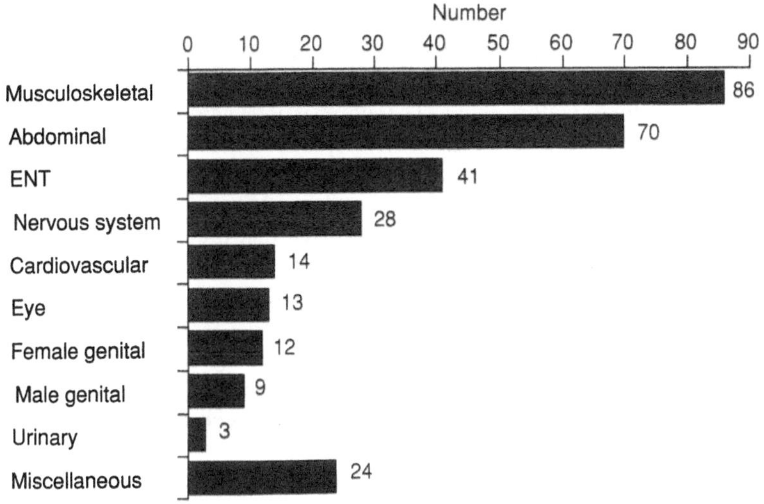

FIG. 4. Types of surgery for 300 f-MH patients. *ENT*, ear-nose-thoracic surgery

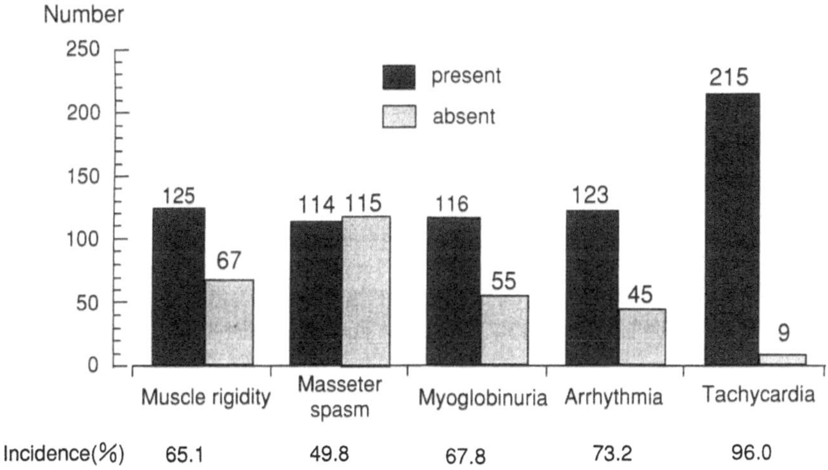

FIG. 5. Incidence of clinical signs in 300 f-MH patients

in the other 7 patients was not accelerated. Simultaneously, 33 a-MH patients were checked with the CICR test; the CICR rate was accelerated in 3 patients and unaccelerated in 30 patients.

Discussion

This retrospective study of MH in Japan reviews published MH cases in Japan from 1961 to 1994. Nakao et al. [9] previously reported that the incidence of f-MH in Japan was 1 in 64 000 general anesthesia events and that the male-to-female ratio was 3.6:1

from 1976 to 1986 in the 5th Workshop on MH and the 6th Myology Calloquim. In the period from 1987 to 1991, the expected incidence of f-MH was 1 per 74 000 general anesthesias in Japan. The demography in these reports is similar to populations in studies by other investigators [4–6]. Therefore, it is suggested that a reasonably representative cross section of the entire MH population has been captured in Japan. In this study, we reported that most f-MH patients (male-to-female ratio = 3.3:1) were male and also 70% of them were younger than 30 years old.

In a recent review, Strazis and Fox [10] reported that the male/female ratio was 2.2:1, that pediatric patients comprised nearly half of MH cases, and that few MH patients were more than 33 years old (both males and females) from the analysis of 503 published MH cases.

In the previous study, Nakao et al. [9] reported that the age distribution of MH was quite different from that of the normal general anesthesia group in Japan (Fig. 6). From this result, we can say that at least in Japan, MH has occurred more frequently in male patients than in female patients.

There are several speculations [11] for the linkage of gender and MH: (1) a higher number of anesthetic episodes in male patients; (2) a higher total muscle volume, which might facilitate triggering of the reaction; (3) possible hormone protection of females; and (4) males might be more susceptible because of sexual hormones. However, it is currently unclear whether there is an obvious linkage of sex and MH.

Previous studies [4,12] have reported orthopedic surgery to comprise about 20% of MH cases. In this study, orthopedic surgery, including musculoskeletal surgery, was related to 28.6% of MH cases. It is suggested that the excess probability of MH in musculoskeletal surgery, otopharyngeal surgery, or eye surgery is usually at least as great for the young adult group as in children [10].

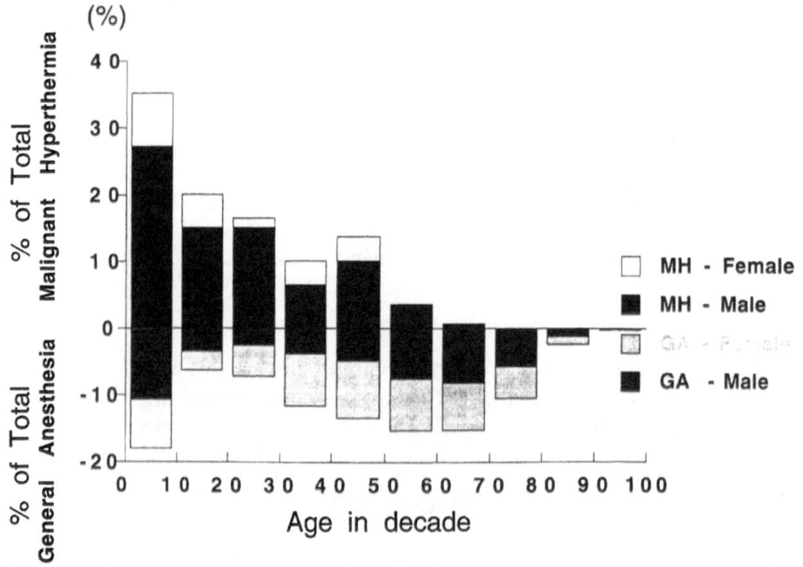

FIG. 6. Age and sex distribution of general anesthesia (*GA*) and malignant hyperthermia (*MH*) cases in Japan in 1976–1987

Halothane and SCC were widely used for induction of anesthesia, especially in children, and this tendency may explain the high incidence of pediatric MH patients in this study. However, in Japan, halothane and SCC have gradually been used less frequently in recent years. In contrast, isoflurane and sevoflurane frequently are used for volatile anesthetics in the clinical field; sevoflurane is used especially for pediatric surgery. Therefore, MH should increase with the use of both of these in Japan.

It is generally accepted that the whole muscle caffeine halothane contracture test is the gold standard as a diagnostic and screening test of MH susceptibility [13,14]. To obtain more definitive information about MH susceptibility, the CICR rate is commonly measured in Japan using single chemically skinned fibers [15,16]. Endo et al [17] reported that the sarcoplasmic reticulum (SR) of MH patients exhibited a sensitivity to Ca 1.8-fold greater and also had a 2-fold higher peak rate of CICR than that of normal patients, whereas the uptake of Ca into the SR and contractile system function were normal. Therefore, we can say it is worthwhile to investigate the function of the SR of human skeletal muscle, especially by direct measurement of the CICR rate, because it provides more definitive information concerning MH susceptibility. From the results of the CICR test for f-MH and a-MH patients in this study, it might be said that disorder in the Ca-release mechanism from SR in skeletal muscle might be one of the main causes of MH.

MH has recently been considered as a molecular disease [18,19]. It is suggested that the abnormality of the skeletal muscle's ryanodine receptor (RYR1) accelerates the CICR rate. Consequently, measurement of the CICR rate might be the best method to detect the abnormality of RYR1. It should be considered, however, that another mechanism might affect disorder of the CICR system because 20% of f-MH patients muscles did not show any accelerated changes with the CICR test. Almost 9% of a-MH patients muscles, however, showed the accelerated changes in the CICR test.

Furthermore, we should consider the renewed clinical criteria of MH including scoring the level of pH, $Pa{CO_2}$, BE, and end-tidal CO_2 (ET-CO_2), which have been shown by Larach et al. [20] as the clinical grading scale of MH.

Conclusion

In conclusion, this study showed the distribution of 300 f-MH cases in Japan from 1961 to 1994 and typical patterns for age, sex, triggering agents, type of surgery, clinical signs, and laboratory findings. We suggest that it is very important to do such a retrospective case study and that the CICR rate measurement is useful to identify MH susceptibility.

References

1. Denborough MA, Lovell RRH (1960) Anesthetic deaths in a family. Lancet 2:45
2. Gronert GA (1980) Malignant hyperthermia. Anesthesiology 53:395–423
3. Kaplan RF (1991) Malignant hyperthermia. In: 42th annual reference course lectures the American Society of Anesthiologists (ASA), p 2311
4. Britt BA, Kalow W (1970) Malignant hyperthermia: statistical review. Can Anesth Soc J 17:293–315
5. Ellis FR, Halsall PJ (1980) Malignant hyperpyrexia. Br J Hosp Med 24:318–327
6. Ørding H (1985) Incidence of malignant hyperthermia in Denmark. Anesth Analg 64:700–704

7. Morio M, Kikuchi H, Yuge O, et al (1987) Incidence of malignant Hyperthermia. In: Proceedings, 7th European congress of anesthesiologists, vol III, pp 82–88
8. Kikuchi H, Matsui I, Morio M (1987) Diagnosis of malignant hyperthermia in Japan by the skinned fiber test. In: Britt BA (ed) Malignant hyperthermia. Nijhoff, Dordrecht, pp 279–294
9. Nakao M, Mukaida K, Kubota M, et al (1991) Variable incidences of fulminant malignant hyperthermia in Japan. In: Proceedings, 14th Japan MH Sympo., pp 83–85
10. Strazis KP, Fox AW (1993) Malignant hyperthermia: a review of published cases. Anesth Analg, 77:297–304
11. Heilinger D, Donner E, Sauberer A, et al (1994) Malignant hyperthermia in Austria 1984–1993. In: 3rd international symposium on malignant hyperthermia, abstract 23
12. Kalow W, Britt BA, Terreau ME, et al (1970) Metabolic error of muscle metabolism after recovery from malignant hyperthermia. Lancet 2:895–898
13. Ellis FR, Harriman DGF, Keaney NP, et al (1971) Halothane-induced muscle contracture as a cause of hyperpyrexia. Br J Anaesth 43:721–722
14. Kalow W, Britt BA, Richter A (1977) The caffeine test of isolated human muscle in relation to malignant hyperthermia. Can Anaesth Soc J 24:678–694
15. Kawana Y, lino M, Horiuti K, et al (1992) Acceleration in calcium-induced calcium release in the biopsied muscle fibers from patients with malignant hyperthermia. Biomed Res 13(4):287–297
16. Matsui K, Fujioka Y, Mukaida K, et al (1989) Comparative study on in vitro diagnosis by the skinned fiber test and clinical diagnosis of malignant hyperthermia. Masui (Jpn J Anesthesiol) 38:195–201
17. Endo M, Yagi S, Ishizuka T, et al (1983) Changes in the Ca-induced Ca release mechanism in the sarcoplasmic reticulum of the muscle from a patient with malignant hyperthermia. Biomed Res 4:83–92
18. McCarthy TV, Healy JMS, Heffron JJA, et al (1990) Localization of the malignant hyperthermia susceptibility locus to human chromosome 19q12–13.2. Nature 343:562–564
19. Gillard EF, Otsu K, Fujii J, et al (1991) A substitution of cysteine for arginine 614 in the ryanodine receptor is potentially causative of human malignant hyperthermia. Genomics 11:751–755
20. Larach MG, Localio AR, Allen GC, et al (1994) A clinical grading scale to predict malignant hyperthermia susceptibility. Anesthesiology 80:771–779

Part 3. Muscular Testing of Malignant Hyperthermia

Testing for Malignant Hyperthermia in North America and Australia

Henry Rosenberg

There are 14 biopsy centers for malignant hyperthermia (MH) in the United States and Canada, and 3 in Australia. In recent years 3 biopsy centers in North America have closed for funding reasons or because the center director changed the emphasis of his/her activities.

The centers in North America have been utilizing, with only small deviations, the biopsy protocol, agreed upon through a series of workshops beginning about 1977. More formalized "biopsy workshops" began in 1987, and the last such workshop was held in 1994.

My information regarding Australian centers derives from Dr. Michael Denborough in Canberra. Centers directed by Dr. Hurse in Perth and Dr. Borton in Sydney follow the European protocol, while Dr. Denborough's protocol utilizes 3% halothane and incremental doses of caffeine. Each group retains its own database. Dr. Denborough's center conducts about 50 biopsies per year. A biopsy center was created in New Zealand several years ago by Dr. Neil Pollock (Table 1).

Methodology of the North American Protocol

The methodology of the protocol for the caffeine halothane contracture test (CHCT) has been well described in several sources, the most complete description being that of Dr.Larach in 1989 [1]. Basically, the test consists of excision of 2–4 g of vastus lateralis muscle (one center uses gracilis), under regional or general anesthesia. Patients should weigh more than 20 kg, preferably be older than 8 years, and not have experienced muscle trauma within 3 months. Testing should be completed within 5 h of biopsy using fascicles of 2–3 cm in length (the longer the better, in terms of repolarization of the cut fiber). After a period of stabilization and the performance of a length tension determination, the muscle is stimulated supramaximally at 0.1–0.2 Hz. Separate fascicles are then tested in triplicate to 3% halothane in the gas phase, which is monitored by gas chromatography, at 37°C. Three other bundles should be tested with incremental doses of caffeine (free base) of 0.5, 1, 2, 4, 8, and 32 m*M*. Each concentration is maintained until the maximum contracture plateau is reached, or for 4 min.

A positive response to halothane is at least 0.5 g, with some laboratories setting a

MCP/Hahnemann University, Philadelphia, PA 19102, USA

TABLE 1. Active muscle biopsy centers in North America, Australia, and New Zealand.

United States	Canada	Australia and New Zealand
Bethesda, MD	Toronto	Canberra
Chicago, IL	Ottowa	Perth
Los Angeles, CA	Winnipeg	Sydney
Philadelphia, PA		Palmerston North (NZ)
Rochester, MN		
Sacramento, CA		
Winston-Salem, NC		

Further information may be obtained from the Malignant Hyperthermia Association of the United States, 32 S. Main St., P.O. Box 1069, Sherburne, NY 13815.

contracture threshold of 0.7 g or greater. It is generally agreed that a contracture of 1 g or more is diagnostic in all laboratories.

A positive response to caffeine is a contracture greater than 0.2 g at 2 m*M* in any one of two or three strips tested. Other variations of the caffeine contracture measurement are the CSC (caffeine-specific concentration, that caffeine concentration which is associated with a 1-g contracture) and the percent of response at 2 m*M* compared to 32 m*M*. The latter two are more controversial. The cutoff of 4 m*M* CSC is clearly too high as 32% of negatives have such a response [1]. A 7% response has been suggested as the cutpoint for the latter test. If any bundle exhibits a positive response, then the patient is believed to be MH susceptible. There are no defined "equivocal" results.

The response to caffeine in the presence of 1% halothane is thought to be an optional test at present, without general agreement on the definition of threshold responses. There are about 60% false-positives with the halothane CSC (HCSC) of 1 m*M*, but a more acceptable incidence with an HCSC of 0.35 m*M*.

Ryanodine has only recently been incorporated into the protocol. The protocol established by the European MH group is being used. Results of the testing protocol are reported to the North American Malignant Hyperthermia Registry for data analysis.

Validation of the Test

Unlike most other tests in medicine, validation of testing for MH is extremely difficult because unequivocal MH-negative and MH-positive patients are difficult if not impossible to identify, even with a negative response to trigger agents. The approach that has been used therefore is to biopsy, for control muscle, patients without known predilection for MH based on their history who are undergoing surgical procedures. Patients believed to have a high likelihood of MH are defined on the basis of the grading scale for MH recently published [2].

The first publication concerning the specificity of the testing procedure revealed an acceptable level of less than 15% if 3% halothane was used and 2 m*M* caffeine was more than 0.2 g [3]. Subsequent studies revealed that a 3% halothane threshold of 0.5 g and a caffeine threshold of 0.3 g to 2 m*M* caffeine produced a sensitivity of 100% (no false-negatives) and a specificity of 78% [4]. Further, two studies show that more than 30 patients who were diagnosed as MH negative (MHN) did not trigger on exposure to trigger anesthetics.

Problem Areas

Despite the foregoing optimistic report, it is clear that this biological test is not without significant problems.

False Positives

First, pooled data have revealed a great deal of center-to-center variation in test results. For example, some centers report a 0% false-positive rate to 3% halothane of 0.7 g and to caffeine at 2 m*M* of more than 0.2 g while others show a 28% false-positive rate. The reasons for this variance may include patient selection, handling of tissue, lack of control of halothane concentration in the bath, presence of occult muscle disease (because most patients labeled as control are undergoing hip surgery, many may have disuse atrophy), and anesthetic regimen at the time of biopsy, among others. In addition, there is often variation in response to the agents among the muscle bundles from each patient.

False Negatives

Of greatest concern is the possibility of false-negative biopsies. Wedel and Nelson have reported four patients who displayed clinical signs compatible with MH and who tested MHN to halothane alone and caffeine alone, but displayed an abnormal HCSC response (<4 m*M*) [5]. One or another parent or sibling tested positive, however, to halothane alone or caffeine alone. Was this a false-negative biopsy response, or was the proband really MHN and the relatives falsely positive, as there is a 32% chance of a false positive? Given our information about variations in biological tests, it would not be surprising to find a few false negatives. Time does not permit a detailed exploration of these two alternatives.

Importance of Selection of Patient for Biopsy

It should also be recalled that sensitivity and specificity are not the only defining characteristics of a successful diagnostic test. Predictive values of the test need to be considered. For a test of low prevalence, such as MH, without careful selection criteria for biopsy the positive predictive value of a test will decrease, exaggerating the false-positive results. In my opinion, the reason that the reported frequency of MH susceptibility after succinylcholine-induced masseter rigidity is 50%, while the true incidence of MH signs after masseter rigidity is about 10% [6], relates to the application of the biopsy test to a population with a low incidence of MH. Another explanation may be that only patients with very exaggerated trismus are referred for biopsy, thereby skewing the results toward MH-susceptible (MHS) patients.

Further complicating the interpretation of the test results is the probability that patients with muscle diseases such as Duchenne muscular dystrophy (DMD), myotonia, and other muscle disorders will display an abnormal response to halothane and caffeine. Are these patients MH susceptible, or do they merely have an abnormal response to halothane and caffeine?

The Future

Resolution of the problems associated with the CHCT will depend on future research into the pathophysiology of MH. Until a complete understanding of the biochemical changes is obtained, it is likely that testing will be less than ideal, but then, what test is ideal?

Great advances in our understanding of the limitations and the meaning of the CHCT have been obtained by the willingness of diagnostic centers to freely share clinical and laboratory data and conform to a basic protocol for testing. Through this cooperation, we have defined many families with MH susceptibility that may serve as a reference standard.

A second promising approach is the addition of ryanodine to the testing protocol, which incidentally was suggested at the second international symposium on MH in 1977 [7]. Another approach is the sharing of biological material between testing centers that are close to one another.

Finally, it is hoped that joint testing by the North American and European protocols on biopsy material will clarify strengths and weaknesses of each approach. It is vital also that the naysayers of the contracture test not discourage those performing this test. The testing protocol has been invaluable in advancing our information about MH and has been of great benefit to many patients and their families.

References

1. Larach MG (1989) Standardization of the caffeine halothane muscle contracture test. Anesth Analg 69:511–515
2. Larach MH, Localio R, Allen GC, et al (1994) A clinical grading scale to predict malignant hyperthermia susceptibility. Anesthesiology 80:771
3. Larach MG, Landis JR, Bunn JS, et al (1992) Prediction of malignant hyperthermia susceptibility in low-risk subjects. An epidemiologic investigation of caffeine halothane contracture rsponses. Anesthesiology 76:16–24
4. Larach MG, Landis JR, Shirk SJ, et al (1992) Prediction of malignant hyperthermia susceptibility in man: improving sensitivity of the caffeine halothane contracture test. Anesthesiology 77:A1052
5. Wedel DJ, Nelson TE (1994) Malignant hyperthermia–diagnostic dilemma: False-negative contracture responses with halothane and caffeine alone. Anesth Analg 78:787–785
6. O'Flynn RP, Shutack JG, Rosenberg H, et al (1994) Masseter muscle rigidity and malignant hyperthermia susceptibility in pediatric patients. Anesthesiology 80:1228–1295
7. Wappler F, Roewer N, Lenzen C, et al (1994) High-purity ryanodine and 9,21-dehydroryanodine for in vitro diagnosis of malignant hyperthermia in man. Br J Anaesth 72:240–242

Further Reading

Fletcher JE (1994) Current laboratory methods for the diagnosis of malignant hyperthermia susceptibility. In: Levitt RC (ed) Temperature regulation during anesthesia. Anesthesiology clinics of North America, vol. 12. Saunders, Philadelphia, pp 553–570

Muscular Testing of Malignant Hyperthermia in Europe

F.R. Ellis

Until 1983, malignant hyperthermia (MH) screening was conducted in a limited number of European Centers in a fairly haphazard and ad hoc way. In 1983, the Banff International Workshop on MH ended in confusion as to how best to screen for MH. At that time various blood tests were still being used or suggested, such as creatine kinase (CK), serum pyrophosphate, osmotic red cell fragility, and various white cell tests. The muscle tests included in vitro adenosine triphosphate (ATP) depletion, intracellular Ca^{2+} release studies, potassium chloride contracture, succinylcholine contracture, and succinylcholine and caffeine contracture. Although all this intense activity was good for learning about the pathoetiology of MH, it was hopeless for diagnostic uniformity. Clearly a disease diagnosed in one center should be diagnosable in the same way as in others, and the testing method should produce the same results.

Following the symposium at Banff, foundation members of the European MH Group put together a protocol [1] that was debated, discussed, and finally accepted. During the first two meetings of the European MH Group in Lund and Leeds, not only the basis of the current in vitro contracture test (IVCT) protocol but also the MH-susceptible (MHS), MH-negative (MHN), and MH-equivocal (MHE) [2] classification of results were established. The current protocol is probably the best that can be achieved using the relatively inexpensive technology we have all had to adopt. Since then there have been a number of refinements to the protocol, especially in terms of quality control, such as measuring drug levels in the tissue bath and setting limits, proof of muscle viability and also of muscle performance, and the number of tests to be performed.

At a recent meeting, the European MH Group [EMHG] addressed the problem of the incidence of the various categories in the different laboratories. In particular, the incidence of the MHE category as a percentage of the total biopsies was examined. Differences between centers could reflect the type of patient referred and preselection criteria rather than technical laboratory differences. Our own preselection criteria are strict and include an anesthetic-induced MH-like episode, a patient aged more than 12 years, and no obvious cause for the reaction.

The recent interest in the potential DNA screening of MH has almost ironically put greater emphasis on the IVCT. If the IVCT is inaccurate, the interpretation of the DNA

University of Leeds, Academic Unit of Anaesthesia, St. James's University Hospital, Beckett St., Leeds LS9 7TF, UK findings and haplotypes will also be inaccurate. The dependence of DNA screening on statistics is such that even one incorrectly categorized MHS or MHN could render a whole pedigree useless.

It is important to recognize the limitations of IVCT. The test is expensive (approximatly £1000 each investigation), it is time consuming, requires special expertise and specially trained technicians, and requires a dedicated laboratory. Technical errors and mistakes are likely to produce false negatives because of failure of muscle viability. All centers have experienced difficulty in obtaining good control results from non-MH families who were biopsied in exactly the same way as the MH group.

There are some less obvious limitations of the IVCTs. First, too many positives are diagnosed. It is easy to overdiagnose MHS, and in a sense this is an accepted part of the protocol so as to avoid false negatives. False positives will result from setting the limits too crudely, by uncritically accepting all MHEs to be MH susceptible, by using poor samples, by using too many samples, and by broadening the scope of the tests, for example, including ryanodine.

There is a tendency to believe that the laboratory findings have absolute diagnostic meaning, but there is a real problem in that false positives may occur in the population more frequently than real positives. This was particularly a problem when CK was used as a screening test because an increased CK is quite a common occurrence. The statistics of uncommon events are complex but could be applied to MH screening. If preconditions are rigorously applied, this latter problem should be a small one. The abnormal results obtained in such a variety of muscle disease conditions underlines the potential crudity or perhaps lack of specificity of the IVCT result. We have to assume that the abnormal IVCT results represents a "final common pathway" for a variety of pathological states. The most exciting prospect for the IVCTs is the contribution they make toward the DNA-based genetics.

In 1990, MacLennan and colleagues [3] and McCarthy and colleagues [4] suggested the same genetic basis for MH. The candidate gene was identified on C19 and was in close association with, or identical with, the RYR gene. The RYR as the site for the MH abnormality is attractive as it allows a rational explanation of the pathoetiology of MH in terms of calcium homeostasis.

It was not long before the new technology was proposed as a screening device, and in 1991 Healy et al. [5] argued that with certain provisos, in large families in which there were no apparent recombinants for the RYRI mutation, the presence of the MH haplotype could be used to infer MHS and the absence could be used to infer MHN. However, subsequent events have seriously challenged their arguments. First, members of the genetic section of the EMHG demonstrated that more than 50% of the European families do not show linkage with the RYRI. Also, in 1992 Levitt and colleagues [6] showed that MH cosegregated with a gene on C17q. This was an important paper as it showed positively that MH was likely to be a heterozygous condition.

Since then there have been an increasing number of papers [7–10] confirming the heterogenous nature of MH and, unlike porcine MH in which a single mutation in the RYR gene is shown to cosegregate with MH across a wide variety of breeds, human MH is diverse. In fact, several mutations have been detected in RYRI. With this diversity one begins to wonder whether there can ever be a DNA test for MH that can be applied generally to a population.

Even more disturbing was the recent revelation by Alastair Stewart [11], at the last EMHG Genetics Section meeting, that there was some evidence that in an individual pedigree there could be two different MH genes, or two separated MH mutations within one gene. Presumably this is inevitable with heterogeneity on the scale we are now detecting. At the same meeting, at least three other families were presented in which the same sort of double genetic defect could have occurred within an individual pedigree. These findings if confirmed make the Healy paper [5] recommendations very dubious indeed.

In the light of these observations, there should perhaps be a reappraisal of the IVCT. Whatever the genetic defect, the IVCT will identify, in a manner that is potentially causally linked with MH, any patient susceptible to MH whether the family defect lies in the ryanodine receptor protein or one of the dihydropyridine protein subunits or elsewhere. Thus in an extended pedigree in which more than one MH gene may be present, MH susceptibility will be identified independently of which MH gene is responsible.

More importantly, however, MH-negative patients will be identified with the same degree of certainty. The MHN group is crucial to justify our continued testing of patients. I do not know of any MHN patient tested using the EMHG protocol who has developed MH when subsequently exposed to trigger agents. This is the strength of the IVCT, which in some ways provides a "gold standard" for MH screening.

References

1. European Malignant Hyperthermia Group (1984) A protocol for the investigation of malignant hyperthermia (MH) susceptibility. Br J Anaesth 56:1267–1269
2. European Malignant Hyperthermia Group (1985) Laboratory diagnosis of malignant hyperthermia susceptibility (MHS). Br J Anaesth 57:1038
3. MacLennan DH, Duff C, Zorato F, Fujii J, Philips MS, Korneluk RG, Frondis W, Britt BA, Worton RG (1990) Ryanodine receptor gene is a candidate for predisposition to malignant hyperthermia. Nature 343:559–561
4. McCarthy TV, Healy JMS, Heffron JJA, Lehane M, Deufel T, Lehmann-Horn F, Farral M, Johnson K (1990) Localisation of the malignant hyperthermia susceptibility locus to human chromosome 19q 12–13.2. Nature 343:562–564
5. Healy SJM, Heffron JJA, Lehane M, Bradley DG, Johnson K, McCarthy TV (1991) Diagnosis of susceptibility to malignant hyperthermia with flanking DNA markers. BMJ (London) 303:1225–1228
6. Levitt RG, Nouri J, Jedika AE, McKusick VA, Marks A, Shutak JG, Fletcher JE, Rosenberg H, Meyers DA (1991) Evidence for the localisation of a malignant hyperthermia susceptible locus (MHS 2) to human chromosome 17q. Genomics 11: 543–547
7. Deufel T, Heitinger T, Golla A, Johnson K, MacLennan DH, Lehmann-Horn F (1991) Two recombinations between malignant hyperthermia susceptibility and human ryanodine receptor in a single family: another gene for MHS? Human Gene Mapping 11: Cytogenet Cell Gene 58:2018–2019
8. Deufel T, Golla A, Iles DE, Meindl A, Meitinger T, Schindelhauer D, DeVries A, Pongratz D, MacLellan DH, Hohnson KJ, Lehmann-Horn F (1992) Evidence for genetic heterogeneity of malignant hyperthermia susceptibility. Am J Hum Genet 50:1151–1161
9. Sudbrak R, Procaccio V, Klausnitzer M, Curran JL, Monsieurs K, Van Broeckhoven C, Ellis FR, Heytens L, Hartung EJ, Kozak-Ribbens G, Heilinger D, Weissenback J, Lehmann-Horn F, Mueller CR, Deufel T, Stewart AD, Lunardi J (1995) Mapping of a further malignant hyperthermia susceptibility (MHS) locus to chromosome 3q13.1. Am J Hum Genet 56:684–691

10. Ball SP, Dorkings HR, Ellis FR, Halsall PJ, Hopkins PM, Mueller RF, Stewart AD (1993) Genetic linkage analysis of chromosome 19 markers in malignant hyperthermia. Br J Anaesth 70:70–75
11. Stewart AD (1993) DNA studies in some Leeds malignant hyperthermia families. Presented at the December Meeting of the Genetics Section of the European Malignant Hyperthermia Group, London 1993

Caffeine Contracture of the Skinned Muscle Fiber in Malignant Hyperthermia and Neuromuscular Diseases

Akio Takagi, Makoto Araki, Susumu Kojima, Masayoshi Ida, and Tomoji Watanabe

Introduction

Caffeine contracture of skinned muscle fiber was analyzed in subjects who had recovered from malignant hyperthermia (MH) and in patients with other neuromuscular diseases. In several muscle specimens, the calcium-induced (Ca-induced) Ca release (CICR) was also studied and compared to the caffeine contracture. The caffeine contracture test was very sensitive in detecting abnormalities, but it seemed less specific for diagnosis of true MH.

Methods

Muscle (usually biceps brachii) was biopsied from 28 subjects who had recovered from episodes of MH or MH-like reactions with their informed consent. Muscle specimens of various neuromuscular diseases, which were biopsied for diagnostic purposes, were also studied. Control muscle was obtained in orthopedic surgery with approval of the patients or from cases in which neuromuscular problems were judged psychogenic after complete studies. MH or MH-like reactions were classed into the fulminant or abortive form by the criteria of Morio et al. [1].

Skinned muscle fibers were prepared and tested as reported elsewhere [2]. Caffeine contracture was analyzed either by the standard (Fig. 1) or by the cumulative method. In the standard method, threshold and concentration of half-maximum contracture were computed from the dose–response curve. When more than one parameter was abnormal, caffeine contracture was judged abnormal. The muscle fiber type of single fibers examined was determined from their response to strontium ion [3]. The rate of CICR was measured according to Endo et al. [4].

The effect of caffeine on the contractile system was analyzed. The effect of caffeine on the pCa-tension relationship was studied in muscle fibers that were treated with 0.1% Brij 58 to abolish the function of the sarcoplasmic reticulum (SR).

Department of Neurology, Toranomon Hospital and Okinaka Memorial Institute for Medical Research, Minato-ku, Tokyo, 105 Japan

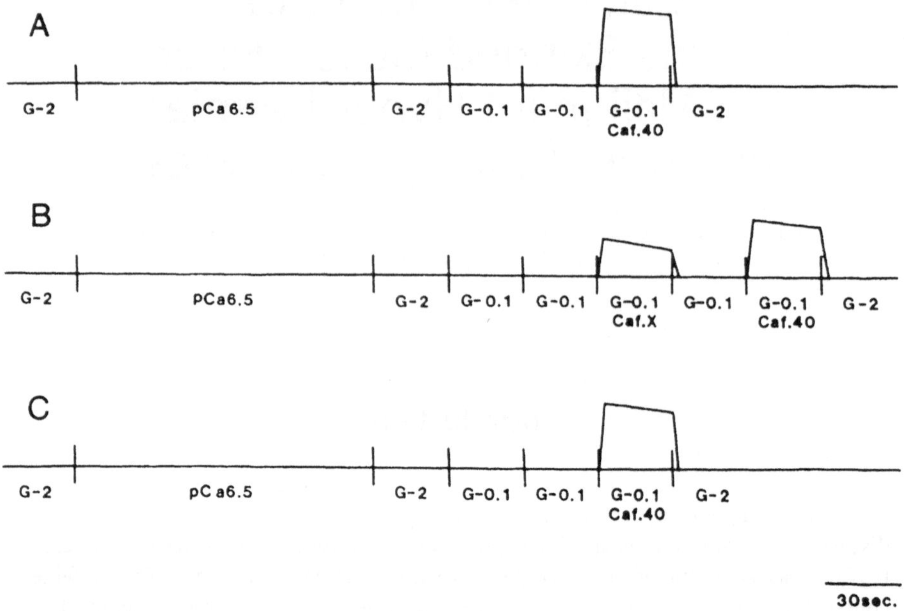

FIG. 1A–C. Quantitative analysis of caffeine contracture of skinned muscle fibers. After the sarcoplasmic reticulum (SR) of the skinned fibers was loaded with Ca ion in solution of pCa 6.5 for 2 min, caffeine contracture was induced by 30 m*M* caffeine (**A**,**C**). In the test, contracture was induced by a variable amount of caffeine (**B**). The peak tension of trace **B** was compared to that of trace **A** and **C**, and relative contracture was calculated. *G-2* or *G-0.1* denotes a relaxing solution containing 2 m*M* or 0.1 m*M* ethylene glycol-bis (β-aminoethyl ether)-*N*,*N*,*N*′,*N*′-tetraacetic acid (EGTA), respectively

Results

Normal Values

In the standard method, threshold and half-maximum activation values were 7.3 + 2.6 and 8.2 + 3.3 m*M* (mean + SD, $n = 20$), respectively, for type 1 muscle fiber and 9.8 + 2.4 and 11.7 + 3.5 m*M* ($n = 26$) for type 2. By the cumulative method, threshold of caffeine was 10.9 + 3.8 m*M* ($n = 8$) for type 1 and 12.4 + 2.1 m*M* ($n = 26$) for type 2 fiber. The threshold value was a little higher using the cumulative method.

Malignant Hyperthermia and Other Neuromuscular Diseases

In 11 of 13 cases with fulminant MH, caffeine contracture was abnormally enhanced. In 7 of 15 cases with abortive MH, the caffeine test was abnormally enhanced. In subclinical hyperCKemia (CK, creatine kinase), Duchenne muscular dystrophy, or neuroleptic malignant syndrome, 71%, 85%, or 75%, respectively, of cases were abnormal (Table 1). In other neuromuscular diseases, the number of cases studied was too small to draw a definite conclusion. Both cases (2) of McArdle disease were abnormal in caffeine contracture.

One of two cases with central-core disease was positive. In exertional heatstroke or myotonic syndrome, there was no constant tendency. Two cases of rippling muscle

TABLE 1. Results of caffeine contracture test in malignant hyperthermia (MH) and other neuromuscular diseases.

Disease	Positive cases/total cases
Malignant hyperthermia	
Fulminant form	11/13
Abortive form	7/15
Duchenne dystrophy	11/13
Subclinical hyperCKemia	12/17
Neuroleptic malignant syndrome	9/12
Becker dystrophy	1/2
Fukuyama congenital dystrophy	1/2
Limbgirdle syndrome	0/1
Central-core disease (typical)	1/2
Myotonic dystrophy	1/4
Myotonia congenita	1/2
Paramyotonia	1/1
Rippling muscle disease	0/2
McArdle disease	2/2
Mitochondria disease	1/2
Exertional heatstroke	1/3
Myoglobinuria	1/3
Scoliosis	1/6
Spinal muscular atrophy	0/4
Peripheral neuropathy	1/3
Guillan-Barre syndrome	1/1

TABLE 2. Variation in response between type 1 and type 2 muscle fibers in the caffeine contracture test[a].

	Abnormal: type 1 and type 2	Abnormal: type 1	Abnormal: type 2	Normal: both types	Total
Fulminant MH	9	0	0	0	9
Abortive MH	2	1	3	6	12
Duchenne dystrophy	3	2	0	1	6
HyperCKemia	0	3	3	5	11
NMS	3	6	0	3	12
Becker dystrophy	0	1	0	1	2
Myotonic dystrophy	0	0	0	1	1
Thomsen myotonia	0	0	1	1	2
Heatstroke	1	0	0	2	3
Central-core disease	0	0	1	0	1

NMS, neuroleptic malignant syndrome.
[a] Results were summarized in cases in which both type 1 and type 2 fibers were analyzed on the same specimen.

disease were normal. In peripheral neuropathy or Guillan-Barre syndrome, some cases were positive.

Variation in Response in Type 1 and Type 2 Muscle Fiber

The difference in caffeine contracture between type 1 and type 2 fibers in various diseases was examined (Table 2). When both type 1 and type 2 fibers from the same

TABLE 3. Calcium-induced calcium release (CICR) test in MH and other neuromuscular diseases.

	Positive/total cases
Malignant hyperthermia	
Fulminant form	3/4
Abortive form	0/1
Duchenne dystrophy	0/4
Becker dystrophy	0/1
HyperCKemia	1/4

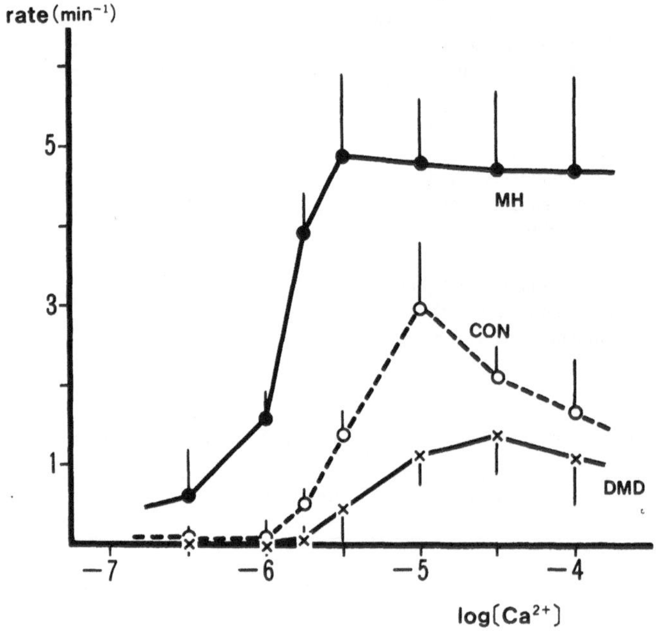

FIG. 2. Examples of the CICR test in malignant hyperthermia (*MH*), Duchenne dystrophy (*DMD*), and control (*CON*). The rate of CICR was plotted for solutions of various Ca ion concentrations (abscissa). The CICR was increased in MH and reduced in DMD. Mean + SD are shown (n = 4 for MH, 5 for con, and 4 for DMD)

sample were analyzed, caffeine contracture was abnormal in both fiber types in all fulminant MH cases examined. In abortive MH, Duchenne dystrophy, or neuroleptic malignant syndrome, both fiber types were abnormal in a few cases. In neuroleptic malignant syndrome, however, only the type 1 fiber was abnormal in 6 cases of 12.

Comparison Between Caffeine Contracture and CICR

The rate of CICR as compared to the caffeine contracture test (Table 3) was abnormally increased in three of four cases of fulminant MH and one of four cases of subclinical hyperCKemia. In abortive MH, the rate of CICR was normal. In Duchenne or Becker dystrophy, the rate of CICR was normal or less than normal (Fig. 2).

Discussion

Kalow et al. [5] first demonstrated that the skeletal muscle of MH patients developed contracture at a lower concentration of caffeine than that of normal persons. We studied the influence of halothane on skeletal muscle and showed that muscle from patients with MH developed contracture from a lower concentration of halothane. Halothane released Ca ion from the SR in a manner similar to caffeine [6]. Because caffeine is easier to apply to in vitro study than halothane, we began to analyze caffeine contracture of skinned muscle fibers.

Caffeine contracture was abnormally enhanced in 85% of cases of fulminant MH; however, it was abnormal in only 47% of cases of abortive MH. Caffeine contracture was abnormally enhanced in 70%–85% of cases of Duchenne muscular dystrophy, subclinical hyperCKemia, or neuroleptic malignant syndrome. In subclinical hyperCKemia, serum CK values are consistently elevated, but neuromuscular abnormality is not detected by physical examination. This condition is naturally heterogeneous, including subclinical polymyositis, preclinical neuromuscular disorders, and so on.

In control muscle fibers, type 1 fibers were more sensitive to caffeine than type 2, developing contracture by a lower dose of caffeine. In the current study, parameters were compared to the same muscle fiber type. In fulminant MH, both type 1 and type 2 fibers were abnormal when both fiber types were analyzed in the same specimens. In neuroleptic malignant syndrome, only type 1 fiber was abnormal in 6 of 12 cases. These observations suggested that increased caffeine response did not arise by only a single mechanism.

In several cases of MH, muscular dystrophy, or hyperCKemia, the rate of CICR was also measured. Increase in the rate of CICR was observed only in fulminant MH and in one case of hyperCKemia. In a case of central-core disease, the CICR was reported abnormal [7]. On the other hand, CICR was normal in abortive MH, Duchenne and Becker dystrophy, and most hyperCKemia cases. The CICR is the main function of the ryanodine receptor or Ca release channel of the SR [8]. Increase of CICR rate might indicate an abnormality of the ryanodine receptor. In some families of MH, a missense mutation of the ryanodine receptor gene has already been described [9,10].

Thus, evidence is accumulating that structural and functional abnormalities of the ryanodine receptor cause part of MH reactions [11]. On the other hand, caffeine contracture was enhanced but CICR was normal in most other neuromuscular diseases such as Duchenne muscular dystrophy. Abnormal CICR or missense mutation of the ryanodine receptor may constitute the true MH or MH myopathy, which might clinically express as the fulminant form of MH. True MH should be separated from MH-like reactions in other neuromuscular diseases. Similar suggestions were proposed by Heytens et al. [12], depending on results of the in vitro muscle test in various neuromuscular diseases.

The incidence of anesthetic complications of Duchenne muscular dystrophy (DMD) ranged from 0 to 28% among various reports [13–15]. In our caffeine contracture test, however, 85% of DMD cases were abnormal. Judging from the results, the caffeine contracture test seemed very sensitive, but might be less specific; 75% of cases of neuroleptic malignant syndrome were abnormal in the caffeine contracture test, but MH reactions were rare in this condition [16]. The main pharmacological action of caffeine on the skeletal muscle is to release Ca ion from the SR by activating the CICR mechanism. Increase in the rate of CICR could be detected by the caffeine contracture method. Caffeine also influences the contractile system [17], producing a larger tension. The enhanced caffeine contracture of DMD might result partly from the response of the contractile system. Skinned fibers of DMD tended to produce a larger tension when 5 m*M* caffeine was added to a solution containing Ca ion, but the difference was statistically insignificant. Dangain and Neering [18] reported similar observations in the dy/dy dystrophic mouse. We observed increased leakage of Ca ion from the SR of the mdx mouse, while the rate of CICR was normal [20,21]. This mechanism might also increase caffeine contracture. Whether causes of MH-like reactions in other neuromuscular diseases are solely the result of the abnormal response of the contractile system still awaits elucidation.

Conclusion

Caffeine contracture of the skinned muscle fiber was abnormally enhanced in 85% of fulminant MH and 47% of abortive MH. In other neuromuscular diseases, it was abnormal in 85% of DMD, 71% of subclinical hyperCKemia, and 75% of neuroleptic malignant syndrome cases. The CICR test was abnormal only in fulminant MH, suggesting that the pathological mechanisms of MH-like reactions in other diseases are different from those of true MH. Causes of MH-like reactions in other neuromuscular diseases may be partly the result of the abnormal response of the contractile system.

Acknowledgments. This work was partly supported by the Research Grant (5A-2) for Nervous and Mental Disorders from the Ministry of Health and Welfare and a grant from Okinaka Memorial Institute for Medical Research. The authors wish to thank Ms. Kazuko Abe and Akiko Murai for technical assistance.

References

1. Morio M, Kikuchi H, Yuge O, et al (1988) Clinical criteria of malignant hyperthermia. Revised criteria. In: Proceedings of the 11th Japanese symposium of malignant hyperthermia, 1988, pp 104–110
2. Takagi A, Sunohara N, Ishihara T, et al (1983) Malignant hyperthermia and related neuromuscular diseases: caffeine contracture of the skinned fibers. Muscle Nerve 6:510–514
3. Takagi A, Yonemoto K, Sugita H (1978) Single human muscle fibers: activation by calcium and strontium. Neurology 28:497–499
4. Endo M, Yagi S, Ishizuka Y, et al (1983) Changes in the Ca-induced Ca release of the sarcoplasmic reticulum of the muscle from a patient with malignant hyperthermia. Biomed Res 4:83–92
5. Kalow W, Britt BA, Terreau ME, et al (1970) Metabolic error of muscle metabolism after recovery from malignant hyperthermia. Lancet 2:895–898
6. Takagi A, Sugita H, Toyokura Y, Endo M (1976) Malignant hyperpyrexia. Effect of halothane on single skinned muscle fibers. Proc Jpn Acad 52:603–606
7. Otsuka H, Komura Y, Mayumi T, et al (1991) Malignant hyperthermia during sevoflurane anesthesia in a child with central core disease. Anesthesiology 75:699–701
8. Lai FA, Erickson HP, Rousseau E, et al (1988) Purification and reconstitution of the calcium release channel from skeletal muscle. Nature 331:315–319
9. Gillard EF, Otsu K, Fujii J, et al (1992) Polymorphisms and deduced amino acid substitutions in the coding sequence of the ryanodine receptor gene in individuals with malignant hyperthermia. Genomics 13:1247–1254

10. Quane KA, Healy JMS, Keating KE, et al (1993) Mutations in the ryanodine receptor gene in central core disease and malignant hyperthermia. Nature Genet 5:51–55
11. Kawana Y, Iino M, Horiuti K, et al (1992) Acceleration in calcium-induced calcium release in the biopsied muscle fibers from patients with malignant hyperthermia. Biomed Res 13:287–297
12. Heytens L, Martin JJ, Van De Kelft E, Bossaert LL (1992) In vitro contracture tests in patients with various neuromuscular diseases. Br J Anaesth 68:72–75
13. Heimann-Patterson TD, Natter HM, Rosenberg HR, et al (1986) Malignant hyperthermia susceptibility in X-linked muscle dystrophies. Pediatr Neurol 2:356–358
14. Sethna NF, Rockoff MA, Worthen M, Rosnow JM (1988) Anesthesia-related complications in children with Duchenne muscular dystrophy. Anesthesiology 68:462–465
15. Larsen UT, Juhl B, Hein-Sorensen O, Olivarius BDF (1989) Complications during anaesthesia in patients with Duchenne's muscular dystrophy (a retrospective study). Can J Anaesth 36:418–22
16. Gronert GA, Fowler W, Cardinet GH, et al (1992) Absence of malignant hyperthermia contractures in Becker-Duchenne dystrophy at age 2. Muscle Nerve 15:52–56
17. Salviati G, Betto R, Ceoldo S, et al (1989) Caffeine sensitivity of sarcoplasmic reticulum of fast and slow fibers from normal and malignant hyperthermia human muscle. Muscle Nerve 12:365–370
18. Dangain J, Neering I (1993) Effect of caffeine and high potassium on normal and dystrophic mouse EDL muscles at various developmental stages. Muscle Nerve 16:33–42
19. Geiduschek J, Cohen SA, Khan A, et al (1988) Repeated anesthesia for a patient with neuroleptic malignant syndrome. Anesthesiology 68:134–137
20. Takagi A, Kojima S, Ida M, Araki M (1992) Increased leakage of calcium ion from the sarcoplasmic reticulum of the mdx mouse. J Neurol Sci 110:160–164
21. Takagi A, Kojima S, Araki M, et al (1989) Clinical implications of enhanced caffeine contracture in malignant hyperthermia and Duchenne muscular dystrophy (in Japanese). Clin Neurol (Tokyo) 29:301–305

10. [illegible] (1991) [illegible] receptor [illegible] hypertrophy [illegible]
[illegible]

[illegible]

11. [illegible]
[illegible] during [illegible] study)

12. [illegible]
[illegible]

13. [illegible]

14. [illegible]

15. [illegible]

Calcium-Induced Calcium Release Test

Makoto Endo

Abstract. We have previously demonstrated [1] that, in skeletal muscle biopsied from a patient with typical malignant hyperthermia (MH) syndrome, Ca-induced Ca release (CICR) in the sarcoplasmic reticulum (SR) was accelerated, while the Ca pump activity of the SR and the Ca sensitivity of the contractile system in the muscle were normal. Therefore, it appeared important to further examine the rate of CICR in muscles from patients in whom MH was suspected. We examined biopsied muscles from 84 subjects, 79 in whom MH was suspected and 5 normal (non-MH-suspected) controls. Skinned fibers were prepared from each specimen and the rates of Ca release were measured at 0, 0.3, 1.0, 3.0, and 10 μM Ca. Since the number of normal control specimens was too small to establish the range of the normal rate of CICR, cluster analysis based on the Ca release rates was made on all the data obtained. As already reported [2], the subjects examined were objectively classified into three groups according to the rate of CICR: unaccelerated (group 1, $n = 61$), moderately accelerated (group 2, $n = 19$), and highly accelerated (group 3, $n = 4$). The normal controls were all in group 1. Almost all the patients with conspicuous fever (≥40.0°C and/or ≥0.5°C rise/15 min) during anesthesia belonged to group two or three. These results suggest that, in the majority of MH patients, the abnormality in the CICR in skeletal muscle SR is involved in the cause of the disease. There could be other causes of the MH syndrome, and for this reason in particular, the CICR test should be fully utilized when genetic studies of the ryanodine receptor are conducted.

References

1. Endo M, Yagi S, Ishizuka T, Horiuti K, Koga Y, Amaha K (1983) Changes in the Ca-induced Ca release mechanism in the sarcoplasmic reticulum of muscle from a patient with malignant hyperthermia. Biomed Res 4:83–92
2. Kawana Y, Iino M, Horiuti K, Matsumura N, Ohta T, Matsui K, Endo M (1992) Acceleration in calcium-induced calcium release in the biopsied muscle fibers from patients with malignant hyperthermia. Biomed Res 13:287–297

Department of Pharmacology, Faculty of Medicine, University of Tokyo, 7-3-1 Hongo, Bunkyo-ku, Tokyo, 113 Japan

Part 4. Genetic Study of Malignant Hyperthermia

Mutations in the Skeletal Muscle Ryanodine Receptor (*RYR1*) Gene Are Linked to Malignant Hyperthermia and Central-Core Disease

David H. MacLennan

The Ca^{2+} Release Channel of Skeletal Muscle Sarcoplasmic Reticulum

Skeletal muscle contraction and relaxation are controlled by Ca^{2+} concentrations within muscle cells [1]. Intracellular Ca^{2+} concentrations are, in turn, controlled by the sarcoplasmic reticulum. The sarcoplasmic reticulum surrounds each muscle fibril like a water jacket. It is subdivided into functional components: the longitudinal sarcoplasmic reticulum and the terminal cisternae, which itself is divided into junctional and nonjunctional terminal cisternae [2]. The longitudinal sarcoplasmic reticulum and the nonjunctional terminal cisternae have, as their major function, the uptake of Ca^{2+} through the activity of the Ca^{2+} pump.

The junctional terminal cisternae is a complex structure. Its major function is the release of Ca^{2+}, but this function requires interaction among proteins in two different membrane systems. The Ca^{2+} release channel in the terminal cisternae, often referred to as the ryanodine receptor, and the slow Ca^{2+} channel in the transverse tubule, often referred to as the dihydropyridine receptor, are physically apposed and are probably functionally connected. It is probable that an interaction between these two protein complexes results in Ca^{2+} release channel opening [3].

Calsequestrin is a luminal protein that binds Ca^{2+} in a matrix which abuts the interior surface of the Ca^{2+} release channel [4]. Thus, Ca^{2+} to be released is concentrated near its site of release. Triadin, a transmembrane protein with a basic cytoplasmic amino acid sequence, may connect the ryanodine receptor and calsequestrin in a functional way [5].

Cloning of cDNA encoding the ryanodine receptor has provided insight into its structure and function [6,7]. The ryanodine receptor from rabbit, human, or pig is composed of about 5035 amino acids and has a mass of about 565 000 daltons. It is predicted to have between 4 and 12 transmembrane sequences, located at the carboxyl-terminal end of the protein. In isolation, the Ca^{2+} release channel is regulated by ligands such as adenosine triphosphate (ATP), Ca^{2+}, Mg^{2+}, and calmodulin. Regulatory domains where these ligands might bind are predicted to lie between amino acid residues 2400 and 3000 [8] and between residues 4250 and 4500 [6].

Banting and Best Department of Medical Research, University of Toronto, Charles H. Best Institute, Toronto, Ontario M5G 1L6, Canada

Malignant Hyperthermia

Malignant hyperthermia (MH) is manifested in humans as an acute hyperthermic reaction, accompanied by skeletal muscle contracture, and usually triggered by a combination of potent inhalational anesthetics and depolarizing muscle relaxants. The corresponding abnormality in swine is referred to as porcine stress syndrome (PSS). Acute hyperthermia and skeletal muscle contracture in swine are most frequently brought on by stress. Both MH and PSS are inherited traits.

Because one of the major features of MH is muscle contracture and contracture can only occur if intracellular Ca^{2+} concentrations are continuously elevated, the underlying basis for MH was predicted to be an alteration in one or more of the elements of the Ca^{2+} regulatory system [9]. If intracellular Ca^{2+} were elevated through an abnormality in the Ca^{2+} regulatory system, then the following reactions would be predicted: muscle contracture would occur with concomitant ATP hydrolysis; glycogen breakdown would be stimulated through Ca^{2+}-dependent phosphorylase kinase activity, resynthesizing ATP; lactic acid, formed from glycolysis, would be metabolized by mitochondria to form CO_2 and ATP, or be excreted into the blood, causing lactic acidosis; and ATP would also be hydrolyzed by the Ca^{2+} pump, responsible for removal of cytosolic Ca^{2+}. The net effect would be heat, acid, and CO_2 production, all of which are features of an MH reaction. In early studies, the Ca^{2+} pump was ruled out as a participant in MH reactions, but comparisons of Ca^{2+} release from preparations of human [10] or porcine [11] sarcoplasmic reticulum have suggested that the Ca^{2+} release channel might be abnormal in MH-susceptible (MHS) individuals.

MH Mutations in the Ryanodine Receptor

We initiated our studies of the genetic basis of MH by cloning the human skeletal muscle ryanodine receptor (*RYR1*) cDNA [7]. We then used human cDNA as a hybridization probe for the detection of polymorphisms in the *RYR1* gene in human genomic DNA that had been cleaved by a series of restriction endonucleases [12]. Since Dr. B.A. Britt had diagnosed the inheritance of MH susceptibility in many MH families using the in vitro caffeine/halothane contracture test (CHCT) [9], we were able to compare the inheritance of MH with the inheritance of one or more polymorphisms in the *RYR1* gene or in markers flanking the *RYR1* gene.

We found that MH and polymorphisms in the *RYR1* gene were inherited through 23 meioses in nine families studied, providing us with a lod score (log of the odds favoring linkage) of 4.2 for a recombinant fraction of 0.0 [12]. The probability, more than 10000 to 1, that MH and *RYR1* were linked, provided us with the incentive to search for mutations in the *RYR1* gene that might cause MH in both humans and swine. In independent studies, McCarthy et al. [13] also linked MH to a region of human chromosome 19q that included the *RYR1* gene. In later studies, others (e.g., [14]) could not link all MH families to chromosome 19q, and proposed that a second MH gene exists.

In our studies of porcine MH, we sequenced *RYR1* cDNAs prepared from skeletal muscle from normal Yorkshire and MH Pietrain animals. We found several nucleotide polymorphisms, but only a single amino acid replacement in the entire group of 5035 amino acids. The replacement in the nucleotide sequence of C1843 with

T led to the replacement of Arg^{615} with Cys [15]. This amino acid change was associated with MH in about 80 animals in five different breeds. In a thorough linkage study, conducted in collaboration with Dr. A. Archibald [16], we were able to demonstrate linkage between the C1843 to T mutation and MH in 338 informative meioses. The lod score of 102 for a recombinant fraction of 0.0, obtained in this study, provided very strong evidence that the mutation was causal of porcine MH.

Our next experiment was to determine whether the corresponding mutation might cause MH in humans. In a survey of 35 MH families, we found a single family that carried an Arg^{614} to Cys mutation (the human *RYR1* cDNA is one amino acid shorter than the porcine cDNA). In this family, the mutation segregated with MH [17]. Thus the causal nature of this mutation is very strongly supported, both by the fact that the corresponding mutation is found in two separate species, where it segregates (in most cases) with MH, and by the extraordinarily high probability that it is linked to MH in swine. Recent biochemical and physiological experiments in which channel properties of the mutant protein have been examined [18,19] also support the causal nature of this mutation.

We also carried out a study of haplotypes in individual swine from five lean, heavily muscled breeds that were either homozygous or heterozygous for the Arg^{615} to Cys mutation [15]. We found that a single haplotype was present in all homozygous animals and that the potential for the MH haplotype was present in all heterozygotes. These data support the view that the mutation came from a founder animal and was propagated or selected for in most lean, heavily muscled breeds. In recent studies, we have shown that the mutation, in a single copy, adds nearly 5% to lean muscle yield, that it redistributes fat from the muscle to the back, and that it improves both carcass index and feed conversion [20]. Thus, it is a beneficial gene for meat production. In the homozygous state, however, these benefits may be lost because of to the increase in stress-induced deaths and the increased incidence of pale, soft, exudative (PSE) meat following slaughter.

We have continued our search for additional MH mutations by sequencing *RYR1* cDNA from MH individuals. In some cases, we have amplified individual exons of the *RYR1* gene and subjected the product to gel electrophoresis under conditions in which single-strand conformational polymorphisms (SSCPs) lead to altered mobility in analytical gels. The first sequencing study led to the discovery of a Gly^{248} to Arg mutation that was linked to MH in a very small family [21]. This mutation has been found in only a single family to date. In independent studies, Quane et al. [22] discovered that the mutation of Gly^{341} to Arg could be linked to MH in a number of European families.

The mutation of Gly^{2433} to Arg was discovered through sequencing of exon 44 from an individual in which an SSCP in exon 44 of *RYR1* genomic DNA had been identified [23]. The mutation was found to be present in 4 of 106 Canadian MH families and to be absent in more than 1000 other chromosomes. In independent studies, Keating et al. [24] found the same mutation in four European families. In our studies, the mutation was present in six individuals in four families who had had an MH reaction, in two obligate carriers, and in ten individuals diagnosed as MH-susceptible by the CHCT. The mutation was linked to MH, as diagnosed by the CHCT, in two families but not in all four. In one of the families in which three discrepancies occurred, the mutation was present in an individual diagnosed as borderline normal by the CHCT, but was absent from another individual diagnosed as borderline MHS by the CHCT. These cases were of little concern because the cutoff points for the MHS diagnosis are arbitrary, although based on a large number of tests [25]. A more difficult problem was posed by an individual in the same family who had a clearly positive CHCT (response to both caffeine and halothane), but did not carry the mutation. The cutoff points for the CHCT are set to provide high sensitivity, and more than 90% sensitivity is obtained in the North American CHCT. However, specificity is only about 75% [25]. Accordingly, we concluded that this patient was an outlier in the normal range of response to the CHCT.

The second family in which a discrepancy occurred presented a more interesting problem. In this family, two brothers had been diagnosed as MHS. Both reacted positively to both caffeine and halothane, but one had a far stronger response to both caffeine and halothane than the other. The stronger reactor carried the Gly^{2433} to Arg mutation, but the weaker reactor did not. It is possible that the CHCT result was inaccurate for the weaker reactor, but it is also possible that the brother with the exceptionally strong CHCT reaction carried two distinct MH mutations, only one of which was detected in our specific test. The brother with the weaker reaction may have carried only the undetected MH mutation. Further analysis of *RYR1* in the weaker reactor will be of great interest.

Future searches for MH mutations in the RYR1 gene are being automated in our laboratory. We have determined all the 105 exon–intron boundaries in the gene, together with flanking intron sequence. Our strategy is to amplify individual exons for rapid, automated sequence analysis. We anticipate that this process will greatly enhance the speed and efficiency of mutant searches starting with genomic DNA. Because genomic DNA is obtained from peripheral blood, the process will be noninvasive and should be cost-effective.

Central-Core Disease

Central-core disease (CCD) is an autosomal dominant, nonprogressive myopathy involving predominantly proximal muscle and presenting in infancy [26]. Diagnosis is made on the basis of the characteristic morphological feature, the presence of myofibrillar cores that are best demonstrated by their lack of oxidative histochemistry. Although other features of the disease are variable, MH susceptibility is a constant feature [27].

On the basis of linkage of MH to the *RYR1* gene on chromosome 19q13.1, the inheritance of CCD was linked to the inheritance of polymorphisms in the *RYR1* gene with a lod score of about 12 for a recombinant fraction of 0.0 [28]. On the basis of this tight linkage, we sequenced *RYR1* cDNA from a CCD patient and found the mutation of Arg^{2434} to His in a single CCD family [29]. This mutation was linked to CCD with a lod score of 4.8 for a recombinant fraction of 0.0 in 16 informative meioses. In independent studies, Quane et al. [30] and Quane et al. [31] also linked CCD to mutations in *RYR1*. These mutations are Arg^{163} to Cys, Ile^{403} to Met, and Tyr^{522} to Ser.

The fact that MH and CCD mutations are found in the same gene suggests that different genotypic variants might result in a spectrum of phenotypic responses of different severity (Fig. 1). At one end of the spectrum, there may be no other phenotypic result of an altered Ca^{2+} release channel than an abnormal response to halothane. A more deleterious mutation, for example, the mutation of Arg^{615} to Cys, might result in spontaneous leakage of Ca^{2+} into the sarcoplasm, which could be

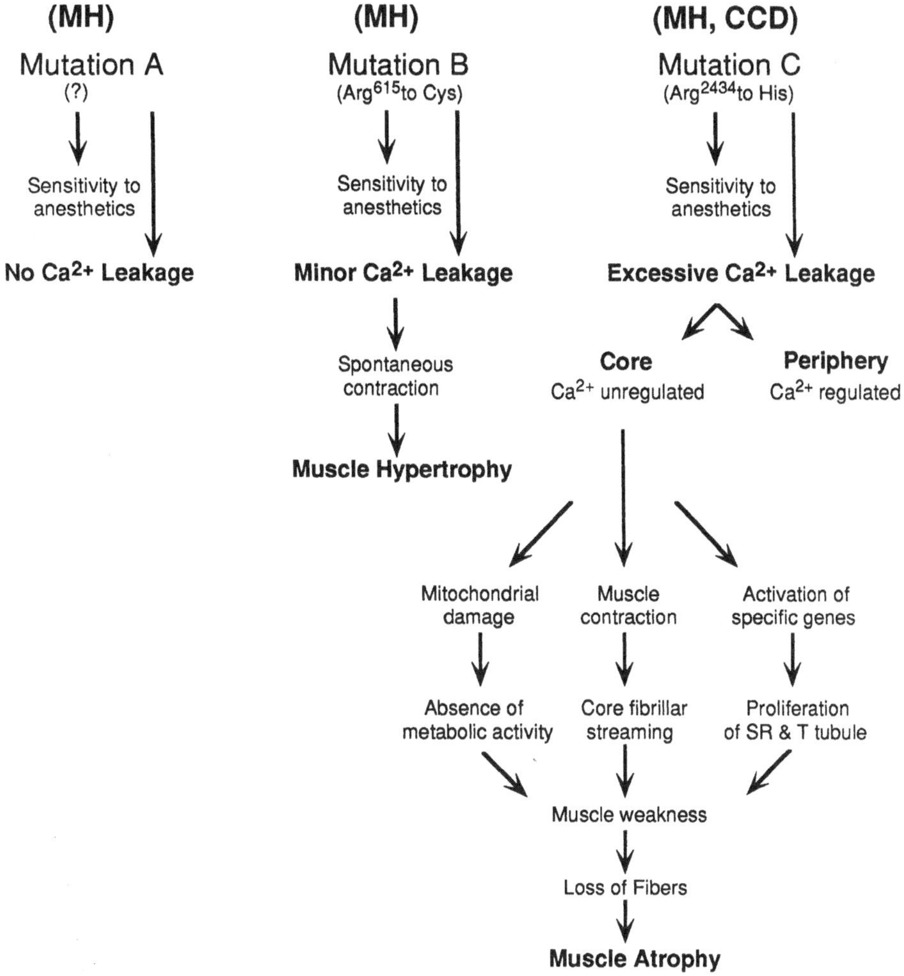

FIG. 1. A proposed mechanism for the differential phenotypic effects of different malignant hyperthermia (*MH*) and central-core disease (*CCD*) mutations. MH mutations in the ryanodine receptor lead to the common phenotype of sensitivity to anesthetics. Some mutations may also lead to spontaneous Ca^{2+} release sufficient to trigger spontaneous contractions. If this trigger Ca^{2+} were readily regulated, the major phenotypic effect would be spontaneous exercise-induced muscle hypertrophy. The Arg^{615} to Cys mutation was selected in swine because it leads to increased lean muscle mass. Mutations leading to excessive spontaneous Ca^{2+} release may have no phenotypic effect on the periphery of the cell, but be deleterious to the central core. Ca^{2+} released from the sarcoplasmic reticulum (SR) can be regulated by four systems, including the organellar sarcoplasmic reticulum and mitochondria and the plasma membrane Ca^{2+} pumps and Na^+/Ca^{2+} exchangers. Under normal circumstances, the bulk of the Ca^{2+} is cycled only through the sarcoplasmic reticulum. If enhanced Ca^{2+} release occurred spontaneously, in CCD muscle, the additional Ca^{2+} regulatory systems might be coopted to regulate Ca^{2+}. The plasma membrane exchangers and pumps would be effective in regulating Ca^{2+} near the periphery, but, in the core, mitochondria may be forced to bear this load, destroying themselves in the process. The degeneration of mitochondria could lead to the degeneration of a central, possibly compartmented, core. The disorganization of the central core might be brought about by higher core levels of Ca^{2+}, which could cause contraction at the core and lead to myofibrillar streaming and membrane disorganization. Elevated Ca^{2+} may stimulate proliferation of internal membrane proteins at the transcriptional level. The phenotypic effects would be the formation of a disorganized, metabolically deficient core, which could lead to cell death and muscle atrophy

regulated throughout the cell by normal Ca^{2+} removal systems. The spontaneous contractions that might result, however, could lead to spontaneous exercise-induced muscle hypertrophy, the phenotype that is selected for in MH swine. A still more deleterious mutation might result in even more leakage of Ca^{2+} into the sarcoplasm. The Ca^{2+} removal systems, including the sarcoplasmic or endoplasmic reticulum (SERCA) type Ca^{2+} pumps, the plasma membrane (PMCA) type Ca^{2+} pumps, Na^+/Ca^{2+} exchangers in the plasma membrane, and mitochondria in the cytoplasm, may have difficulty in removing the excess cytoplasmic Ca^{2+} and the problem may be severe in the core of the cell where only SERCA pumps and mitochondria operate [29]. If the mitochondria became overloaded with Ca^{2+} they could become inactivated [32], accounting for the loss of oxidative activity in the core.

The presence of a high concentration of Ca^{2+} in the core and a lower Ca^{2+} concentration at the periphery might also result in differential contraction of those fibers that are located at the periphery and those located in the core of the cell. This could result in the sarcomeric streaming that is seen in the cores. The abnormal Ca^{2+} concentrations may signal protein synthetic systems, which might result in proliferation of sarcoplasmic reticulum and transverse tubules in core. The deterioration of the core might result in muscle weakness and ultimately in fiber loss, leading to the weakness and atrophy that is associated with CCD. Thus a spectrum from potentially beneficial muscle hypertrophy to potentially deleterious muscle atrophy might result from the same basic cause, a spontaneous Ca^{2+} leak from an abnormal Ca^{2+} release channel.

MH Regulatory Domains in the Ryanodine Receptor

An interesting feature of all the MH and CCD mutations discovered to date is that they lie in two distinct regions of the ryanodine receptor. The first of these regions lies between amino acid residues 163 and 615 and the second includes amino acids 2433 and 2434. Otsu et al. [8] predicted that a regulatory domain might exist in the ryanodine receptor between amino acids 2400 and 3000, because predicted phosphorylation sites and sequences forming potential binding sites for ATP and calmodulin were present in this sequence. This prediction was strengthened by studies of Chen et al. [33], which showed that the sequence between amino acids 2400 and 3000 is bracketed by tryptic cleavage sites, in line with the existence of this sequence as a unique domain. The sequence between amino acids 163 and 615 has not been predicted to contain ligand-binding sites, but it may well form a very important domain, perhaps one that interacts with the dihydropyridine receptor. Studies of tryptic cleavage [33] showed the presence of a tryptic cleavage site in the middle of this sequence; perhaps the sequence forms two interacting regulatory domains. Further analysis of these two interesting domains should lead to new insights into the regulatory features of the skeletal muscle Ca^{2+} release channel.

Acknowledgments. I thank my many colleagues for advice and discussion in the preparation of this review. Research grants to D.H.M., supporting original work from our laboratory, were from the Medical Research Council of Canada (MRCC), the Muscular Dystrophy Association of Canada (MDAC), the Heart and Stroke Foundation of Ontario (HSFO), and the Canadian Genetic Diseases Network of Centers of Excellence.

References

1. Ebashi S, Endo M, Ohtsuki I (1969) Control of muscle contraction. Q Rev Biophys 2:351–384
2. Fleischer S, Inui M (1989) Biochemistry and biophysics of excitation-contraction coupling. Annu Rev Biophys Chem 18:333–364
3. Catterall WA (1991) Excitation-contraction coupling in vertebrate skeletal muscle: a tale of two calcium channels. Cell 64:871–874
4. MacLennan, DH, Campbell KP, Reithmeier RAF (1983) Calsequestrin. In: Cheung WY (ed) Calcium and cell function. Academic, New York, pp 151–173
5. Knudson CM, Stang KK, Moomaw CR, Slaughter CA (1993) Primary structure and topological analysis of a skeletal muscle-specific junctional sarcoplasmic reticulum glycoprotein (triadin). J Biol Chem 268:12646–12654
6. Takeshima H, Nishimura S, Matsumoto T, Ishida H, Kangawa K, Minamino N, Matsuo H, Ueda M, Hanaoka M, Hirose T, Numa S (1989) Primary structure and expression from complementary DNA of skeletal muscle ryanodine receptor. Nature 339:439–445
7. Zorzato F, Fujii J, Otsu K, Phillips M, Green NM, Lai FA, Meissner G, MacLennan DH (1990) Molecular cloning of cDNA encoding human and rabbit forms of the Ca^{2+} release channel (ryanodine receptor) of skeletal muscle sarcoplasmic reticulum. J Biol Chem 265:2244–2256
8. Otsu K, Willard HF, Khanna VK, Zorzato F, Green NM, MacLennan DH (1990) Molecular cloning of cDNA encoding the Ca^{2+} release channel (ryanodine receptor) of rabbit cardiac muscle sarcoplasmic reticulum. J Biol Chem 265:13472–13483
9. Kalow WB, Britt A, Terreau ME, Haist C (1970) Metabolic error of muscle metabolism after recovery from malignant hyperthermia. Lancet ii:895–898
10. Endo M, Yagi S, Ishizuka T, Horiuti K, Koga Y, Amaha K (1983) Changes in the Ca-induced Ca release mechanism in sarcoplasmic reticulum from a patient with malignant hyperthermia. Biomed Res 4:83–92
11. Ohnishi ST, Taylor S, Gronert GA (1983) Calcium-induced Ca^{2+} release from sarcoplasmic reticulum of pigs susceptible to malignant hyperthermia. The effects of halothane and dantrolene. FEBS Lett 161:103–107
12. MacLennan DH, Duff C, Zorzato F, Fujii J, Phillips M, Korneluk RG, Frodis W, Britt BA, Worton RG (1990) Ryanodine receptor gene is a candidate for predisposition to malignant hyperthermia. Nature 343:559–561
13. McCarthy TV, Healy JMS, Heffron JJA, Lehane M, Deufel T, Lehmann-Horn F, Faralli M, Johnson K (1990) Localization of the malignant hyperthermia susceptibility locus to human chromosome 19q12–13.2. Nature 343:562–564
14. Deufel T, Golla A, Iles D, Meindl A, Meitinger T, Schindelhauer D, DeVries A, Pongratz D, MacLennan DH, Johnson KJ, Lehmann-Horn F (1992) Evidence for genetic heterogeneity of malignant hyperthermia susceptibility. Am J Hum Genet 50:1151–1161
15. Fujii J, Otsu K, Zorzato F, deLeon S, Khanna VK, Weiler J, O'Brien PJ, MacLennan DH (1991) Identification of a mutation in porcine ryanodine receptor associated with malignant hyperthermia. Science 253:448–451
16. Otsu K, Khanna VK, Archibald AL, MacLennan DH (1991) Co-segregation of porcine malignant hyperthermia and a probable causal mutation in the skeletal muscle ryanodine receptor gene in backcross families. Genomics 11:744–750
17. Gillard EF, Otsu K, Fujii J, Khanna VK, de Leon S, Derdemezi J, Britt BA, Duff CL, Worton RG, MacLennan DH (1991) A substitution of cysteine for arginine-614 in the ryanodine receptor is potentially causative of human malignant hyperthermia. Genomics 11:751–755
18. Fill M, Coronado R, Mickelson JR, Vilven J, Ma J, Jacobson BA, Louis CF (1990) Abnormal ryanodine receptor channels in malignant hyperthermia. Biophys J 50:471–475
19. Otsu K, Nishida K, Kimura Y, Kuzuya T, Hori M, Kamada T, Tada M (1994) The point mutation Arg615 to Cys in the Ca^{2+} release channel of skeletal sarcoplasmic reticulum

is responsible for hypersensitivity to caffeine and halothane in malignant hyperthermia. J Biol Chem 269:9413–9415
20. O'Brien PJ, Ball RO, MacLennan DH (1994) Effects of heterozygosity for the mutation causing porcine stress syndrome on carcass quality and live performance characteristics. In: Proceedings, 13th international pig veterinary society congress, Bangkok, p 481
21. Gillard EF, Otsu K, Fujii J, Duff CL, de Leon S, Khanna VK, Britt BA, Worton RG, MacLennan DH (1992) Polymorphisms and deduced amino acid substitutions in the coding sequence of the ryanodine receptor (RYR1) gene in individuals with malignant hyperthermia. Genomics 13:1247–1254
22. Quane KA, Keating KE, Manning BM, Healy JMS, Monsieurs K, Heffron JJA, Lehane M, Heytens L, Krivosic-Harber R, Adnet P, Ellis FR, Monnier N, Lumardi J, McCarthy TV (1994) Detection of a novel common mutation in the ryanodine receptor gene in malignant hyperthermia: implications for diagnosis and heterogeneity studies. Hum Mol Genet 3:471–476
23. Phillips MS, Khanna VK, de Leon S, Frodis W, Britt BA, MacLennan DH (1994) The substitution of Arg for Gly2433 in the human skeletal muscle ryanodine receptor is associated with malignant hyperthermia. Hum Mol Genet 3:2181–2186
24. Keating KE, Quane KA, Manning BM, Lehane M, Hartung E, Censier K, Urwyler A, Klausnitzer M, Muller CR, Heffron JJA, McCarthy TV (1994) Detection of a novel RYR1 mutation in four malignant hyperthermia pedigrees. Hum Mol Genet 3:1855–1858
25. Larach MG, Landis JR, Shirk BS, Diaz M (1992) The North American Malignant Hyperthermia Registry. Prediction of malignant hyperthermia susceptibility in man: improving sensitivity of the caffeine halothane contracture test. Anesthesiology 77A: 1052
26. Shy GM, Magee KR (1956) A new congenital non-progressive myopathy. Brain 79:610–621
27. Shuaib A, Paasuke IY, Brownell KW (1987) Central core disease: clinical features in 13 patients. Medicine (Baltimore) 66:389–396
28. Mulley JC, Kozman HM, Phillips HA, Gedeon AK, McCure JA, Iles DE, Gregg RG, Hogan K, Couch FJ, Weber JL, MacLennan DH, Haan EA (1993) Refined genetic localization for central core disease. Am J Hum Genet 52:398–405
29. Zhang Y, Chen HS, Khanna VK, de Leon S, Phillips MS, Schappert K, Britt BA, Brownell AKW, MacLennan DH (1993) Identification of a mutation in human ryanodine receptor associated with central core disease. Nature Genet 5:61–65
30. Quane KA, Healy JMS, Keating KE, Manning BM, Couch FJ, Palmucci LM, Doriguzzi C, Fagerlund TH, Berg K, Ording H, Bendixen D, Mortier W, Linz U, Muller CR, McCarthey TV (1993) Mutations in the ryanodine receptor gene in central core disease and malignant hyperthermia. Nature Genet 5:51–55
31. Quane KA, Keating KE, Healy JMS, Manning BM, Krivosic-Harbey R, Krivosic I, Monnier N, Lunardi J, McCarthy TV (1994) Mutation screening of the RYR1 gene in malignant hyperthermia: detection of a novel Tyr to Ser mutation in a pedigree with associated central cores. Genomics 23:236–239
32. Wrogemann K, Pena SDJ (1976) Mitochondrial calcium overload: a general mechanism for cell necrosis in muscle diseases. Lancet 1:672–673
33. Chen SRW, Airey JA, MacLennan DH (1993) Positioning of major tryptic fragments in the Ca^{2+} release channel (ryanodine receptor) of rabbit skeletal muscle sarcoplasmic reticulum. J Biol Chem 268:22642–22649

Chromosome 17 Candidate Gene Analysis in a Population Referred Because of Suspected Susceptibility to Malignant Hyperthermia

Gary M. Vita[1], Antonel Olckers[1,2], Anne E. Jedlicka[1], Terry Heiman-Patterson[3], Alfred L. George[4], Jeffery E. Fletcher[5], Henry Rosenberg[5], and Roy C. Levitt[1]

Introduction

Malignant hyperthermia susceptibility (MHS) is an autosomal dominant disorder of muscle metabolism that is triggered by potent inhalational anesthetics and depolarizing muscle relaxants. The signs and symptoms associated with an episode of malignant hyperthermia can include generalized muscle rigidity, masseter muscle spasm, rhabdomyolosis, acidosis, and hyperthermia. However, the clinical expression of malignant hyperthermia varies greatly from patient to patient. Moreover, the diagnosis is not easily established on the basis of clinical evidence alone, because the signs and symptoms of malignant hyperthermia are nonspecific. Thus, a clinical diagnosis must be sought by performing an in vitro contracture test (IVCT) to halothane or caffeine. Because the IVCT is costly and invasive, numerous investigators are searching for a noninvasive diagnostic test using molecular genetic approaches [1]. However, the development of noninvasive molecular genetic tests for MHS will depend on the identification of the genes responsible for this disorder.

Two molecular strategies are commonly employed to identify the chromosomal localization of a disease gene using genetic linkage analysis. First, a genomic search can be performed with highly informative polymorphic genetic markers whose location is known. Linkage is confirmed when markers in a particular chromosomal region are consistently coinherited with MHS [1]. Alternatively, a candidate gene approach can be used. A gene candidate may be chosen because it is logically related to the pathogenesis of the disorder or it is mapped to the same chromosomal region as the disorder. For example, genes involved in muscle metabolism and excitation contraction coupling might be logically related to the pathogenesis of MHS. In this chapter, we briefly describe our efforts focused on the evaluation of gene candidates on chromosome 17q that may be associated with an abnormal IVCT in patients suspected of MHS.

[1] Department of Anesthesiology and Critical Care Medicine, The Johns Hopkins Medical Institutions, Baltimore, MD 21287-7834, USA
[2] Department of Human Genetics and Developmental Biology, University of Pretoria, Pretoria, South Africa
[3] Thomas Jefferson University, Philadelphia, PA 19102, USA
[4] Vanderbilt University Medical Center, Nashville, TN 37232-2372, USA
[5] Department of Anesthesiology, Hahnemann University, Philadelphia, PA 19102, USA

Genetic Heterogeneity of Malignant Hyperthermia Susceptibility

It is generally accepted that calcium is the primary regulator of skeletal muscle metabolism and contraction, and that MHS is caused by a genetic defect that alters calcium regulation. It has recently been reported that at least some forms of MHS are associated with DNA polymorphisms in the calcium release channel gene (ryanodine receptor, RYR1) located on chromosome 19q13.1 [2]. Seven putative mutations in RYR1 have been described so far [3]. Although these reports suggest the calcium release channel may play a critical role in certain patients with MHS, we and others have demonstrated evidence for significant genetic heterogeneity in families affected with MHS [4–8]. Most MHS families studied by linkage thus far appear unlinked to chromosome 19q13.1–q13.2 [4–8]. Based on our results we predicted that genetic loci other than on chromosome 19q are important in causing malignant hyperthermia susceptibility [5]. Thus, in contrast to other autosomal dominant disorders in which a single genetic mutation in a single gene is responsible for the disease, MHS is heterogenous. These data suggest that an abnormal IVCT result in any individual with suspected MHS may be caused by a genetic mutation in any one of at least three different genes.

The Association of MHS with Other Neuromuscular Disorders

A number of neuromuscular disorders have been associated with both clinical signs of MHS and an abnormal IVCT used to diagnose this disorder [9,10]. Central-core disease (CCD) is an autosomal dominant muscle disorder that has also been mapped to chromosome 19q13.1, in the vicinity of the ryanodine receptor [11]. Recently, mutations in RYR1 have been described for CCD [12], suggesting that this disorder represents another form of MHS. MHS is also associated with the King–Denborough syndrome, carnitine palmitoyl transferase deficiency, Duchenne muscular dystrophy, and the myotonias, including the Schwartz–Jampel syndrome [10]. Because these disorders most likely result from defects in a wide variety of genes, mutations in these genes could account for some of the biological variability observed in the IVCT and perhaps MHS.

Search for Additional MHS Loci

Candidate Genes

Recent studies have identified additional loci associated with MHS. Candidate genes for these loci include the dihydropyridine receptor (DHPR) gene and the adult sodium channel α-subunit (SCN4A) gene. DHPR is an integral part of the skeletal muscle junctional triad composed of DHPR, RYR1, and the sarcolemmal membrane. Its importance in muscle function as well as the close association with the ryanodine receptor support its role as an MHS gene candidate. Linkage has now been reported between a single MHS family and markers located on chromosome 7q11.23–q21.1 near the DHPR α_2/δ subunits [13]. Thus, further investigation of these genes may uncover a role for them in certain forms of MHS [13].

Mutations have been described in SCN4A that are associated with various forms of generalized nondystrophic myotonia and hyperkalemic periodic paralysis [14]. Malignant hyperthermia-like reactions to anesthetic agents, including masseter muscle rigidity, have been reported in myotonic patients associated with an abnormal IVCT result [15]. Reactions resembling malignant hyperthermia have also been reported along with elevated myoplasmic calcium and abnormal sodium channel gating in the Schwartz–Jampel syndrome [16,17]. The relationship between MHS, masseter muscle rigidity, and these sodium channel-associated myopathies has remained obscure until recently. Weiland et al. demonstrated abnormal sodium channel function in muscle from individuals with an abnormal IVCT result who were suspected of having MHS. These data lend support to the hypothesis that mutation(s) of the SCN4A gene may cause an abnormal IVCT response and malignant hyperthermia-like syndrome [18]. Thus, there appears to be a relationship between sodium channel dysfunction, myotonia, and an abnormal IVCT. Because of these relationships we tested for genetic linkage between an abnormal IVCT result and the inheritance of polymorphic markers on chromosome 17q in the vicinity of the SCN4A gene candidate.

Our studies provided evidence for genetic linkage between an abnormal IVCT result in families suspected of having MHS and polymorphic markers in the vicinity of the α-subunit of the adult skeletal muscle sodium channel on chromosome 17q [5,19]. These studies have now been extended by examining the SCN4A gene candidate for mutations. We tested the hypothesis that specific mutations in the SCN4A gene may account for an abnormal IVCT in patients referred for a diagnosis of suspected MHS.

Examination of the SCN4A Gene Candidate

Utilizing single-strand conformational polymorphism (SSCP) analysis [20,21], we screened 238 unrelated individuals for mutations in the SCN4A gene who had been referred for MHS biopsy because of a family history or suspected episode of MH. We began by examining those exons of the SCN4A gene in which known mutations were associated with abnormal sodium channel function. From these analyses, polymorphisms were detected in two individuals in exon 22 (Fig. 1). Each of these individuals had either a family history or personal experience with masseter muscle rigidity (MMR) and generalized muscle spasm after exposure to succinylcholine. Direct DNA sequence analyses confirmed that these individuals carry a C-to-T mutation at nucleotide position 3938 described previously [22]. As a result of this polymorphism, a methionine is substituted for a threonine at amino acid position 1313. Interestingly, the methionine substitution at amino acid 1313 causes a more severe form of generalized myotonia, termed paramyotonia [22]. Although both individuals were suspected of having MHS by history, only one developed an abnormally large contracture on testing in vitro to halothane (0.9-g contracture on one muscle strip exposed to 3% halothane).

Using this same approach, we examined the SCN4A gene for known mutations in five families (US1, US2, US3, US5, and SA1) that were potentially linked to this gene on chromosome 17q. A single DNA polymorphism was detected by SSCP in exon 22 of SCN4A in family US5. Two probands from this family experienced MMR and whole-body rigidity after exposure to succinylcholine. Subsequent muscle biopsies and IVCT results suggested susceptibility to MH in these two family members [15,23].

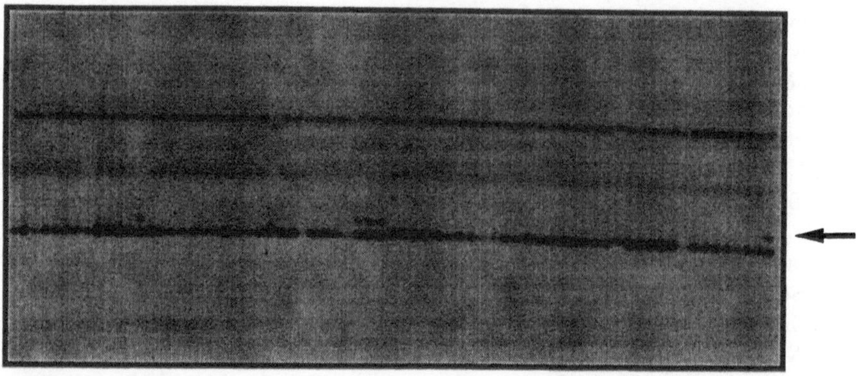

FIG. 1. Examination for DNA polymorphisms in 238 unrelated individuals referred for in vitro contracture testing for malignant hyperthermia using single-stranded conformational polymorphism (SSCP) analyses. *Arrow* points to abnormal conformers representing a single base change (mutation) in the DNA sequence. Primers for the polymerase chain reaction (PCR) were synthesized and selected from a unique sequence from the adult skeletal muscle sodium channel α-subunit (SCN4A) gene surrounding each exon, using OLIGO 4.0 [27]. SSCP was carried out largely as described elsewhere [20,21] with modifications [28]

The polymorphism detected occurred only in individuals diagnosed with myotonia and MHS, or in those who were affected with myotonia but were not tested by IVCT. Direct DNA sequencing of this exon revealed a single base substitution at nucleotide position 3917 in these individuals (Fig. 2). As a result of this polymorphism, an alanine is substituted for a glycine at amino acid position 1306 within the cytoplasmic loop connecting repeats III and IV of SCN4A.

The SCN4A mutation, $Gly^{1306}Ala$, has been previously described and predisposes patients to myotonia fluctuans [24]. This form of myotonia is characterized by very mild daily fluctuations of muscle stiffness that are aggravated by exercise [25]. Because this mutation is associated with subtle signs and mild symptoms, this disorder is likely to be missed on routine preoperative screening, as it was in two individuals from this family. Moreover, the administration of succinylcholine to patients with myotonia fluctuans may result in increased morbidity and mortality during anesthesia.

A restriction fragment length polymorphism (RFLP) is produced by this G-to-C mutation at nucleotide 3917 of the SCN4A gene. This base substitution produces the addition of an *AciI* restriction enzyme site. Therefore, we were able to develop a molecular diagnostic test to detect the $Gly^{1306}Ala$ mutation. When the 199-base polymerase chain reaction (PCR) product that includes this site is digested with *AciI* in affected individuals, it is cleaved into two fragments of 139 and 60 bases in length which are easily detected by gel electrophoresis. This diagnostic RFLP assay represents a rapid, sensitive, specific, and inexpensive noninvasive preoperative method of screening for this disorder [23].

On the basis of our findings, it is reasonable to speculate that other mutations in the 22 remaining exons of the SCN4A gene, or in the β-subunit gene of the sodium channel mapped to chromosome 19q13.1–q13.2 [26], may produce abnormal IVCT, MMR, and potentially other malignant hyperthermia-like reactions to anesthetics.

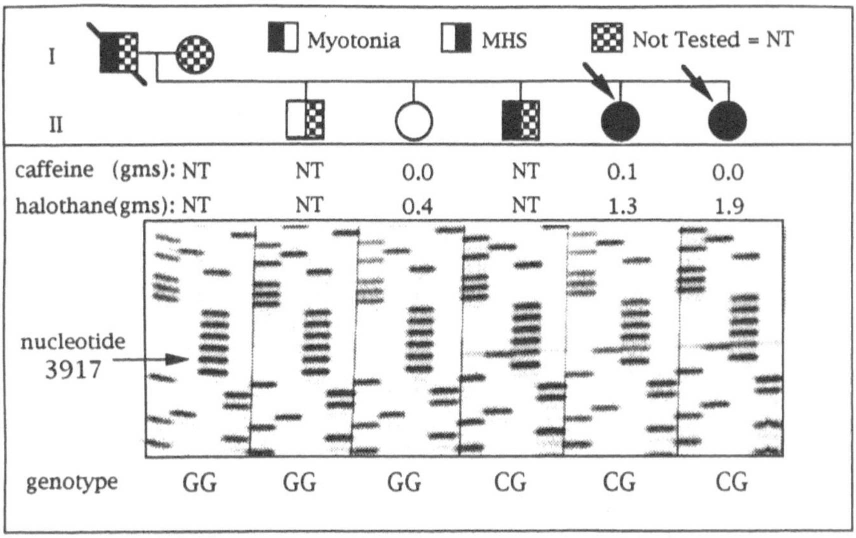

FIG. 2. Nuclear family shows relationship between malignant hyperthermia susceptibility (*MHS*)/masseter muscle rigidity (MMR), myotonia, the caffeine halothane in vitro contracture test (IVCT) result, and a segment of the DNA sequence from exon 22 of the adult skeletal muscle sodium channel α-subunit (SCN4A) gene. Note the G-to-C substitutions at position 3917 (*lighter arrow*) in exon 22 of the SCN4A gene in individuals II-2, II-3, and II-4. An alanine is substituted for a glycine at position 1306 as a result of this mutation. The *bold arrows* indicate individuals who experienced MMR and were referred for evaluation because of suspected MHS. History or clinical examination of individuals I-1, II-4, II-3, and II-2 revealed evidence of myotonia. Individuals who did not harbor this polymorphism displayed no evidence of myotonia by history or neurological exam. Electromyography (EMG) data confirm these diagnoses for individuals II-6, II-3, II-22, and III-2. Established protocols were used as a standardized test to determine the MHS phenotype and are based on skeletal muscle IVCT to halothane or caffeine [29]. Conditions for PCR, SSCP, and direct DNA sequencing were carried out as described elsewhere [23]. (This figure has been reproduced with permission from J.B. Lippincott Company, Philadelphia, PA, (from [23])

Summary

In this chapter we present evidence for an association between mutations in the SCN4A gene, masseter muscle rigidity (MMR), and an abnormal IVCT used to diagnose MHS. In one family examined, two members experienced MMR on exposure to succinylcholine. These individuals were shown to have a very mild atypical form of nondystrophic generalized myotonia, confirmed by electrornyoglaphy (EMG), and were considered susceptible to malignant hyperthermia because of an abnormal IVCT result. Subsequently, we have characterized their disease on a molecular level as myotonia fluctuans. Two other unrelated individuals referred for a diagnosis of MHS were found to harbor a different mutation in the SCN4A gene candidate associated with MMR that was not consistently identified with an abnormal IVCT. Thus, specific mutations in the SCN4A gene may predispose certain patients to an abnormal IVCT and MMR in response to succinylcholine administration. These data advance our understanding of the molecular pathogenesis of MMR. Moreover, these findings provide new insights into the relationship between biological variability in the IVCT and MMR, which is closely associated with MHS and an important clinical complication during anesthesia.

Acknowledgments. We thank the families and individuals who participated in this study. Dr. Levitt was supported by the Hahnemann Anesthesia Research Foundation, MDA, MHAUS, FAER, and National Institutes of Health grant GM47145. A. Olckers was supported by the MDA of South Africa, and the Department of Human Genetics and Developmental Biology, University of Pretoria. Dr. George is a Lucille P. Markey Scholar and is supported by the National Institutes of Health grant AR01862.

References

1. Levitt RC (1992) Prospects for the diagnosis of malignant hyperthermia susceptibility using molecular genetic approaches. Anesthesiology 76:1039–1048
2. Fujii J, Otsu K, Zorrato F, Leon S, Khanna K, Weiler JE, O'Brien PJ, MacLennan DH (1991) Identification of a mutation in porcine ryanodine receptor associated with malignant hyperthermia. Science 253:448–451
3. Quane KA, Keating KE, Healy JMS, Manning BM, Krivosic-Horber R, Krivosic I, Monnier N, Lunardi J, McCarthy TV (1994) Mutation screening of the RYR1 gene in malignant hyperthermia: detection of a novel Tyr to Ser mutation in a pedigree with associated central cores. Genomics 23:236–239
4. Levitt RC, Nouri N, Jedlicka AE, McKusick VA, Marks A, Shutack JG, Fletcher JE, Rosenberg H, Meyers DA (1991) Evidence for genetic heterogeneity in malignant hyperthermia susceptibility. Genomics 11:534–547
5. Levitt RC, Olckers A, Meyers S, Fletcher JE, Rosenberg H, Isaacs H, Meyers DA (1992) Evidence for the localization of a malignant hyperthermia susceptibility locus to human chromosome 17q. Genomics 14:562–566
6. Fagerlund T, Islander G, Ranklev E, Harbitz I, Hauge JG, Mokleby E, Berg K (1992) Genetic recombination between malignant hyperthermia and calcium release channel in skeletal muscle. Clin Genet 41:270–272
7. Deufel T, Golla A, Iles D, Meindl A, Meitinger T, Schindelhaer D, DeVries A, Pongratz D, MacLennan DH, Johnson KJ, Lehmann-Horn F (1992) Evidence for genetic heterogeneity of malignant hyperthermia susceptibility. Am J Hum Genet 50:1151–1161
8. Iles DE, Segers B, Heytens L, Sengers RC, Wieringa B (1992) High-resolution physical mapping of four microsatellite repeat markers near the RYR1 locus on chromosome 19q13.1 and apparent exclusion of the MHS locus from this region in two malignant hyperthermia-susceptible families. Genomics 14:749–754
9. Heiman-Patterson TD, Rosenberg H, Fletcher JE, Tahmoush AJ (1988) Halothane-caffeine contracture testing in neuromuscular diseases. Muscle & Nerve 11:453–457
10. Allen GC (1994) Malignant hyperthermia susceptibility. In: Temperature regulation during anesthesia. Anesthesiology clinics of North America, vol 12. Saunders, Philadelphia, pp 513–536
11. Mulley JC, Kozman HM, Phillips HA, Gedeon AK, McCure JA, Iles DE, Gregg RG, Hogan K, Couch FJ, MacLennan DH, Haan EA (1993) Refined genetic localization for central core disease. Am J Genet 52:398–405
12. Quane KA, Healy JMS, Keating KE, Manning BM, Couch FJ, Palmucci LM, Doriguzzi C, Fagerlund TH, Berg K, Ording H, Bendixon D, Mortier W, Linz U, Muller CR, McCarthy TV (1993) Mutations in the ryanodine receptor gene in central core disease and malignant hyperthermia. Nature Genet 5:51–55
13. Iles DE, Lehmann-Horn F, Scherer SW, Tsui L-C, Olde Weghuis D, Suijkerbuijk R, Heytens L, Mikala G, Schwartz A, Ellis FR, Stewart AD, Deufel T, Wieringa B (1994) Localization of the gene encoding the α_2/δ-subunits of the L-type voltage-

dependent calcium channel to chromosome 7q and analysis of the segregation of flanking markers in malignant hyperthermia susceptibile families. Hum Mol Genet 3:969–975
14. Rudel R, Ricker K, Lehmann-Horn F (1993) Genotype-phenotype correlations in human skeletal muscle sodium channel diseases. Arch Neurol 50:1241–1248
15. Heiman-Patterson T, Martino C, Rosenberg H, Fletcher J, Tahmoush A (1988) Malignant hyperthermia in myotonia congenita. Neurology 38:810–812
16. Seay AR, Ziter FA (1978) Malignant hyperthermia in a patient with Schwartz-Jampel syndrome. J Pediatr 93:82–87
17. Lehmann-Horn F, Iaizzo PA, Franke C, Hatt H, Spaans F (1991) Schwartz-Jampel syndrome: II. Na+ channel defect causes myotonia. Muscle & Nervel 13:528–535
18. Weiland SJ, Fletcher JE, Rosenberg H, Gong Q-H (1989) Malignant hyperthermia: slow sodium current in cultured human muscle cells. Am J Physiol 257:C759–C765
19. Olckers A, Meyers DA, Meyers S, Taylor EW, Fletcher JW, Rosenberg H, Isaacs H, Levitt RC (1992) Adult muscle sodium channel α-subunit gene candidate for malignant hyperthermia susceptibility. Genomics 14:829–831
20. Orita M, Suzuki Y, Sekiya T, Hayashi K (1989) Rapid and sensitive detection of point mutations and DNA polymorphisms using the polymerase chain reaction. Genomics 5:874–879
21. Orita M, Iwahana H, Kanazawa H, Hayashi K, Sekiya T (1989) Detection of polymorphisms of human DNA by gel electrophoresis as single strand conformation polymorphisms. Proc Natl Acad Sci USA 86:2766–2770
22. McClatchey AI, Van den Bergh P, Pericak-Vance MA, Raskind W, Verellen C, McKenna-Yasek D, Rao K, Haines JL, Bird T, Brown RH, Gusella JF (1992) Temperature sensitive mutations in the III–IV cytoplasmic loop region of the skeletal muscle sodium gene in paramyotonia congenita. Cell 68:769–774
23. Vita GM, Olckers A, Jedlicka A, George AL, Fletcher JE, Rosenberg H, Heiman-Patterson T, Levitt RC (1995) Masseter muscle rigidity associated with glycine^{1306} to alanine mutation in the adult muscle sodium channel α-subunit. Anesthesiology 82:1097–1103
24. Lerche H, Heine R, Pika U, George AL, Mitrovic N, Browatzki M, Weib T, Rivet-Bastide M, Franke C, Lomonaco M, Ricker K, Lehmann-Horn F (1993) Human sodium channel inactivation due to substitutions for a glycine within the III-IV linker. J Physiol (Lond) 470:13–22
25. Ricker K, Lehmann-Horn F, Moxley R (1990) Myotonia fluctuans. Arch Neurol 47: 268–272
26. Makita N, Sloan-Brown K, Weghuis DO, Ropers HH, George AL Jr (1994) Genomic organization and chromosomal assignment of the human voltage-gated Na+ channel β-subunit gene (SCN1B). Genomics 23:628–634
27. George AL, Iyer GS, Kleinfield R, Kallen RG, Barchi RL (1993) Genomic organization of the human skeletal sodium channel gene. Genomics 15:598–606
28. Schwengel DA, Nouri N, Meyers DA, Levitt RC (1993) Linkage mapping of the human thromboxane A_2 receptor (TBXA2R) to chromosome 19q13.3 using transcribed 3′ untranslated DNA polymorphisms. Genomics 18:212–215
29. Fletcher JE (1994) Current laboratory methods for the diagnosis of malignant hyperthermia susceptibility. In: Temperature regulation during anesthesia. Anesthesiology clinics of North America, vol 12. Saunders, Philadelphia, pp 553–570

Disease Expression and Gene Expression Can Be Quite Distinct: Modulation of Skeletal Muscle Sodium Channels by the Human Myotonin Dystrophy Protein Kinase (HMPK)*

J. Paul Mounsey[1], Puting Xu[2], J. Edward John III[1], John Gilbert[2], Allen D. Roses[2], and J. Randall Moorman[1]

Current Status of Myotonic Dystrophy

The genetic mutation responsible for the autosomal dominant inheritance of myotonic dystrophy (DM) is a variable triplet repeat (CTG) that is located in the 3′ untranslated region of a gene with protein kinase domains named myotonin protein kinase [1–4]. The CTG repeat segregates in the normal population and varies from 5 to 40 repeats, while in asymptomatic DM gene carriers and patients it varies from 50 to several thousand repeats. The role of the human myotonin protein kinase gene (HMPK) and the effects of the triplet repeat are unknown [2,5–7].

Because of the characteristic myotonia, reduced muscle resting membrane potential, and the wide clinical involvement of many organ systems including the lens and the cardiac conduction pathways, the biochemical involvement of cell membranes has been postulated for many years [8–10]. There are mutations of the chloride and sodium channels in other diseases in which myotonia or paramyotonia are signs, as in congenital myotonia [11,12] and hyperkalemic periodic paralysis [13–15].

The mechanism by which the genetic mutation causes symptoms and signs in DM is unknown. It is similar to fragile X syndrome in that the expanded repeat can become very large with several thousand repeats [2,16]. This differs from other triplet repeat diseases such as Kennedy's disease [17] or Huntington's disease in which the expansion of the repeat is more modest, with a two- to threefold increase [18]. Studies of the expression of HMPK mRNA in skeletal muscle have demonstrated conflicting results [19–21]. The normal HMPK mRNA is expressed in muscle. There is an abnormally large mRNA expressed in congenital DM [6,22]. There is yet to appear an adequate hypothesis to explain the variable effect on phenotypic expression based on the coding potential of the HMPK gene or the protein kinase enzymatic activity in the presence of at least one normal allele.

The pathogenic effect of the unstable triplet repeat could be at the level of gene regulation, rather than an effect limited to the coding capacity of the HMPK gene

---

* This work was published in part in the Journal of Clinical Investigation (95:2379–2384) 1995

[1] Departments of Internal Medicine (Cardiovascular Division), Molecular Physiology, and Biological Physics, University of Virginia, Charlottsville, VA 22908, USA

[2] Departments of Medicine (Neurology) and Neurobiology, Duke University Medical Center, Durham, NC 27710, USA

[19,23]. Simple tandem repeat sequences are binding sites for specific nuclear proteins. The binding of nuclear regulatory proteins may have a greater effect and extend to genes other than the HMPK. Thus, while the unstable triplet repeat is the inherited locus, the mechanism of disease expression may be related to its effects on regulation of other nuclear mechanisms. Studies in DM are in progress, extending the pioneering work of Richards and Sutherland [24,25] in fragile X syndrome.

In this chapter, we demonstrate a phenotypic effect resembling that found in DM skeletal muscle by the coexpression of muscle Na^+ channels with the HMPK gene with an expanded triplet repeat. These coexpression data are consistent with the demonstration that both the normal and DM HMPK alleles are expressed in DM [26] and that phenotypic expression may be caused by an additional unique effect of the DM HMPK.

Results

Methods are described and cited in Mounsey et al. [27].

Figure 1 illustrates the Na^+ currents in a *Xenopus* oocyte expressing the alpha subunit of the rat skeletal muscle Na^+ channel alone (Fig. 1a) and another oocyte from

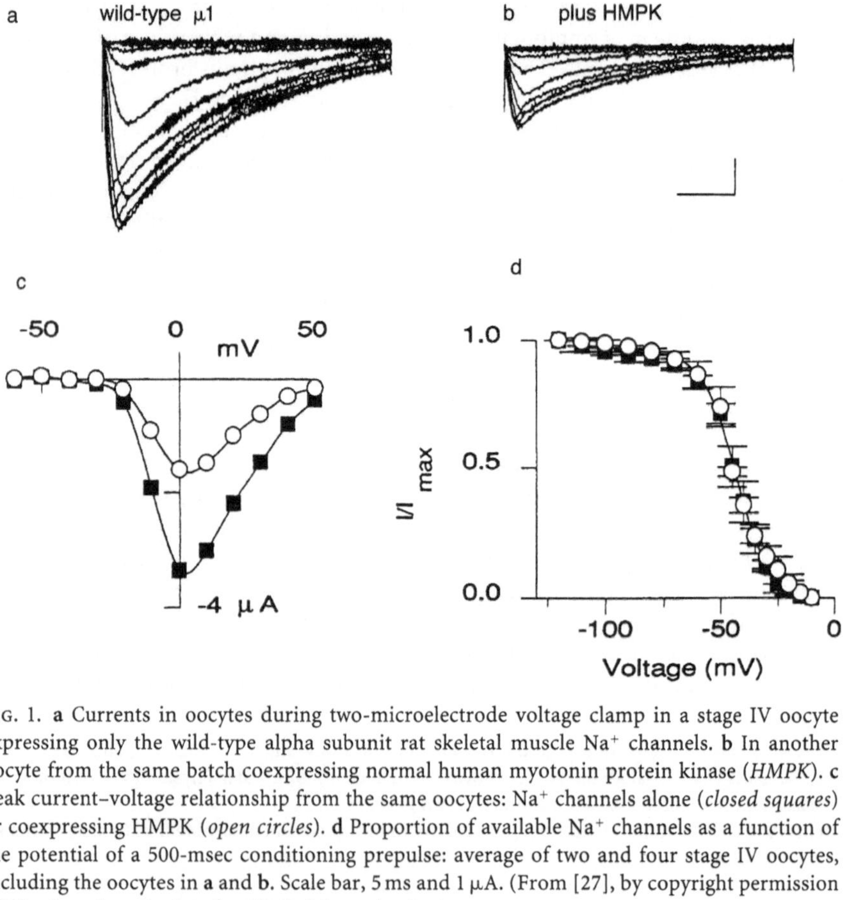

FIG. 1. **a** Currents in oocytes during two-microelectrode voltage clamp in a stage IV oocyte expressing only the wild-type alpha subunit rat skeletal muscle Na^+ channels. **b** In another oocyte from the same batch coexpressing normal human myotonin protein kinase (*HMPK*). **c** Peak current–voltage relationship from the same oocytes: Na^+ channels alone (*closed squares*) or coexpressing HMPK (*open circles*). **d** Proportion of available Na^+ channels as a function of the potential of a 500-msec conditioning prepulse: average of two and four stage IV oocytes, including the oocytes in **a** and **b**. Scale bar, 5 ms and 1 μA. (From [27], by copyright permission of The American Society for Clinical Investigation)

the same batch coexpressing a HMPK with a 5-mer triplet repeat (Fig. 1b). The currents were smaller when the HMPK was coexpressed, but the voltage-dependent activation and steady-state inactivation of the currents were unchanged.

When the Na^+ channel is mutated at a potential phosphorylation site (S1321) in the cytoplasmic linker between domains III and IV so that it cannot be phosphorylated when expressed, the effect of HMPK is much smaller (Fig. 2a,b). HMPK with a 5-mer repeat reduced the current through wild-type Na^+ channels, but had no effect on the Na^+ channels mutated at the phosphorylation site [phos(-)] (Fig. 2c). This linking region plays a major role in Na^+ channel inactivation [28–33]. When a HMPK constructed to contain a 75-mer triplet repeat was coexpressed with the Na^+ channels, the reduction in current amplitude was reduced by 9% compared to a 48% reduction by 5-mer HMPK (Fig. 2d).

To better resolve the peak and the kinetics of fast currents, patch-clamp techniques were used. Patches of oocyte membranes expressing only Na^+ channels had larger currents than those coexpressing either normal (5-mer) or DM (75-mer) HMPK. Figure 3 shows the averages of 50–100 traces in oocytes expressing muscle Na^+ channels with the 5-mer HMPK, with the DM HMPK, or alone. With either form of HMPK there were fewer channels than when the Na^+ channel mRNA was expressed alone. The change in current amplitude could not be attributed to a change in the voltage dependence of activation gating; in patches with very large currents, peak current–voltage relationships showed no shift along the voltage axis (Fig. 3d).

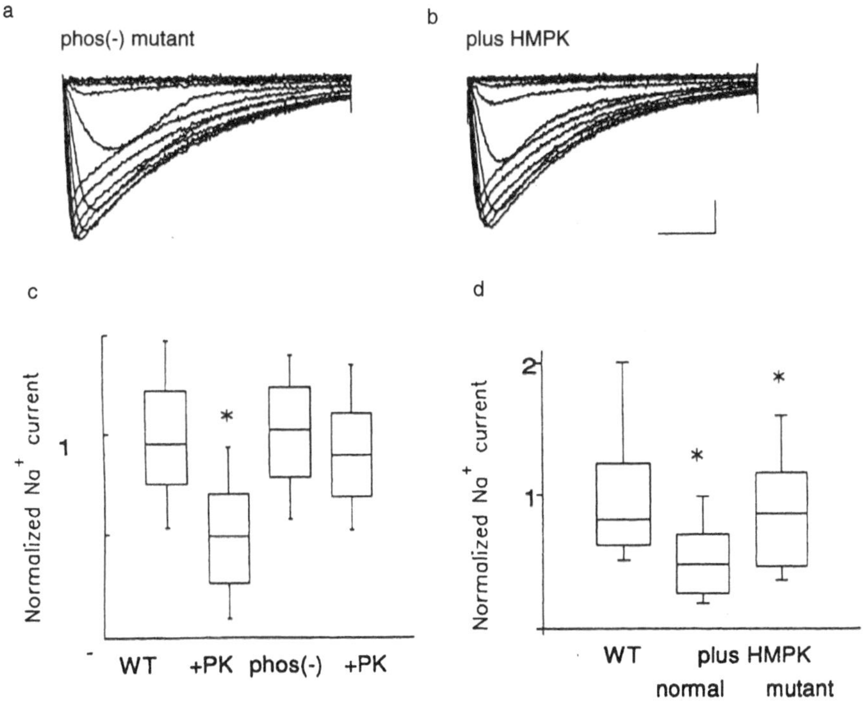

FIG. 2. **a** Currents in oocytes expressing the phos(-) mutated Na^+ channels alone and **b** coexpressing HMPK. **c** Box plots of peak Na current for the wild-type Na^+ channel alone and **d** coexpressing HMPK. Scale; 5 ms and 1 μA. (From [27], by copyright permission of The American Society for Clinical Investigation)

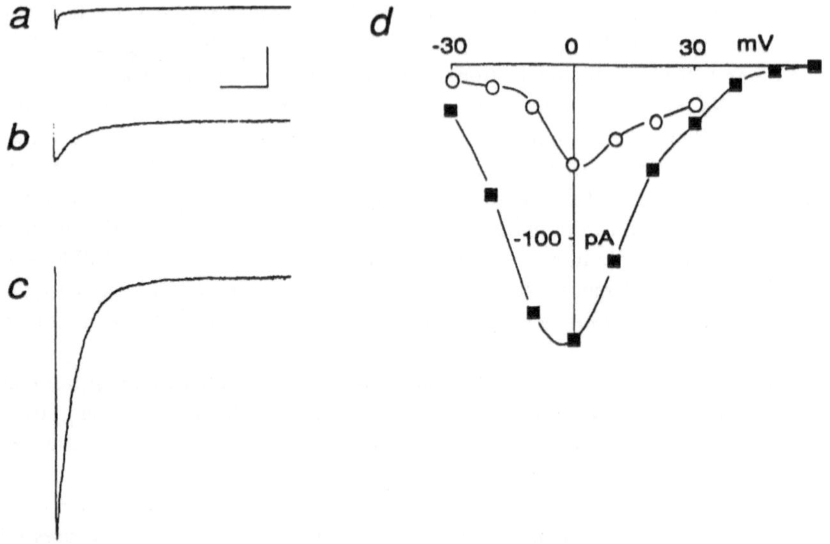

FIG. 3a–d. Patch currents. **a** Average of 50–100 traces in oocytes expressing wild-type rat skeletal muscle Na^+ channels alone, **b** with the myotonic dystrophy (DM) mutant HMPK, or **c** with the normal HMPK. Scale bar, 20 msec and 10 pA. **d** Peak current–voltage relationship for two patches, one in an oocyte expressing only Na^+ channels (*closed squares*) and in another oocyte coexpressing HMPK (*open circles*). (From [27], by copyright permission of The American Society for Clinical Investigation)

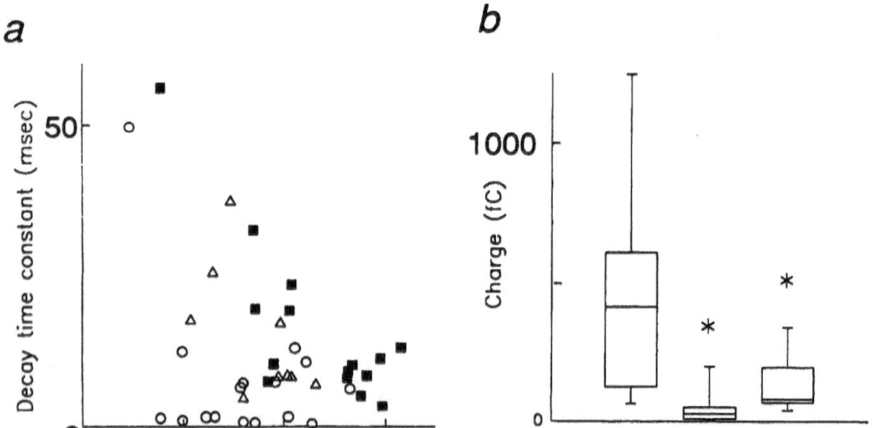

FIG. 4. **a** Time to half-decay as a function of peak amplitude for patch currents in oocytes expressing only Na^+ channels (*closed squares*), and oocytes coexpressing normal (*open circles*) and DM mutant HMPK (*open triangles*). **b** Box plots of transmembrane charge movement. *Asterisks* indicate $P < 0.005$ for any pair of groups; Mann-Whitney U test. (From [27], by copyright permission of The American Society for Clinical Investigation).

In addition to reducing Na^+ current amplitude, coexpression of HMPK changed the current decay rate. Figure 4a shows the decay rate as a function of current amplitude. Currents decayed more slowly in oocytes expressing only the Na^+ channels. At all amplitudes, the time constants of Na^+ currents were significantly less when the normal HMPK was coexpressed ($P < 0.005$, Mann–Whitney U test). When the DM HMPK was coexpressed, there were no significant differences in the time constants of Na^+ currents.

The current waveform was integrated to compare the total charge carried across each patch. The median charge in patches expressing the Na^+ channels alone was 15-fold larger than in oocytes coexpressing the normal HMPK, and 5-fold larger in oocytes coexpressing the DM HMPK.

Conclusions

These data suggest the HMPK does have a protein kinase function and exerts a biological effect by modulating the Na^+ channel. A target phosphorylation site is in the inactivation mechanism of the skeletal muscle Na^+ channel. The effect of expressing the normal HMPK is likely to be dampened excitivity; the diminished activity of the DM HMPK may help explain the hyperexcitability of muscle in DM. It is likely that this effect operates on other channels with similar inactivation mechanisms. One of the disease-specific effects of expression of the DM HMPK may be to abnormally modulate intrinsic channels in many membranes and initiate a chronic cascade of events leading to the manifestations of the disease. The view of DM as a generalized membrane disease associated with abnormal intrinsic membrane protein phosphorylation was recognized two decades before the molecular tools for these experiments were available [8,9,34].

Acknowledgments. This work by supported by NIH-NS19999 (ADR), the Piton Foundation, Denver, CO (ADR), and the Muscular Dystrophy Association (JG, ADR).

References

1. Aslanidis C, Jansen G, Amemiya C, Shutler G, Mahadevan M, Tsilfidis C, Chen C, Alleman J, Wormskamp NG, Vooijs M, Buxton J, Johnson K, Smeets HJM, Lennon GG, Carrano AV, Korneluk RG, Wieringa B, de Jong PJ (1992) Cloning of the essential myotonic dystrophy region and mapping of the putative defect. Nature 355:548–551
2. Brook JD, McCurrach ME, Harley HG, Buckler AJ, Church D, Aburatani H, Hunter K, Stanton VP, Thirion JP, Hudson T, Sohn R, Zemelman B, Snell RG, Rundle S, Crow S, Davies J, Shelbourne P, Buxton J, Jones C, Juvonen V, Johnson K, Harper PS, Shaw D, Housman DE (1992) Molecular basis of myotonic dystrophy: expansion of a trinucleotide (CTG) repeat at the 3′ end of a transcript encoding a protein kinase family member. Cell 68:799–808
3. Mahadevan M, Tsilfidis C, Sabourin L, Shutler G, Amemiya C, Jansen G, Neville C, Narang M, Barcelo J, O'Hoy K, Leblond S, Earle-MacDonald J, de Jong PJ, Wieringa B, Korneluk RG (1992) Myotonic dystrophy mutation: an unstable CTG repeat in the 3′untranslated region of the gene. Science 255:1253–1255
4. Fu YH, Pizzuto A, Fenwick RG Jr, King J, Rajnarayan S, Dunne PW, Dubel J, Nasser GA, Ashizawa T, de Jong P, Wieringa B, Korneluk R, Perryman MB, Epstein HF, Caskey CT (1992) An unstable triplet repeat in a gene related to myotonic muscular dystrophy. Science 255:1256–1258

5. Shaw DJ, Harper PS (1992) Myotonic dystrophy: advances in molecular genetics. Neuromuscular Disorder 2:241–243
6. Roses AD (1992) Myotonic dystrophy. Trends Genet 8:254–255
7. Brunner HG, Nillesen W, van Oost BA, Jansen G, Wieringa B, Ropers HH, Smeets H (1992) Presymptomatic diagnosis of myotonic dystrophy. J Med Genet 29:780–784
8. Roses AD, Appel SH (1974) Muscle membrane protein kinase in myotonic muscular dystrophy. Nature 250:245–247
9. Roses AD, Appel SH (1975) Phosphorylation of component "a" of the human erythrocyte membrane in myotonic dystrophy. J Membr Biol 20:51–58
10. Wong P, Roses AD (1977) Altered component "a" phosphorylation in erythrocyte membrane in myotonic muscular dystrophy. J Membr Biol 45:145–166
11. Koch MC, Steinmeyer K, Lorenz C, Kicker K, Wolf F, Otto M, Zoll B, Lehmann-Horn F, Grzeschik KH, Jentsch TJ (1992) The skeletal muscle chloride channel in dominant and recessive human myotonia. Science 257:797–800
12. George AL Jr, Crackower MA, Abdalla JA, Hudson AJ, Ebers GC (1993) Molecular basis of Thomsen's disease (autosomal dominant myotonia congenita). Nature Genet 3: 305–310
13. Fontaine B, Khurana TS, Hoffman EP, Bruns GA, Haines JL, Trofatter JA, Hanson MP, Rich J, McFarlane H, Yasek DM, Romano D, Gusella JF, Brown RH Jr (1990) Hyperkalemic periodic paralysis and the adult muscle sodium channel alpha subunit gene. Science 250:1000–1002
14. Rojas CV, Wang JZ, Schwartz LS, Hoffman EP, Powell BR, Brown RH Jr (1991) A Met-to-Val mutation in the skeletal muscle Na^+ channel alpha-subunit in hyperkalaemic periodic paralysis. Nature 354:387–389
15. Ptacek LJ, George AL Jr, Griggs RC, Tawil R, Kallen RG, Barchi RL, Robertson M, Leppert MF (1991) Identification of a mutation in the gene causing hyperkalemic periodic paralysis. Cell 67:1021–1027
16. Kremer EJ, Pritchard M, Lynch M, Yu S, Holman K, Baker E, Warren ST, Schlessinger D, Sutherland GR, Richards RI (1991) Mapping of the DNA instability at the fragile X to a trinucleotide repeat sequence p(CCG)n. Science 252:1711–1714.
17. LaSpada AR, Wilson EM, Lubahn DB, Harding AE, Fischbeck KH (1991) Androgen receptor gene mutations in X-linked spinal and bulbar muscular atrophy. Nature 352:77–79
18. MacDonald ME, Ambrose CM, Duyao MP, Myers RH, Lin C, Srinidhi L, Barnes G, Taylor SA, James M, Groot N, MacFarlane H, Jenkins B, Anderson MA, Wexler NS, Gusella JF (1993) A novel gene containing a trinucleotide repeat that is expanded and unstable on Huntington's disease chromosomes. Cell 72:971–983
19. Gilbert JR, Taylor H, Schwartzbach C, Roses AD (1995) Myotonic dystrophy. In: Kakulas BA, Howell J McC, Roses AD (eds) Duchenne muscular dystrophy: animal and genetic manipulation. Raven Press, New York (in press)
20. Fu YH, Friedman DL, Richards S, Pearlman JA, Gibbs RA, Pizzuti A, Ashizawa T, Perryman MB, Scarlato G, Fenwick RG, et al (1993) Decreased expression of myotonin protein kinase messenger RNA and protein in adult form of myotonic dystrophy. Science 260:235–238
21. Sabouri LA, Mahadevan MS, Narang M, Lee DS, Surh LC, Korneluk RG (1993) Effect of the myotonic dystrophy (DM) mutation on mRNA levels of the DM gene. Nature Genet 4:233–238
22. Roses AD (1993) Myotonic dystrophy. In: Rosenberg RN, Prusiner SB, DiMauro S, Barchi RL, Kunkel LP (eds) The molecular and genetic basis of neurological disease. Butterworth-Heineman, Stoneham, MA, pp 633–646
23. Richards RI, Holman K, Yu S, Sutherland GR (1993) Fragile X syndrome unstable element, p(CCG)n, and other simple tandem repeat sequences are binding sites for specific nuclear proteins. Hum Mol Genet 2:1429–1435
24. Richards RI, Sutherland GR (1992) Dynamic mutations: a new class of mutations causing human disease. Cell 70:709–712

25. Richards RI, Sutherland GR (1992) Heritable unstable DNA sequences. Nature Genet 1:7–9
26. Krahe R, Ashizawa T, Funanage V, et al (1994) Decreased steady-state levels of myotonic dystrophy protein kinase mutant hnRNA and mRNA in muscle of myotonic dystrophy patients (Abstract). In: MDA symposium on molecular mechanisms of neuromuscular disease, Tucson, AZ, January 1994, and VIIIth International Congress on Neuromuscular Diseases, Kyoto, Japan, July 1994
27. Mounsey JP, Xu P, John JE III, Gilbert J, Roses AD, Moorman JR (1995) Modulation of skeletal muscle sodium channels by the human myotonin protein kinase. J Clin Invest 95:2379–2384
28. Vassilev PM, Scheuer T, Catterall WA (1988) Identification of an intracellular peptide segment involved in sodium channel inactivation. Science 241:1658–1661
29. Stuhmer W, Conti F, Suzuki H, Wang X, Noda M, Yahagi N, Kubo H, Numa S (1989) Structural parts involved in activation and inactivation of the sodium channel. Nature 339:597–603
30. Moorman JR, Kirsch GE, Brown AM, Joho RH (1990) Changes in sodium channel gating produced by point mutations in a cytoplasmic linker. Science 250:688–691
31. Moorman JR, Kirsch GE, Van Dongen AMJ, Joho RH, Brown AM (1990) Fast and slow gating of sodium channels encoded by a single mRNA. Neuron 4:243–252
32. West JW, Numann R, Murphy BJ, Scheuer T, Catterall WA (1991) A phosphorylation site in the Na^+ channel required for modulation by protein kinase C. Science 254:866–868
33. West JW, Patton DE, Scheuer T, Wang Y, Goldin AL, Catterall WA (1992) A cluster of hydrophobic amino acid residues required for fast Na(+)-channel inactivation. Proc Natl Acad Sci USA 89:10910–10914
34. Roses AD, Harper P, Bossen E (1979) Myotonic muscular dystrophy (dystrophia myotonica, myotonia atrophy). In: Vinken PJ, Bruyn CW (eds) Handbook of clinical neurology. Wiley, New York, pp 485–532

Part 5. Biological Study of Malignant Hyperthermia

Intracellular Calcium Release Channels in Muscles Related to Excitation-Contraction-Coupling and Malignant Hyperthermia

MAKOTO ENDO

Abstract. A plant alkaloid, ryanodine, was found to act specifically on intracellular Ca release channels, fixing them in the open state, only when the channels were open [1]. Utilizing the specific action of ryanodine, the channel protein was isolated, purified [2–4], and sequenced [5]. The purified channel protein showed all the properties of the Ca-induced Ca release (CICR) channel [3,4,6]; activation by Ca, acceleration of opening by ATP, inhibition of opening by Mg, and so on. Morphologically, the channel appeared to be very similar to the foot structure at the triad junction [2,4], which is considered to be the physiological Ca release channel. Curiously enough, however, CICR does not seem to be the mechanism of physiological Ca release. It is likely that the Ca release channel in skeletal muscle operates in different modes when stimulated through the physiological pathway (depolarization of the T-tubule membrane, probably through changes in voltage sensor molecules) and when stimulated by increased Ca ion concentration (CICR). Although the CICR opening mode does not seem to be important physiologically, it is important pathophysiologically, as in malignant hyperthermia (MH), and pharmacologically, as in caffeine contracture. These two opening modes are pharmacologically different; procaine and adenine inhibit the CICR mode, but not the physiological mode [7]. An interesting agent is dantrolene; it inhibits both modes to an almost equal extent at 37°C, but at 20°C the agent does not inhibit the CICR mode at all, whereas it still effectively inhibits the physiological mode [8].

References

1. Fleischer S, Ogunbunmi EM, Dixon MC, Fleer EA (1985) Localization of Ca^{2+} release channels with ryanodine in junctional terminal cisternae of sarcoplasmic reticulum of fast skeletal muscle. Proc Natl Acad Sci USA 82:7256–7259
2. Inui M, Saito A, Fleischer S (1987) Purification of the ryanodine receptor and identity with feet structures of junctional terminal cisternae of sarcoplasmic reticulum from fast skeletal muscle. J Biol Chem 262:1740–1747
3. Imagawa T, Smith JS, Coronado R, Campbell KP (1987) Purified ryanodine receptor from skeletal muscle sarcoplasmic reticulum is the Ca^{2+}-permeable pore of the calcium release channel. J Biol Chem 262:16636–16643

Department of Pharmacology, Faculty of Medicine, University of Tokyo, 7-3-1 Hongo, Bunkyo-ku, Tokyo, 113 Japan

4. Lai FA, Erickson HP, Rousseau E, Liu Q-Y, Meissner G (1988) Purification and reconstitution of the calcium release channel from skeletal muscle. Nature 331:315–319
5. Takeshima H, Nishimura S, Matsumoto T, Ishida H, Kangawa K, Minamino N, Matsuo H, Ueda M, Hanaoka M, Hirose T, Numa S (1989) Primary structure and expression from complementary DNA of skeletal muscle ryanodine receptor. Nature 339:439–445
6. Hymel L, Inui M, Fleischer S, Schindler HG (1988) Purified ryanodine receptor of skeletal muscle forms Ca^{2+}-activated oligomeric Ca^{2+} channels in planar bilayers. Proc Natl Acad Sci USA 85:441–445
7. Endo M (1985) Calcium release from sarcoplasmic reticulum. Curr Top Membr Transp 25:181–230
8. Kobayashi T, Endo M (1988) Temperature-dependent inhibition of caffeine contracture of mammalian skeletal muscle by dantrolene. Proc Japan Acad 64:76–79

Metabolism of Inositol Polyphosphates by Malignant Hyperthermia-Susceptible and Control Porcine Skeletal Muscle

PAUL S. FOSTER and SIMON P. HOGAN

Introduction

Malignant hyperthermia-susceptible (MH-susceptible, MHS) skeletal muscle cells have an inherited abnormality in the processes that are intimately involved in excitation-contraction coupling (EC coupling) or myoplasmic Ca^{2+} regulation [1,2]. MH-susceptible muscle cells have an elevated level of myoplasmic Ca^{2+}, and a rapid and sustained rise in myoplasmic Ca^{2+} is also central to the etiology of fulminant anesthetic-induced MH. The mechanism of signal transduction across the triadic junction during EC coupling of skeletal muscle is unknown [3]. However, mechanical coupling between highly specialized triadic proteins has been proposed as the primary mechanism for voltage-activated generation of sarcoplasmic reticulum (SR) Ca^{2+} signals and subsequent construction. The dihydropyridine-sensitive Ca^{2+} channel (voltage sensor) of the transverse tubule (T-tubule) membrane and the ryanodine receptor or Ca^{2+} release channel of the SR are key proteins involved in this process [3].

In the pig, the molecular defect predisposing to MH susceptibility is associated with the ryanodine receptor [4–6]. A mutation in the ryanodine receptor gene leads to a substitution of Cys for Arg615 in the Ca^{2+} release channel. While mutations in the ryanodine receptor gene have been observed in some MH-susceptible families [7], the primary cause of human MH is unknown. Recent investigations in swine genotyped as homozygous or heterozygous for the ryanodine receptor mutation indicate that a second factor or modulator is required for the consistent expression of the MH syndrome [8]. The modulating factor may be associated with processes that regulate SR Ca^{2+} transients and EC coupling. Dysfunctions in the phosphoinositide cycle and fatty acid metabolism have also been associated with MH-susceptible muscle [8–11].

Calcium signals generated by activation of the phosphoinositide pathway control many cellular processes [12]. D-*myo*-Inositol 1,4,5-trisphosphate ($Ins(1,4,5)P_3$) is intimately involved in the generation of endoplasmic reticulum-derived Ca^{2+} signals and the regulation of the complex spatiotemporal patterns of Ca^{2+} waves and oscillations that are coupled to inositol lipid hydrolysis. $Ins(1,4,5)P_3$ is deactivated by rapid dephosphorylation to D-*myo*-inositol 1,4-bisphosphate and by phosphorylation to D-*myo*-inositol 1,3,4,5-tetrakisphosphate ($Ins(1,3,4,5)P_4$) [12]. Dephosphorylation is as-

Division of Biochemistry and Molecular Biology, The John Curtin School of Medical Research, Australian National University, Canberra, ACT 2601, Australia

sociated with both membrane and soluble extracts from cells, while phosphorylation occurs predominantly in the cytosol. The conversion of Ins(1,4,5)P_3 to Ins(1,3,4,5)P_4 may be a secondary pathway for Ins(1,4,5)P_3 metabolism, with the main function of Ins(1,4,5)P_3 3-kinase being to regulate the relationship between Ins(1,4,5)P_3 and Ins(1,3,4,5)P_4, whose roles in Ca^{2+} signaling may be closely associated. Ins(1,3,4,5)P_4 may activate plasma membrane Ca^{2+} influx associated with the activation of the phosphoinositide signaling pathway. Ins(1,3,4,5)P_4 is also dephosphorylated by soluble and membrane-associated 5-phosphate specific phosphatases.

The pivotal role of D-*myo*-inositol 1,4,5-trisphosphate (Ins(1,4,5)P_3) in the generation of intracellular Ca^{2+} signals in many cell types has led to investigations of the function of the phosphoinositide signaling pathway in skeletal muscle Ca^{2+} regulation. In skeletal muscle, the sensitivity of the intracellular Ca^{2+} release mechanism to Ins(1,4,5)P_3 is highly dependent on experimental conditions, and this has led to the conflicting views on the role of the phosphoinositide pathway in muscle [14]. However, while the physiological role of Ins(1,4,5)P_3 in skeletal muscle Ca^{2+} homeostasis is yet to be defined, recent investigations suggest that the inositol polyphosphate may have a modulating function on SR Ca^{2+} transients and Ca^{2+}-dependent processes of muscle metabolism that are associated with the contractile cycle [14]. Ins(1,4,5)P_3 can activate channels in the SR membrane that display rapid kinetics for Ca^{2+} release and are distinct from ryanodine-sensitive Ca^{2+} channels. The enzymes required for the synthesis of phosphatidylinositol 4,5-bisphosphate and the generation of Ins(1,4,5)P_3 are found in the T-tubules of muscle cells, and high concentrations of Ins(1,4,5)P_3 are found bound to triadic structures. The pathways for the rapid clearance of Ins(1,4,5)P_3/Ins(1,3,4,5)P_4 to other inositol polyphosphates and for their sequential metabolisms to inositol are also present in skeletal muscle [15].

Understanding signaling pathways that are involved in the regulation of myoplasmic Ca^{2+} will be central to defining the mechanisms of expression and pathogenesis of human MH. Phosphoinositide metabolism in porcine and human MH-susceptible muscle cells is altered, resulting in elevated levels of myoplasmic Ins(1,4,5)P_3 and increased sensitivity of the SR Ca^{2+} release mechanism to this inositol polyphosphate [9–11]. In this investigation we have characterized the kinetics of Ins(1,4,5)P_3 and Ins(1,3,4,5)P_4 catabolism by purified soluble and particulate inositol (1,4,5)P_3/(1,3,4,5)P_4-polyphosphate 5-phosphatase (Ins(1,(3)4,5)PP 5-phosphatase or inositol polyphosphate 5-phosphatase) and Ins(1,4,5)P_3 3-kinase from MH-susceptible and control skeletal muscle. These enzymes play a pivotal role in regulating the intracellular concentrations of these inositol polyphosphates. The pathways for the dephosphorylation of Ins(1,4,5)P_3 and Ins(1,3,4,5)P_4 to inositol in extracts from MH-susceptible and control muscle have also been characterized.

Experimental Procedures

Isolation of Soluble and Particulate Extracts

The biceps femoris and semitendinosis muscles were removed from pigs as previously described [15]. Isolated muscle (200 g) was minced finely, resuspended in 4 volumes of ice-cold 20 m*M* Tris/HCl (pH 8.3)/0.25 *M* sucrose/12 m*M* 2-mercaptoethanol/ 0.5 m*M* ethylene diaminetetraacetic acid (EDTA)/0.5 m*M* ethylene glycol-bis(β-aminoethyl ether)-*N,N,N′,N′*-tetraacetic acid (EGTA) containing 0.4 m*M* phenylmethylsulphonyl fluoride, 5 m*M* leupeptin, and 1 m*M* benzamamide (protease inhibitors) (buffer A) and stirred for 10 min at 4°C. The suspension was homogenized in 3 × 30-s bursts (Waring blender London, UK) separated by 30-s intervals, before centrifugation at 3000 *g* for 30 min. The supernatant was then filtered and centrifuged at 10 000 *g* for 30 min, and the resulting supernatant was further centrifuged at 110 000 *g* for 30 min. The supernatant (soluble extract) was retained and the pellet from the 110 000 *g* spin (particulate extract) was resuspended in 0.6 *M* KCl in buffer A for 1 h and then centrifuged at 110 000 *g* for 30 min. The particulate pellet was then resuspended (5–10 mg/ml) in 50 m*M* *N*-2-hydroxy ethylpiperazine-*N*′-2-ethane sulfonic acid/HCl (HEPES/HCl) (pH 7.2)/2 m*M* dithiothreitol (DTT) and homogenized or solubilized. The particulate extracts consisted of fragmented SR and T-tubule (sarcotubular) membranes. All procedures were carried out at 4°C, and extracts for storage were snap-frozen in liquid nitrogen and rapidly thawed before use.

Preparation of Detergent-Solubilized Particulate Inositol Polyphosphate 5-Phosphatase

The pellet from the 0.6 *M* KCl wash was resuspended in 20 m*M* Tris/HCl (pH 8.0)/1 m*M* EDTA/1.0% (v/v) glycerol/2 m*M* DTT and protease inhibitors (buffer B) and then centrifuged at 110 000 *g* for 30 min. The particulate extract was then resuspended in buffer B and homogenized to give a final protein concentration of 15 mg/ml. Particulate extracts were solubilized in buffer B containing polyoxyethylene 10-tridecyl ether (P10-TE) for 2 hr at 4°C at a detergent to protein ratio of 4, and at a final protein concentration of 5 mg/ml. Solubilized proteins were obtained after centrifugation at 110 000 *g* for 1 h and were then diluted into buffer B to give a final P10-TE concentration of 0.5% (v/v).

Purification of Particulate and Soluble Inositol Polyphosphate 5-Phosphatase and Ins(1,4,5)P_3 3-Kinase

Soluble type I and particulate inositol polyphosphate 5-phosphatase and Ins(1,4,5)P_3 3-kinase were purified by a combination of diethylamino ethanol (DEAE) sephacel, blue sepharose, heparin agarose, and structural analog affinity (resin #204) chromatography as previously described [15].

Assay of Inositol Polyphosphate 5-Phosphatase and Ins(1,4,5)P_3 3-Kinase Activities

Inositol polyphosphate 5-phosphatase activity was assayed at 37°C, in 100 μl of 50 m*M* HEPES/HCl (pH 7.5)/5 m*M* $MgCl_2$ (buffer C) with 20 nCi [2-^{3}H] Ins(1,4,5)P_3/10 μ*M* Ins(1,4,5)P_3 or 20 nCi [2-^{3}H] Ins(1,3,4,5)P_4/2 μ*M* Ins(1,3,4,5)P_4. Reactions were initiated by the addition of 100 μg of protein from soluble or particulate extracts. Ins(1,4,5)P_3 3-kinase activity in soluble extracts was assayed at 37°C in 100 μl of 50 m*M* HEPES/HCl (pH 7.5)/2 m*M* DTT/1 μ*M* free $CaCl_2$/3 μ*M* calmodulin (CaM)/5 m*M* adenosine triphosphate (ATP)/7 m*M* $MgCl_2$ (concentrations that induced maximal activity) (buffer D) and 20 nCi [2-^{3}H]Ins(1,4,5)P_3/2 μ*M* Ins(1,4,5)P_3. In some experiments, attempts were made to convert [2-^{3}H]Ins(1,3,4)P_3 to [2-^{3}H]Ins(1,3,4,6)P_4. [2-^{3}H]Ins(1,3,4)P_3 was prepared by incubating 20 nCi [2-^{3}H]Ins(1,3,4,5)P_4/2 μ*M* Ins(1,3,4,5)P_4 with soluble inositol polyphosphatase 5-phosphatase in buffer C for

30 min. D-2,3-Bisphosphoglyceric acid (10 mM), 5 mM ATP, and another 100 μg of protein from soluble extracts were then added to the reaction medium. The reaction was then allowed to continue for another hour. All reactions were terminated by adding 1 ml of 0.1 M ammonium formate/0.1 M formic acid. The [^{3}H]inositol phosphates were then separated by high pressure liquid chromatography (HPLC) and quantified by liquid scintillation spectrometry.

Kinetic Characterization of Particulate and Soluble Inositol Polyphosphate 5-Phosphatase and Ins(1,4,5)P_3 3-Kinase Activities

Substrate–velocity relationships for inositol polyphosphate 5-phosphatases were determined using 2.5–80 μM (soluble type I) or 10–200 μM (particulate) Ins(1,4,5)P_3/20 nCi [2-^{3}H]Ins(1,4,5)P_3 or 0.5–16 μM (soluble type I) or 0.5–4 μM (particulate) Ins(1,3,4,5)P_4/20 nCi [2-^{3}H]Ins(1,3,4,5)P_4 in buffer C containing 2 mM $MgCl_2$. Kinetic investigations on Ins(1,4,5)P_3 3-kinase were performed in buffer D containing 0.05–4 μM Ins(1,4,5)P_3/20 nCi [2-^{3}H]Ins(1,4,5)P_3 and 5 mM $MgCl_2$. The products of Ins(1,4,5)P_3 and Ins(1,3,4,5)P_4 metabolism were separated by HPLC. In some investigations, [^{3}H]inositol phosphates were separated on 2 ml Dowex 1 or 1 ml (100 mg of resin) Amprep (Amersham, UK) strong-anion-exchange (SAX) columns. Amprep SAX columns were washed with 10 ml of 1.0 M ammonium formate/0.1 M formic acid, followed by 100 ml of deionized water. The sample was then applied and the column washed with 3 × 2 ml of deionized water. Inositol phosphates were separated by isocratic batch elution using various concentrations of ammonium formate [15]. Between 5% and 20% of substrate was metabolized in all kinetics assays.

High Pressure Liquid Chromatography of [^{3}H]Inositol Phosphates

Inositol phosphates were separated and analyzed by a modification of a method previously described [15]. Inositol phosphates were eluted from a Partisil 10 SAX column (4.6 × 250 mm) with a linear gradient from 0 to 1.4 M $NH_4H_2PO_4$/orthophosphoric acid (pH 3.7); gradients were developed over various periods as required, followed by a high salt wash, at a flow rate of 1.0 ml/min. [^{3}H]Inositol phosphates were automatically detected by an online continuous flow scintillation counter (Packard Instruments, Illinios, IL, USA). Standard retention times for ^{3}H-labeled inositol phosphates were obtained by eluting authentic inositol phosphates under identical separation conditions. A mixture of ATP, ADP and AMP was included in all samples as an internal control of gradient stability.

Other Methods

Protein determinations were carried out using the coomassie or BCA protein estimation kits (Pierce, Rockford, IL, USA). The free Ca^{2+} concentration in Ins(1,4,5)P_3 3-kinase assays was calculated by using an apparent affinity constant of EGTA for Ca^{2+} of $3.17 \times 10^{-6} M$ at pH 7.5 [15].

Results and Discussion

Phosphoinositide metabolism in porcine and human MH-susceptible muscle cells is altered [9–11]. In soluble extracts from MH-susceptible porcine and human skeletal muscle, elevated levels of Ins(1,4,5)P_3 have been observed. Recent investigations have shown that microinjection of Ins(1,4,5)P_3 into intact human skeletal muscle induces Ca^{2+} release from intracellular stores that is greater in MH-susceptible fibers, than in control fibers. The reason for increased Ins(1,4,5)P_3 sensitivity in MH-susceptible muscle may be due to the elevated resting Ins(1,4,5)P_3 levels. An abnormality in the metabolism of Ins(1,4,5)P_3 may explain the high intracellular Ins(1,4,5)P_3 concentration, the more sustained effects of Ins(1,4,5)P_3 on the Ca^{2+} transient, and the high resting Ca^{2+} levels observed in isolated MH-susceptible muscle. Alternatively, elevated levels of myoplasmic Ins(1,4,5)P_3 may be an adaptation to altered Ca^{2+} homeostasis of MH-susceptible myocytes. Ins(1,4,5)P_3 may also modulate SR Ca^{2+} transients and thus play a role in the expression of the MH syndrome.

HPLC Chromatograms of Ins(1,4,5)P_3 and Ins(1,3,4,5)P_4 Metabolism

The HPLC chromatograms in Fig. 1 show the retention times of known inositol phosphate standards relative to the [^{3}H]-metabolite profiles of Ins(1,4,5)P_3 (Fig. 1a) and Ins(1,3,4,5)P_4 (Fig. 1b) that were observed after incubation of the inositol polyphosphates in soluble and particulate extracts from muscle. Metabolite profiles were selected at incubation times that show the presence of all metabolic products. Unknown [^{3}H]inositol phosphates were identified by running known standards under identical conditions and comparing retention times. Unknown [^{3}H]inositol phosphates were also identified by comparing elution profiles (retention times) with those previously reported for known [^{3}H]inositol phosphates run under similar conditions [15].

Metabolism of Ins(1,4,5)P_3/Ins(1,3,4,5)P_4 in Particulate and Soluble Fractions

In many cell extracts, both Ins(1,4,5)P_3 and Ins(1,3,4,5)P_4 are sequentially dephosphorylated to free inositol. The metabolite profiles of the sequential dephosphorylation of Ins(1,4,5)P_3 and Ins(1,3,4,5)P_4 to inositol (Fig. 2) by soluble and particulate extracts from MH-susceptible and control skeletal muscle were similar. Ins(1,4,5)P_3 and Ins(1,3,4,5)P_4 are rapidly and predominantly metabolized to Ins(1,4)P_2 and Ins(1,3,4)P_3, respectively. The rapid appearance of Ins(1,3,4)P_3 in amounts that were in direct proportion to the amount of Ins(1,3,4,5)P_4 metabolized, together with the absence of detectable higher inositol polyphosphates (penta- or hexakisphosphates), suggests that Ins(1,3,4,5)P_4 is the primary intermediate in the conversion of Ins(1,4,5)P_3 to Ins(1,3,4)P_3. In soluble extracts, some phosphorylation of Ins(1,4,5)P_3 to Ins(1,3,4,5)P_4 was observed in the absence of 2,3-bisphosphoglycerate and in the presence of ATP. In both extracts from MH-susceptible and control muscle, Ins(1,4)P_2 was sequentially dephosphorylated to Ins(4)P and inositol. In both MH-susceptible and control muscle, Ins(1,3,4)P_3 was further dephosphorylated to Ins(3,4)P_2 or Ins(1,3)P_2 and then to free inositol in soluble and particulate extracts, respectively. Ins(1,3,4)P_3 is also metabolized to Ins(3,4)P_2 in pancreatic islets; in parotid and liver tissue, both Ins(1,3,)P_2 and Ins(3,4)P_2 are formed, with the predominant metabolite in homogenates being Ins(3,4)P_2. Inositol monophosphates derived from Ins(1,3,4,5)P_4 metabolism were not observed in all extracts from MH-susceptible or control muscle; however, in some soluble extracts Ins(3)P was formed, indicating the presence of Ins(3,4)P_2 4-phosphatase. In all the MH-susceptible and control muscle cell extracts, both Ins(1,4,5)P_3 and Ins(1,3,4,5)P_4

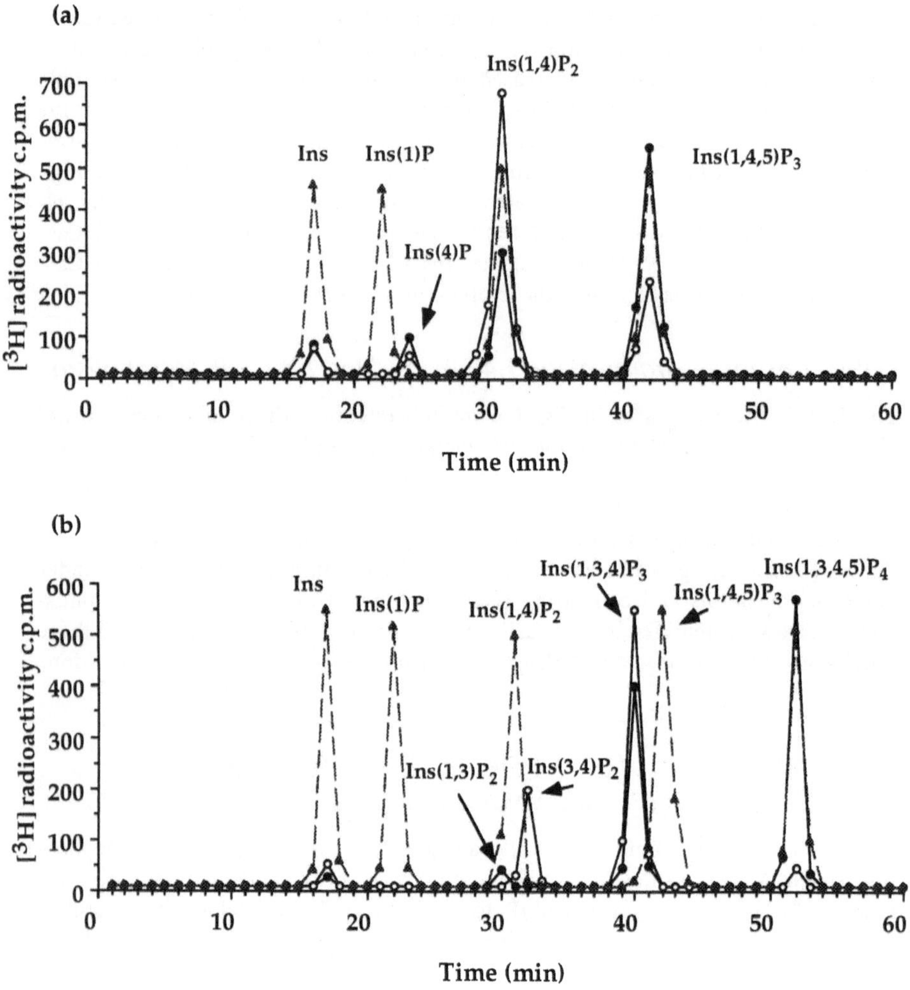

FIG. 1**a,b.** High pressure liquid chromatography (HPLC) profiles of inositol polyphosphate metabolism in soluble and particulate extracts of muscle represent the time-dependent conversion of the respective ^{3}H-labeled inositol polyphosphates into their various metabolites. Incubations were performed in buffer E (see experimental procedures) containing 20 nCi [2-^{3}H] Ins(1,4,5)P$_3$/10 μ*M* Ins(1,4,5)P$_3$ (**a**) or 20 nCi [2-^{3}H] Ins(1,3,4,5)P$_4$/2 μ*M* Ins(1,3,4,5)P$_4$ (**b**) and 100 μg of protein from the respective extract. The data shown represent a period of incubation that shows all the metabolites for the respective inositol polyphosphate in soluble (*solid circles*) and particulate (*open circles*) extracts. Similar results were obtained in six soluble and particulate extracts from six pigs. The retention times of known inositol phosphate standards (*solid triangles*) relative to the metabolites are also shown. [^{3}H]Inositol phosphates were identified and separated by HPLC as described in the experimental procedures

were rapidly dephosphorylated to inositol, and the metabolite profiles were the same for the respective inositol polyphosphates.

The dephosphorylation pathway of Ins(1,3,4,5)P$_4$ not only returns the inositol polyphosphate to free inositol for its incorporation into the membrane but also

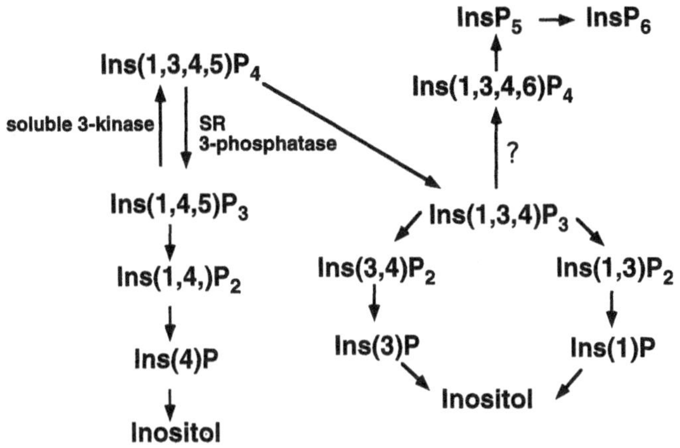

FIG. 2. The metabolism of $Ins(1,4,5)P_3$ and $Ins(1,3,4,5)P_4$ to inositol in malignant hyperthermia-susceptible (MH-susceptible) porcine muscle. Similar metabolite profiles are observed in control muscle extracts

provides a pathway for the formation of $Ins(1,3,4,6)P_4$ and eventually $Ins(1,3,4,5,6)P_5$ [16]. Mass measurements of inositol phosphates in various types of isolated skeletal muscles from *Xenopus laevis* (frog) and rat, at rest, indicate the presence of $Ins(1,4)P_2$ (3.5–6.9 μM), $Ins(1,4,5)P_3$ (1.2–2.5 μM), $Ins(1,4,5,6)P_4$/$Ins(3,4,5,6)P_4$ (0.2–0.9 μM), $Ins(1,3,4,5,6)P_5$ (1 μM), and $InsP_6$ (3.2–4.6 μM) [14]. $Ins(1,3,4)P_3$ prepared from the dephosphorylation of $Ins(1,3,4,5)P_4$ could not be phosphorylated in soluble extracts from muscle. In two experiments, attempts were made to detect $Ins(1,3,4)P_3$ 6-kinase activity in soluble extracts from control muscle. When soluble extracts (100 μg) were incubated with $[^3H]Ins(1,3,4,5)P_4$ for 30 min, more than 75% of the dephosphorylated product was $[^3H]Ins(1,3,4)P_3$. After inhibiting the phosphatase activity, then adding ATP and another 100 μg of protein and incubating for a further hour, the only ^3H metabolite detected was $[^3H]Ins(1,3,4)P_3$. Furthermore, $Ins(1,4,5)P_3$ was only phosphorylated to $Ins(1,3,4,5)P_4$, and the tetrakisphosphate was not phosphorylated to higher inositol polyphosphates. Further experiments are required to characterize muscle inositol polyphosphate kinases and to define the pathway for the generation of $Ins(1,4,5,6)P_4$/$Ins(3,4,5,6)P_4$ and the inositol penta- and hexa-kisphosphates described in muscle.

Muscle $Ins(1,3,4,5)P_4$ 3-phosphatase (3-phosphatase) activity was also detected in solubilized particulate extracts after fractionation by DEAE-Sephacel [15]. Crude cell extracts are known to contain endogenous inhibitors of 3-phosphatase activity that mask its presence [16]. $Ins(1,3,4,5)P_4$ is metabolized by a 3-phosphatase in extracts from a number of cells and may function to prolong the $Ins(1,4,5)P_3$ signal. However, $Ins(1,3,4,5,6)P_5$ and $InsP_6$ are potent inhibitors of $Ins(1,3,4,5)P_4$ dephosphorylation by hepatic 3-phosphatase and may be the preferred substrates in vivo. The physiological role of the 3-phosphatase in the turnover of inositol polyphosphates is unknown, and conjecture is complicated by the recent observation that hepatic 3-phosphatase activity is almost exclusively restricted to the endoplasmic reticulum and appears to be predominantly intraluminal [15,16].

Kinetics of Ins(1,4,5)P_3 and Ins(1,3,4,5)P_4 Metabolism by Ins(1,4,5)P_3 3-Kinase and Particulate and Soluble Type I Inositol Polyphosphate 5-Phosphatase

In cells in which the phosphoinositol lipid cycle is functional, intracellular Ca^{2+} mobilization is terminated by the rapid metabolism of Ins(1,4,5)P_3 and Ins(1,3,4,5)P_4 by soluble and membrane-bound 5-phosphate-specific phosphatases. A number of forms of 5-phosphatase exist in cells, with different substrate specificities for Ins(1,4,5)P_3 and Ins(1,3,4,5)P_4. The presence of a type or form of 5-phosphatase may be dependent on the metabolic function of the cell and the role of the phosphoinositol lipid cycle in the Ca^{2+} signaling of that cell.

Inositol polyphosphate 5-phosphatases were found in membranes that are intimately involved in EC coupling and in the myoplasm. The dephosphorylation of Ins(1,4,5)P_3 and Ins(1,3,4,5)P_4 by soluble type 1 and particulate 5-phosphatases was Mg^{2+} dependent (with maximal activity at 2 m*M* $MgCl_2$), linear with time, and inhibited by 1–10 m*M* D-2.3-bisphosphoglyceric acid (results not shown). Before kinetic analysis, 5-phosphatase activities were shown to be 5-phosphate specific and display Michaelis–Menten behavior. Substrate concentrations in the ranges of 2.5–200 μ*M* Ins(1,4,5)P_3 and 0.5–16 μ*M* Ins(1,3,4,5)P_4 were used to generate substrate–velocity plots. Kinetic constants for the dephosphorylation of Ins(1,4,5)P_3 and Ins(1,3,4,5)P_4 by soluble type 1 and particulate 5-phosphatases are shown in Tables 1 and 2. The

TABLE 1. Kinetic characteristics of Ins(1,4,5)P_3 metabolism by malignant hyperthermia-susceptible (MHS) and control porcine skeletal muscle.

Inositol polyphosphate 5-phosphatase	K_m (μ*M*)	V_{max} (μmol/min per mg)
Soluble		
Control	8.90 ± 0.55	2.44 ± 2.32
MHS	8.87 ± 2.23	3.55 ± 1.24
Particulate		
Control	46.3 ± 1.3	115.2 ± 11.0
MHS	42.6 ± 3.1	112.8 ± 7.8

Observations were made in duplicate or triplicate from any 1 preparation and the mean used to generate K_m and V_{max} values from Lineweaver–Burke plots. Data shown are the means of all K_m and V_{max} values obtained from 4–6 individual preparations. No more than 2 preparations were isolated from any one pig.

TABLE 2. Kinetic characteristics of Ins(1,3,4,5)P_4 metabolism by MH-susceptible and control porcine skeletal muscle.

Inositol polyphosphate 5-phosphatase	K_m (μ*M*)	V_{max} (μmol/min per mg)
Soluble		
Control	1.053 ± 0.348	0.127 ± 0.059
MHS	0.985 ± 0.225	0.330 ± 0.355
Particulate		
Control	2.10 ± 0.25	0.00533 ± 0.00001
MHS	1.89 ± 0.08	0.00458 ± 0.00045

Observations were made in duplicate or triplicate from any 1 preparation and the mean used to generate K_m and V_{max} values from Lineweaver–Burke plots. Data shown are the means of all K_m and V_{max} values obtained from 4–6 individual preparations. No more than 2 preparations were isolated from any one pig.

apparent K_m (46.3 μM) for Ins(1,4,5)P_3 metabolism by purified particulate 5-phosphatase from porcine skeletal muscle is higher than that reported in rabbit muscle (15–18 μM) and other tissues (3–14 μM). In contrast, purified soluble type 1 5-phosphatase had an apparent K_m of 8.9 and 1.0 μM for Ins(1,4,5)P_3 and Ins(1,3,4,5)P_4, respectively. The apparent K_m values for Ins(1,3,4,5)P_4 metabolism by particulate (2.1 μM) and soluble type 1 5-phosphatases are similar to those reported for other tissues [13–15]. Soluble type 1 5-phosphatases from human platelets and bovine and rat brain have apparent K_m values between 3 and 10 μM and between 0.5 and 1 μM for Ins(1,4,5)P_3 and Ins(1,3,4,5)P_4, respectively. Muscle particulate 5-phosphatase is much more active (115 μmol of Ins(1,4,5)P_3 hydrolyzed/min per mg of protein) than the soluble type 1 enzyme from muscle (2.44 μmol of Ins(1,4,5)P_3 hydrolyzed/min per mg of protein) and the type 1 cytosolic forms from rat brain and platelets (1–5 μmol of Ins(1,4,5)P_3 hydrolyzed/min per mg of protein). As in other cells, muscle 5-phosphatases have a lower affinity, but a higher capacity to metabolize Ins(1,4,5)P_3, than Ins(1,3,4,5)P_4. However, unlike some cells, particulate and soluble type 1 5-phosphatases do not have similar kinetic profiles [14]. Soluble type 1 5-phosphatase may have a functional role in the metabolism of both inositol polyphosphates while particulate 5-phosphate-specific dephosphorylation may be the primary route for Ins(1,4,5)P_3 clearance once soluble pathways have become saturated. Particulate and soluble inositol polyphosphate 5-phosphatases have similar kinetic constants (K_m and V_{max} values [see Tables 1 and 2]) for the same form of enzyme, from MH-susceptible and control muscle.

It was recently observed that T tubules and triadic elements contain significant amounts of bound Ins(1,4,5)P_3 (300–400 pmol/mg protein), although SR vesicles do not. The nature of this binding site is not clear, but an association with the putative voltage-sensor protein is possible. During depolarization, Ins(1,4,5)P_3 may be released from T-tubule membranes and triadic structures and induce Ca^{2+} transients. The presence of a particulate 5-phosphatase with low affinity but which is very active would ensure that Ins(1,4,5)P_3 released from, or proximal to, these membrane elements would be rapidly metabolized after the generation of Ca^{2+} signals. Muscle also possesses a soluble type 2 5-phosphatase [17] that has substrate specificity and physical constants that differ from soluble type 1 5-phosphatase. As in other cells, there are multimeric forms of inositol polyphosphate 5-phosphatases in muscle.

Muscle Ins(1,4,5)P_3 3-kinase activity was found only in soluble extracts and is a high-affinity, low-capacity clearance pathway for Ins(1,4,5)P_3 (Table 3). Purified

TABLE 3. Kinetic characteristics of Ins(1,4,5)P_3 metabolism by MH-susceptible and control porcine skeletal muscle.

Inositol (1,4,5)P_3 3-kinase	K_m (μM)	V_{max} (μmol/min per mg)
Soluble		
Control	0.35 ± 0.03	1.90 ± 0.20
MHS	0.51 ± 0.06	2.19 ± 0.70

Observations were made in duplicate or triplicate from any 1 preparation and the mean used to generate K_m and V_{max} values from Lineweaver–Burke plots. Data shown are the means of all K_m and V_{max} values obtained from 4–6 individual preparations. No more than 2 preparations were isolated from any one pig.

$Ins(1,4,5)P_3$ 3-kinase activity has an absolute requirement for Mg^{2+} and ATP and is stimulated by CaM (0.1–3 μ*M*) and Ca^{2+} (0.01–1 μ*M*). Optimal $Ins(1,4,5)P_3$ 3-kinase activity was achieved at 1 μ*M* free Ca^{2+}, 3 μ*M* CaM, 5 m*M* Mg^{2+}, and 1 m*M* ATP, and these concentrations were used in all kinetic experiments. $Ins(1,4,5)P_3$ 3-kinase activity displayed Michaelis–Menten behavior and was linear with time. The apparent K_m (0.35 μ*M*) of the purified muscle $Ins(1,4,5)P_3$ 3-kinase for $Ins(1,4,5)P_3$ is similar to that reported in other tissues (0.21–0.70 μ*M*) [14,15]. MH-susceptible and control $Ins(1,4,5)P_3$ 3-kinase activity was also found to have similar kinetic constants for metabolism of $Ins(1,4,5)P_3$ (see Table 3). The conversion of $Ins(1,4,5)P_3$ to $Ins(1,3,4,5)P_4$ may be a secondary pathway for $Ins(1,4,5)P_3$ metabolism, with the main function of $Ins(1,4,5)P_3$ 3-kinase being to regulate the relationship between $Ins(1,4,5)P_3$ and $Ins(1,3,4,5)P_4$, whose roles in Ca^{2+} signaling may be closely associated.

Conclusions

Elevated levels of $Ins(1,4,5)P_3$ in MH-susceptible porcine skeletal muscle are not directly caused by a dysfunction in particulate or soluble inositol polyphosphate 5-phosphatases or $Ins(1,4,5)P_3$ 3-kinase activities. Abnormalities in other processes that regulate myoplasmic $Ins(1,4,5)P_3$ concentrations predispose to the elevated levels of $Ins(1,4,5)P_3$ in intact MH-susceptible muscle cells. These processes may either directly or indirectly regulate the primary clearance pathways (5-phosphatase activities) for $Ins(1,4,5)P_3$. In muscle, like other cells, inositol polyphosphate metabolism is complex. $Ins(1,3,4,5)P_4$ is also metabolized by an $Ins(1,3,4,5)P_3$ 3-phosphatase that may function to prolong $Ins(1,4,5)P_3$ levels and the Ca^{2+} signal. Crude extracts from cells are also known to contain endogenous inhibitors of inositol polyphosphate metabolism. $Ins(1,3,4,5,6)P_5$ and $InsP_6$ are potent inhibitors of $Ins(1,3,4,5)P_4$ dephosphorylation by hepatic 3-phosphatase and may be the preferred substrates in vivo. Recently, $Ins(1,4,5)P_3$ has been located bound at high concentrations to the T-tubule and triadic elements [14]. The nature of this binding site is unknown. Production of $Ins(1,4,5)P_3$ is controlled by the hydrolysis of phosphotidylinositol (4,5)-bisphosphate by phospholipase C in many cells, and hydrolysis may be stimulated by an increase in cytosolic Ca^{2+} [13]. It is likely that altered phosphoinositide metabolism in MH-susceptible porcine muscle cells is an adaptation to changes in Ca^{2+} homeostasis induced by the mutation observed in the Ca^{2+} release channel of SR.

The physiological roles of $Ins(1,4,5)P_3$/$Ins(1,3,4,5)P_4$ in regulating the Ca^{2+} transients of muscle cells is yet to be defined. However, the presence of enzymes in muscle that are involved in the regulation of inositol polyphosphates is consistent with a role for the inositol lipid signaling pathway in myocyte Ca^{2+} homeostasis.

References

1. Foster PS (1990) Malignant hyperthermia. Int J Biochem 22:1217–1222
2. MacLennan DM, Philips MS (1992) Malignant hyperthermia. Science 256:789–794
3. Dulhunty A (1992) The voltage-activation of contraction in skeletal muscle. Prog Biophys Mol 57:181–223
4. MacLennan DH, Duff C, Zorzato F, Fujii J, Phillips M, Korneluk R, Frodis W, Britt BA, Worton RG (1990) Ryanodine receptor gene is a candidate for predisposition to malignant hyperthermia. Nature 343:559–561

5. Fujii J, Otsu K, Zorzato F, de Leon S, Khanna VJ, Weiler J, O'Brien PJ, MacLennan DH (1991) Identification of a mutation in porcine ryanodine receptor associated with malignant hyperthermia. Science 253:448–451
6. Otsu K, Nishida K, Kimura Y, Kuzuya T, Hori M, Kamada T, Tada M (1994) The point mutation $Arg^{615} \rightarrow$ Cys in the Ca^{2+} release channel of skeletal sarcoplasmic reticulum is responsible for hypersensitivity to caffeine and halothane in malignant hyperthermia. J Biol Chem 265:13472–13483
7. Quane KA, Keating KE, Manning BM, Healy JMS, Monsieurs K, Heffron JJA, Lehane M, Heytens L, Krisovic-Horber R, Adnet P, et al (1994) Detection of a novel common mutation in the ryanodine receptor gene in malignant hyperthermia: implications for diagnosis of heterogeneity studies. Hum Mol Genet 3:471–476
8. Fletcher JE, Calvo PA, Rosenberg H (1993) Phenotypes associated with malignant hyperthermia susceptibility in swine genotyped as homozygous or heterozygous for the ryanodine receptor mutation. Br J Anaesth 71:410–417
9. Foster PS, Gesini E, Claudianos C, Hopkinson KC, Denborough MA (1989) Inositol 1,4,5-trisphosphate phosphatase deficiency and malignant hyperpyrexia in swine. Lancet 2:124–127
10. Lopez JR, Perez C, Alfonzo M, Cordovez G, Linares N, Allen PD (1993) Changes in intracellular Ca^{2+} concenration induced by inositol 1,4,5-trisphoshate in human malignant hyperthermia skeletal muscle. Biophys (Life Sci Adv) 12:131–140
11. Scholz J, Troll U, Schulte AE, Hartung E, Patten M, Sandig P, Schmitz W (1991) Inositol-1,4,5-trisphosphate and malignant hyperthermia. Lancet 337:1361–1362
12. Berridge MJ (1993) Inositol trisphosphate and calcium signalling. Nature 361:315–325
13. Berridge MJ, Irvine RF (1989) Inositol phosphates and cell signalling. Nature 341:197–205
14. Foster PS (1994) The role of phosphoinsositide metabolism in Ca^{2+} signalling of skeletal muscle cells. Int J Biochem 26:449–468
15. Foster PS, Hogan SP, Hansbro PM, O'brien R, Potter BVL, Ozaki S, Denborough MA (1994) The metabolism of D-*myo*-inositol 1,4,5-trisphosphate and D-*myo*-inositol 1,3,4,5-tetrakisphosphate by porcine skeletal muscle. Eur J Biochem 222:955–964
16. Mennati FS, Oliver KG, Putney JW Jr, Shears SB (1993) Inositol phosphates and cell signaling: new views of $InsP_5$ and InsP. Trends Biol Sci 18:53–56
17. Hansbro PM, Foster PS, Hogan SP, Ozaki S, Denborough MA (1994) Purification and characterization of D-*myo*-inositol (1,4,5)/(1,3,4,5)-polyphosphate 5-phosphatase from skeletal muscle. Arch Biochem Biophys 311:47–54

Fatty Acids: Potentially Crucial Modulators of the Malignant Hyperthermia Syndrome

JEFFREY E. FLETCHER[1] and STEVEN J. WIELAND[2]

The Malignant Hyperthermia Syndrome: Role of a Crucial Modulator

Need to Invoke the Modulator

While defects in the ryanodine receptor may in some cases be necessary to impart the potential for malignant hyperthermia (MH) susceptibility, these defects are not sufficient to account for the MH syndrome. Obvious examples include swine homozygous for the proposed ryanodine receptor arginine to cysteine #615 MH mutation that do not exhibit an MH reaction at a young age [1] and those that do not consistently exhibit a reaction even as an adult, despite the administration of more than adequate amounts of triggering agents [2].

Modulator as a Target for Prophylaxis

An analogy can be drawn between MH inheritance, the MH syndrome, and firecrackers. Although all firecrackers contain the proper ingredients for explosion, there are occasional firecrackers that do not explode. Just as factors such as disconnecting the fuse from the explosive, or dampening the fuse or explosive, can prevent the explosion in the case of a firecracker, there is a powerful modulator of the MH syndrome that would be a highly desirable target for therapy and prophylaxis. That is, perhaps we can downregulate the modulator and prevent the MH syndrome, regardless of the actual primary defect in MH. This is a concept that is becoming more important as our knowledge of the complexities of MH expands.

Ion Channels Altered in MH: Ryanodine Receptor and Sodium Channel

General

The functions of several proteins are altered in MH skeletal muscle, most notably the ryanodine receptor and sodium channels (Fig. 1). There are interesting similarities in

[1] Department of Anesthesiology and [2] Department of Anatomy, Hahnemann University, Philadelphia, PA 19102, USA

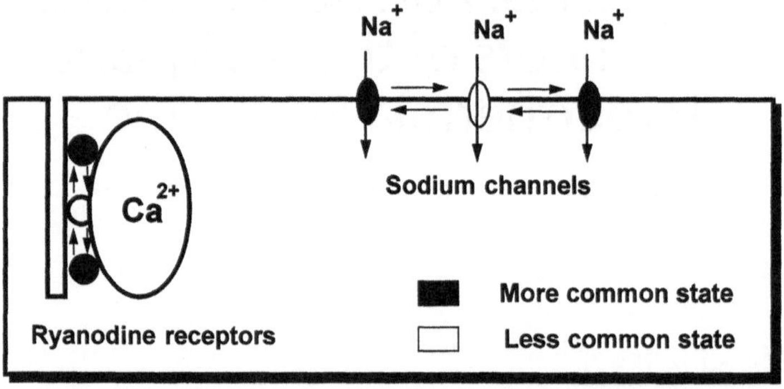

FIG. 1. Interconvertible states of ryanodine and sodium channels. Both the ryanodine receptors and sodium channels have multiple states; *solid symbols*, more common; *open symbols*, less common. In the case of the ryanodine receptor this can be observed by the threshold of calcium-induced calcium (Ca^{2+}) release or the K_D of ryanodine binding. In the case of the sodium (Na^+) current, this can be observed by the rate of inactivation. The factors converting the channels from one state to another are not known

the function of these two membrane-bound ion channels. Both the ryanodine receptor [3,4] and the sodium channel [5,6] can exist in multiple interconvertible functional states. However, in normal muscle there is one state that usually predominates. The MH defect appears to shift a larger percentage of channels into a less common state.

Ryanodine Receptor

The abnormal ryanodine receptor function in MH muscle has been sufficiently addressed by other investigators. However, it is important to note that while the function of the ryanodine receptor appears to be abnormal in muscle from MH swine in the absence of anesthetics or caffeine [7–12], the function of the ryanodine receptor appears to be normal in MH humans under these conditions [4,13] and, as demonstrated by Dr. Nelson and collaborators, requires the presence of caffeine [13] or halothane [14] to reveal the functional abnormalities that occur in about 50% of the channels.

Sodium Channel

Unlike that of the ryanodine receptor, the function of the sodium channels is altered in human MH cell cultures in the absence of anesthetics [15,16]. Similar studies have not been conducted in swine. The sodium currents are normally carried by embryonic tetrodotoxin-insensitive currents in primary cultures of human skeletal muscle [5]. There are actually two types of abnormalities observed in primary cultures of human MH skeletal muscle (Fig. 2). The first, a kinetic abnormality, is manifested as a slow inactivation of the sodium currents [15]. The second abnormality results in about 70% of the sodium currents being carried by adult tetrodotoxin-sensitive channels instead of the embryonic tetrodotoxin-insensitive currents typical of normal primary cultures [16].

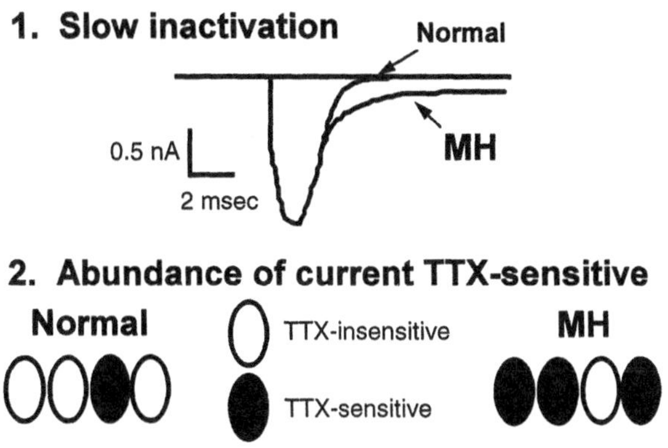

FIG. 2. The two functional abnormalities of the sodium channel in human malignant hyperthermia (*MH*) muscle. The sodium currents in primary cultures of human MH muscle have an overall slower inactivation rate than normal and are considerably more tetrodotoxin (*TTX*) sensitive. The abnormal tetrodotoxin sensitivity of the sodium current is apparently caused by the presence of relatively more active adult tetrodotoxin sodium channels than embryonic tetrodotoxin-insensitive channels. These two forms of the sodium channel are encoded by different genes

Relationship Between Ion Channel Defects, Contracture Response, and MH Syndome

Ryanodine Receptor

We have demonstrated that the altered ryanodine receptor function in swine appears to be directly linked to the abnormal contracture response to halothane [2]. However, in agreement with other investigators, we found that there is no direct cause-and-effect relationship between the altered ryanodine receptor function and the MH syndrome and that it is necessary to invoke a modulator to account for the MH syndrome [2].

Sodium Channel

The halothane-induced inhibition of sodium channel currents has not been exhaustively examined as regards MH, but appears to be similar in MH and normal muscle [17,18], suggesting that unlike the ryanodine receptor, this abnormality in function does not lead directly either to an abnormal contracture response or to the MH syndrome. Instead, a modulator must be invoked to account for the action of halothane in either case.

Free Fatty Acids as a Model for the Crucial Modulator

Interplay of Fatty Acids, Phospholipase C, and the Antioxidant Abnormality

Keep in mind that there are a number of potential modulators in MH and there is considerable interplay among a particular group of processes that makes dissecting

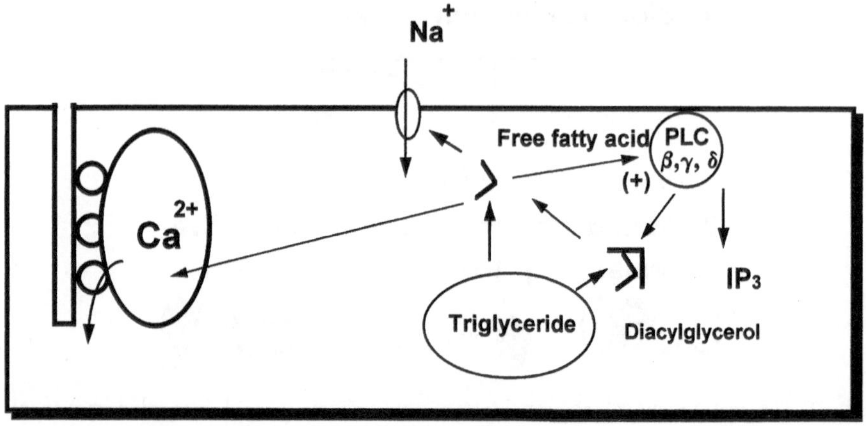

FIG. 3. Free fatty acid as a modulator of the functions of the sodium channel and ryanodine receptor. Fatty acids derived from skeletal muscle triglyceride stores or phospholipase C (*PLC*, isoforms β, γ, δ) activity interact with sodium channels and ryanodine receptors. PLC also releases 1,4,5-inositol trisphosphate (*IP_3*)

cause-and-effect relationships extremely difficult (Fig. 3). For example, free fatty acid production [1,19,20] and phospholipase C activity [21–23] have both been shown by at least two different groups of investigators to be elevated in MH. Phospholipase C activity can increase the levels of free fatty acids through deacylation of diacylglycerol. Also, free fatty acids stimulate phospholipase C activity. A third area of research, the antioxidant abnormality [24], could be the result of excess unsaturated fatty acids. Conversely, an antioxidant abnormality would stimulate the activity of lipid-metabolizing enzymes, causing liberation of free fatty acids and 1,4,5-inositol triphosphate (IP_3).

Source and Types of Free Fatty Acids

The source of excess free fatty acids had originally been proposed to be phospholipase A_2 activity, based on several logical assumptions [1,19]. However, it now appears that either triglyceride-associated free fatty acid metabolism [25,26] or elevated phospholipase C activity [21–23] or both are the most likely potential sources. There are two basic types of fatty acids, saturated and unsaturated, and each has very different effects on protein function. Most of the enzymes liberating fatty acids from the storage forms increase the relative proportion of unsaturated fatty acids. The unsaturated fatty acids are more susceptible than saturated fatty acids to lipid peroxidation, and this may be the link to the antioxidant abnormality in MH.

Effects of Elevated Free Fatty Acids in vivo in the Presence of Halothane

Free fatty acids are a normal energy source in skeletal muscle, whether derived from dietary fatty acids or muscle triglyceride stores (Fig. 4). A high free fatty acid flux would not be a noticeable problem in the absence of anesthesia, but could be detected by nuclearmagnetic resonance (NMR) [27]. Indeed, large changes in fatty acid β-oxidation and muscle triglyceride utilization can occur with little to no noticeable

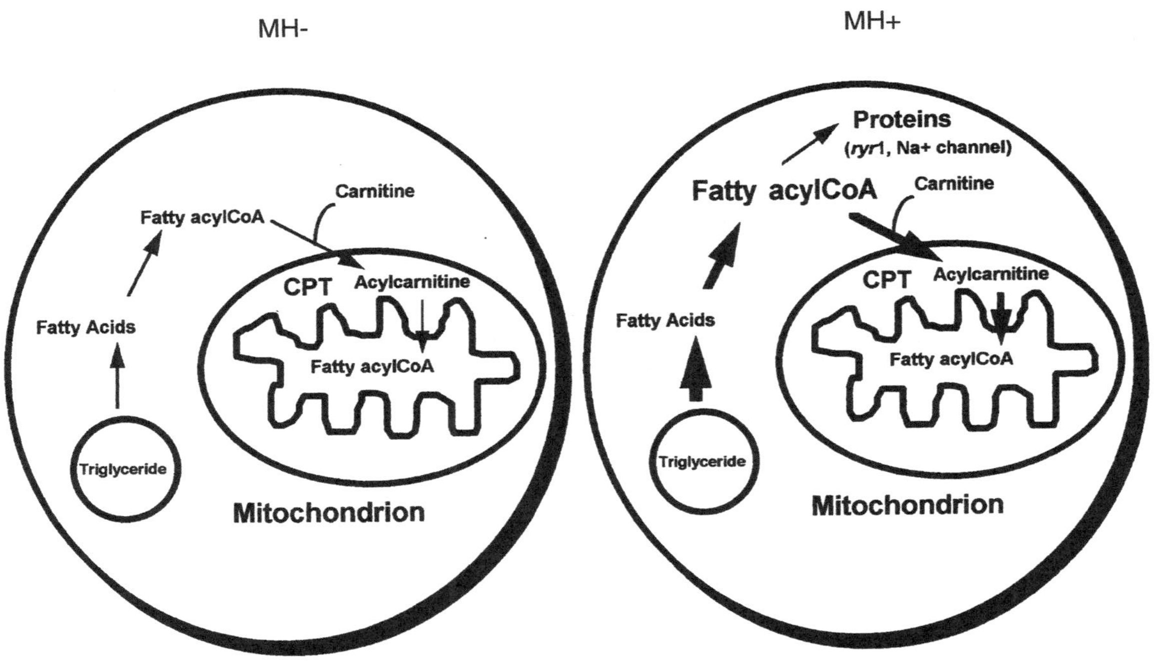

FIG. 4. Fatty acid metabolism in normal (*MH−*) and MH-susceptible (*MH+*) skeletal muscle. In normal muscle, fatty acids derived either from the diet or from skeletal muscle triglyceride stores are converted to acyl-CoAs, which are then transported to the mitochondria to produce energy through β-oxidation. The fatty acids are transported across the inner mitochondrial membrane by conversion to acylcarnitine by carnitine palmitoyl transferase (*CPT*). The acylcarnitines are then converted back to acyl-CoAs, which are subject to β-oxidation. The acyl-CoA is the "activated" form of fatty acid that can be used in lipid synthesis and covalent attachment to proteins through acylation mediated by acyltransferases. In MH+ muscle, the flux of fatty acids through β-oxidation is increased, but the levels of fatty acids at any point in time are unchanged. The acyl-CoAs may increase (this has not been tested). The excess of acyl-CoA may abnormally acylate proteins, such as the ryanodine receptor (*ryr*1) and sodium (Na^+) channel

change in overall muscle function. Whether it is the supply of free fatty acids or the inhibition of β-oxidation of fatty acids that becomes important during anesthesia has not been determined. An unusually large number of MH patients have low carnitine palmitoyl transferase (CPT) activity [28]. This would reduce β-oxidation and allow fatty acids to accumulate, making this a potential problem in at least some cases of MH. Also, the most likely forms that the fatty acids take when they increase have not been identified. The most probable is the acyl-CoA form, because the free form is at normal [29] to low [26] values and the acylcarnitines should not build up unless the mitochondrial translocase activity is somehow blocked. The acyl-CoAs have been demonstrated to increase calcium release [30]. The elevated production of free fatty acids or metabolites may be a secondary compensatory response to a primary mutation.

Requirements for a Modulator

Requirements

The modulator is the system that can be downregulated to abolish MH. To qualify as a modulator, a particular process should minimally (1) correlate quantitatively with the presentation of the MH syndrome; (2) greatly affect only calcium release in the presence of anesthetic; (3) be able to be downregulated without greatly affecting skeletal muscle function; (4) regulate the function of at least the sodium and calcium release channels; and (5) exhibit the same temperature dependence as does the contracture response to halothane.

Elevated Free Fatty Acids Correlate with the MH Syndrome

Elevated free fatty acids had originally been suggested by Khay and Anne Cheah to be a causative factor in MH because of their close association with MH susceptibility in swine [1,19]. However, with a primary defect in the ryanodine receptor as the possible cause of porcine MH [31], free fatty acids perhaps should now be viewed as an ideal candidate for the crucial modulator of the MH syndrome in at least some forms of MH [2]. In agreement with this proposal, Khay and Anne Cheah [1] have demonstrated that free fatty acid production is normal in young MH-susceptible swine at the time when they cannot be triggered and increases with age in parallel with increasing MH susceptibility. Fatty acid *production* was shown by our group to be elevated in humans testing positive for MH [20]. The actual *levels* of free fatty acid in intact tissue unexposed to halothane are normal [29] to low [26]. However, the low triglyceride levels in MH-susceptible swine [26] and humans [25] suggest that the *flux* of free fatty acids through mitochondrial β-oxidation is high in MH muscle. Elevated fatty acid production would not be noticeable in the absence of anesthetics, but could account for the lean meat associated with MH swine.

Effects of Fatty Acids on the Ryanodine Receptor and Sodium Channel

Ryanodine Receptor

There are three types of effects of free fatty acids on Ca^{2+} release from the sarcoplasmic reticulum. First, there is an immediate free fatty acid-induced Ca^{2+}

release that is not mediated through the ryanodine receptor [32,33]. Second, there is an effect of free fatty acids on the threshold of Ca^{2+}-induced Ca^{2+} release in the absence of anesthetic [26] that is similar to one abnormality observed in porcine, but not human, MH. Third, there is a 30-fold enhancement of halothane-induced Ca^{2+} release [11] that is blocked by ruthenium red [4], suggesting involvement of the ryanodine receptor. This latter effect is actually a net efflux of Ca^{2+} in the presence of the Ca^{2+}-pumping activity. This interaction of free fatty acids with halothane is by far the most pronounced of the three fatty acid effects and may be the only relevant effect, considering that in most cases the excess free fatty acids would be utilized as a source of energy through mitochondrial β-oxidation. It should be noted that the cysteine site created by the porcine mutation [31] is a potential fatty acid-binding site [34]. The 30-fold increase in sensitivity of halothane-induced calcium release by fatty acids is the only temperature-dependent interaction described so far that could explain the temperature dependence of contractures induced by halothane for the diagnosis of MH [4].

Sodium Channel

In primary cultures of human skeletal muscle, two isoforms of the sodium channel are expressed: the tetrodotoxin-sensitive adult form and a tetrodotoxin-insensitive embryonic form [5]. The relative proportion of functional sodium channels of each form is usually related to the state of differentiation of the cell. In normal muscle, elevation of myoplasmic fatty acids turn on latent tetrodotoxin-sensitive sodium currents ([35]; Fig. 5). In MH muscle the bulk of tetrodotoxin-sensitive sodium currents are already activated and fatty acids have no further effect [16]. This finding provides further support for the hypothesis of elevated fatty acid production in MH muscle.

While it is difficult to envision a direct cause–effect relationship between a sodium channel defect and disrupted fatty acid metabolism, at least one muscle disorder (hyperkalemic periodic paralysis) with a known mutation in the sodium channel, also exhibits altered triglyceride-associated fatty acid metabolism [36].

The alteration in sodium channel kinetics (a delayed inactivation of sodium currents) is similar to the kinetics of the α-subunit of the sodium channel in the absence of the β-subunit [37,38]. It may come as a surprise to fans of the ryanodine receptor that the β-subunit of the sodium channel is encoded on chromosome 19 [39], suggesting that a mutation in the β-subunit of the sodium channel may be the cause of some forms of MH.

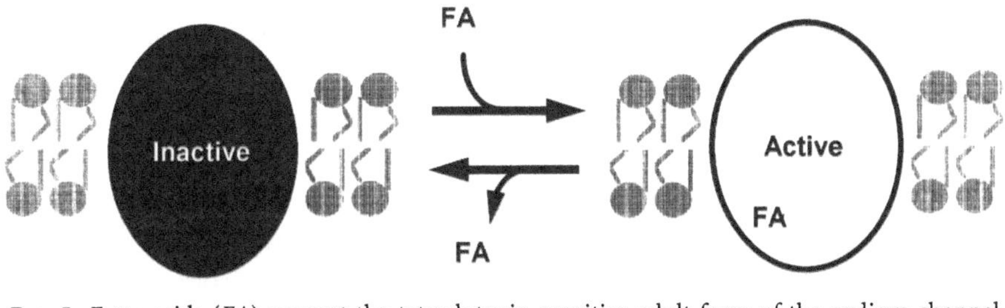

FIG. 5. Fatty acids (*FA*) convert the tetrodotoxin-sensitive adult form of the sodium channel from a functionally inactive to an active state. The reversibility of this reaction has not been tested, but it is assumed to be reversible

Conclusion

Free fatty acid, or an intermediate form such as acyl-CoA, is a likely candidate and good model for the crucial modulator of the MH syndrome because the effects of fatty acids have been well characterized. Many of the questions unanswerable by considering only a mutation in the ryanodine receptor can be explained by invoking a modulator, such as free fatty acids. Note that other modulators, such as diacylglycerol or IP_3 derived from phospholipase C activity, could have actions similar to those reported for free fatty acids, especially because fatty acids and diacylglycerol are believed to bind to similar sites and fatty acids can be derived from diacylglycerol. As all the mutations causing MH may never be identified and a blood test for MH may never be developed, a rational approach to the problem is to prophylactically downregulate the modulator of the MH syndrome for all patients undergoing anesthesia.

Acknowledgments. Support for this research was obtained from the Hahnemann Anesthesia Research Foundation.

References

1. Cheah KS, Cheah AM, Waring JC (1986) Phospholipase A_2 activity, calmodulin, Ca^{2+} and meat quality in young and adult halothane-sensitive and halothane-insensitive British Landrace pigs. Meat Sci 17:37–53
2. Fletcher JE, Calvo PA, Rosenberg H (1993) Phenotypes associated with malignant hyperthermia susceptibility in swine genotyped as homozygous or heterozygous for the ryanodine receptor mutation. Br J Anaesth 71:410–417
3. Hawkes MJ, Nelson TE, Hamilton SL (1992) [^{3}H]Ryanodine as a probe of changes in the functional state of the Ca^{2+}-release channel in malignant hyperthermia. J Biol Chem 267:6702–6709
4. Fletcher JE, Tripolitis L, Rosenberg H, Beech J (1993) Malignant hyperthermia: halothane-and calcium-induced calcium release in skeletal muscle. Biochem Mol Biol Int 29:763–772
5. Catterall WA (1992) Cellular and molecular biology of voltage-gated sodium channels. Physiol Rev 72:S15–S48
6. Vital Brazil O, Fontana MD (1983) Review article—Toxins as tools in the study of sodium channel distribution in the muscle fibre membrane. Toxicon 31:1085–1098
7. Nelson TE (1983) Abnormality in calcium release from skeletal sarcoplasmic reticulum of pigs susceptible to malignant hyperthermia. J Clin Invest 72:862–870
8. Ohnishi ST, Taylor S, Gronert GA (1983) Calcium-induced Ca^{2+} release from sarcoplasmic reticulum of pigs susceptible to malignant hyperthermia: the effects of halothane and dantrolene. FEBS Lett 161:103–107
9. Kim DH, Sreter FA, Ohnishi ST, Ryan JF, Roberts J, Allen PD, Meszaros LG, Antoniu B, Ikemoto N (1984) Kinetic studies of Ca^{2+} release from sarcoplasmic reticulum of normal and malignant hyperthermia-susceptible pig muscles. Biochim Biophys Acta 775:320–327
10. Mickelson JR, Ross JA, Reed BK, Louis CF (1986) Enhanced Ca^{2+}-induced calcium release by isolated sarcoplasmic reticulum vesicles from malignant hyperthermia-susceptible pig muscle. Biochim Biophys Acta 862:318–328
11. Fletcher JE, Mayerberger S, Tripolitis L, Yudkowsky M, Rosenberg H (1991) Fatty acids markedly lower the threshold for halothane-induced calcium release from the terminal cisternae in human and porcine normal and malignant hyperthermia-susceptible skeletal muscle. Life Sci 49:1651–1657

12. Nelson TE, Lin M, Volpe P (1991) Evidence for intraluminal Ca^{++} regulatory site defect in sarcoplasmic reticulum from malignant hyperthermia pig muscle. J Pharmacol Exp Ther 256:645–649
13. Fill M, Stefani E, Nelson TE (1991) Abnormal human sarcoplasmic reticulum Ca^{2+} release channels in malignant hyperthermic skeletal muscle. Biophys J 59:1085–1090
14. Nelson TE (1992) Halothane effects on human malignant hyperthermia skeletal muscle single calcium-release channels in planar lipid bilayers. Anesthesiology 76: 588–595
15. Wieland SJ, Fletcher JE, Rosenberg H, Gong QH (1989) Malignant hyperthermia: slow sodium current in cultured human muscle cells. Am J Physiol 257:C759–C765
16. Wieland SJ, Gong Q-H, Fletcher JE, Rosenberg H (1992) Fatty acid activation of silent sodium channels in cultured human skeletal muscle. Anesthesiology 77:A761
17. Wieland SJ, Fletcher JE, Gong Q-H, Rosenberg H (1991) Effects of lipid-soluble agents on sodium channel function in normal and MH-susceptible skeletal muscle cultures. In: Blanck TJJ, Wheeler DM (eds) Mechanisms of anesthetic action in muscle. Plenum, New York, pp 9–19
18. Ruppersberg JP, Rudel R (1988) Differential effects of halothane on adult and juvenile sodium channels in human muscle. Pflügers Arch 412:17–21
19. Cheah KS, Cheah AM (1981) Skeletal muscle mitochondrial phospholipase A_2 and the interaction of mitochondria and sarcoplasmic reticulum in porcine malignant hyperthermia. Biochim Biophys Acta 638:40–49
20. Fletcher JE, Rosenberg H (1986) In vitro muscle contractures induced by halothane and suxamethonium: II. Human skeletal muscle from normal and malignant hyperthermia-susceptible patients. Br J Anaesth 58:1433–1439
21. Foster PS, Gesini E, Claudianos C, Hopkinson KC, Denborough MA (1989) Inositol 1,4,5,-trisphosphate phosphatase deficiency and malignant hyperpyrexia in swine. Lancet 1:124–126
22. Scholz J, Roewer N, Rum U, Schmitz W, Scholz H, Schulte am Esch J (1991) Possible involvement of inositol-lipid metabolism in malignant hyperthermia. Br J Anaesth 66:692–696
23. Scholz J, Troll U, Schulte am Esch J, Hartung E, Patten M, Sandig P, Schmitz W (1991) Inositol-1,4,5-trisphosphate and malignant hyperthermia. Lancet 337:1361
24. Duthie GG, Arthur JR (1993) Free radicals and calcium homeostasis: relevance to malignant hyperthermia. Free Radical Biol Med 14:435–442
25. Fletcher JE, Rosenberg H, Michaux K, Tripolitis L, Lizzo FH (1989) Triglycerides, not phospholipids, are the source of elevated free fatty acids in muscle from patients susceptible to malignant hyperthermia. Eur J Anaesth 6:355–362
26. Fletcher JE, Tripolitis L, Erwin K, Hanson S, Rosenberg H, Conti PA, Beech J (1990) Fatty acids modulate calcium-induced calcium release from skeletal muscle heavy sarcoplasmic reticulum fractions: implications for malignant hyperthermia. Biochem Cell Biol 68:1195–1201
27. Olgin J, Rosenberg H, Allen G, Seestedt R, Chance B (1991) A blinded comparison of noninvasive, in vivo phosphorus nuclear magnetic resonance spectroscopy and the in vitro halothane/caffeine contracture test in the evaluation of malignant hyperthermia susceptibility. Anesth Analg 72:36–47
28. Vladutiu GD, Hogan K, Saponara I, Tassini L, Conroy J (1993) Carnitine palmitoyl transferase deficiency in malignant hyperthermia. Muscle Nerve 16:485–491
29. Fletcher JE, Rosenberg H, Michaux K, Cheah KS, Cheah AM (1988) Lipid analysis of skeletal muscle from pigs susceptible to malignant hyperthermia. Biochem Cell Biol 66:917–921
30. Fulceri R, Nori A, Gamberucci A, Volpe P, Giunti R, Benedetti A (1994) Fatty acyl-CoA esters induce calcium release from terminal cisternae of skeletal muscle. Cell Calcium 15:109–116
31. Fujii J, Otsu K, Zorzato F, de Leon S, Khanna VK, Weiler JE, O'Brien PJ, MacLennan DH (1991) Identification of a mutation in porcine ryanodine receptor associated with malignant hyperthermia. Science 253:448–451

32. Cheah AM (1981) Effect of long chain unsaturated fatty acids on the calcium transport of sarcoplasmic reticulum. Biochim Biophys Acta 648:113–119
33. Dettbarn C, Palade P (1993) Arachidonic acid-induced Ca^{2+} release from isolated sarcoplasmic reticulum. Biochem Pharmacol 45:1301–1309
34. Grand RJA (1989) Acylation of viral and eukaryotic proteins. Biochem J 258:625–638
35. Wieland SJ, Fletcher JE, Gong Q-H (1992) Differential modulation of a sodium conductance in skeletal muscle by intracellular and extracellular fatty acids. Am J Physiol 263:C308–C312
36. Fletcher JE, Erwin K, Beech J (1993) Phenytoin increases specific triacylglycerol fatty esters in skeletal muscle from horses with hyperkalemic periodic paralysis. Biochim Biophys Acta 1168:292–298
37. Bennett PB Jr, Makita N, George AL Jr (1993) A molecular basis for gating mode transitions in human skeletal muscle Na^+ channels. FEBS Lett 326:21–24
38. Chahine M, Bennett PB, George AL Jr, Horn R (1994) Functional expression and properties of the human skeletal muscle sodium channel. Pflügers Arch 427:136–142
39. Makita N, Bennett PB Jr, George AL Jr (1994) Voltage-gated Na^+ channel β_1 subunit mRNA expressed in adult human skeletal muscle, heart, and brain is encoded by a single gene. J Biol Chem 269:7571–7578

Mitochondrial Disorders in Malignant Hyperthermia

GENEVIÈVE KOZAK-RIBBENS

Introduction

Some muscle disorders are known to be associated with malignant hyperthermia (MH), including King-Denborough syndrome and central-core disease (CCD), and occasionally Duchenne muscular dystrophy and myotonias. Among them, mitochondrial myopathies (MM) have been occasionally reported [1,2]. In a recent publication by Harriman concerning 1400 patients investigated by the Leeds MH unit (R. Ellis), no histological sign of MM such as ragged red fibers or negative staining of cytochrome *c* oxidase was described [3]. Our results are in agreement with those of Harriman: whatever the in vitro contracture test results, pathological findings were heterogenous [4]. Therefore, our observation of mitochondrial abnormalities in 25 of 234 patients (10.6%) led us to investigate these patients further by additional adequate tests.

Among patients who suffered from exertional heatstroke (EHS), exercise rhabdomyolysis (ER), and some forms of pharmacological-induced rhabdomyolysis (PIR), some demonstrated abnormal responses to caffeine or halothane when they were investigated according to the European Malignant Hyperthermia Group (EMHG) protocol [5].

A similar MH syndrome occurs in many breeds of pigs. However, porcine MH is triggered by physical and stress factors in addition to inhalational anesthetics and succinylcholine [6]. Abnomalies of mitochondria have been described in the muscles of MHS pigs.

Most experimental findings in human and porcine MH indicate that the regulation of the intracellular free calcium concentration in skeletal muscle is defective. Investigation of the primary lesion in MH led to revealing a defect of the sarcoplasmic reticulum calcium channel, the ryanodine receptor (RYR), induced by mutations of the RYR 1 gene localized on chromosome 6 in pigs and 19 in humans [7,8]. For 60% of the French families who are now undergoing genetic research, the RYR locus is involved. For the remaining 40%, different loci are now being searched.

Thirty years ago, several investigators proposed that the primary defect of MH was located in the mitochondrion. They postulated that triggering agents uncoupled

Centre de Résonance Magnétique et Médicale (CRMBM), Centre National de la Recherche Scientifique (CNRS) Unité de Recherche Associée (URA) 1186, Laboratoire Associé à l'Assistance Publique à Marseille, Faculté de Médecine, 27 boulevard Jean Moulin, 13005 Marseille, France

oxidative phosphorylation thus accelerating heat, CO_2, and lactate formation and reducing adenosine triphosphate (ATP) production. This hypothesis was rapidly discarded [9,10]. The role of mitochondria in MH has been intensively debated. Ca^{2+} efflux from mitochondria of MH porcine muscle has largely been studied by K.S. Cheah and his group, who concluded first a role for mitochondria and second an altered phospholipase A2 activity in the defective Ca^{2+} regulation in MH muscle [11]. Fletcher et al. [12] disproved these conclusions but delegated to free fatty acids a role in modulating the calcium channel defect.

In 1979, Gronert and Heffron [13] had stated that "the reduced respiratory and calcium binding activities in mitochondria from MHS swine support the diagnosis of a myopathy, but that these do not account for the functional and biochemical derangement observed in clinical MH." It is now generally admitted that mitochondrial defects could be secondary to the increased concentration of cytosolic Ca^{2+} [14].

We present data obtained from 254 patients investigated in Marseille since 1990. Mitochondrial abnormalities have been observed in 24 of 254 biopsied patients, none of whom exhibited clinical symptoms of mitochondrial defects. Studies on some families for which a transmission of mitochondrial defects was observed are also discussed here. Another defect of mitochondrial transport was observed: a reduced activity of carnitine palmytoyltransferase (CPT) activity. Vladutiu et al. [15] reported the association of MH and CPT deficiency in 7 MHS patients. We have observed a similar CPT deficiency in 51 of the 138 patients investigated for this enzyme activity.

Methods

Halothane/Caffeine Contracture Tests

Patients were classified into groups according to their pathology: the MH group, patients who suffered from a suspected MH crisis (probands), and members of known MH families (135); the EHS group, patients who suffered from exertional heatstroke (84); the ER group, patients who suffered from exertional rhabdomyolysis with a core temperature less than 38°C (22); and the PIR group, patients who suffered from rhabdomyolysis (13).

According to the EMHG protocol, contracture of at least 0.2 g at a concentration of 2% halothane or less and 2 m*M* caffeine or less is required to diagnose MH susceptibility [6]. A halothane + caffeine test was generally added according to the procedure used by B.A. Britt. The muscle was exposed to 1% halothane for 10 min. Incremented doses of caffeine were then applied (0.125, 0.25, 0.35, 0.5, 1 and 32 m*M*). The K phenotype was recognized if a 1-g contracture value was recorded for 0.5 m*M* caffeine or less.

Morphological and Histoenzymological Examinations

The samples were frozen in isopentane cooled by liquid nitrogen for histochemical explorations that were carried out on serial sections; in each case, the cytochrome *c* oxidase (COX) and the adenosine monophosphate (AMP) deaminase were highlighted. For conventional histology, samples were placed in Boin's fixative then embedded in paraffin; for electron microscopy, they were placed in 4% gluraraldehyde and embedded in araldite.

Biochemical Tests

Activity of Respiratory Chain Complexes

Muscle samples were immediately frozen in isopentane and stored at −80°C until enzymatic tests were performed. The method used for preparation of mitochondrial fractions and the determination of each respiratory chain enzyme complex have been previously described [16]. All optical measurements were performed using a Hitachi 557 (Tokyo, Japan) dual wavelength spectrophotometer.

Carnitine Palmytoyl Transferase Activity

CPT activity was measured from muscle samples by isotopic exchange technique. The normal range of values was up to 26 nanomoles min^{-1} mg $protein^{-1}$, corresponding to a value greater than 50% of the mean value of control subjects.

P-31 Nuclear Magnetic Resonance (NMR) Spectroscopy (P-31 MRS)

P-31 MRS is a noninvasive technique to study in vivo muscular energetic metabolism. NMR spectra were recorded at 4.5 tesla (T) on a Bruker 4730 Biospec system (Wissembourg, Germany) equipped with a horizontal superconducting magnet (bore diameter, 30 cm). A 50-mm-diameter surface coil was positioned over the flexor digitorum superficialis muscle. Spectra were time averaged during 1 min and sequentially recorded during 3 min of rest, 3 min of exercise, and 20 min of recovery. The exercise protocol consisted of finger flexions every 1.5 s for 3 min lifting a 6-kg weight. After 1 h of rest, the protocol was applied a second time under ischemic conditions, inflating a cuff above the maximum arterial pressure during the 3-min exercise only. Relative concentrations of phosphorylated metabolites, creatine phosphate (PCr), inorganic phosphate (Pi), ATP, and monophosphoesters (sugar phosphates) were determined by integration of the respective resonances. The intracellular pH (pH_i) was calculated from the chemical shift of inorganic phosphate relative to creatine phosphate [17].

Screening of mtDNA Deletions

Search of mtDNA deletion was performed on 12 of the 24 patients belonging to the MH, EHS, and ER groups. For southern blot analysis, 5 mg of muscle total DNA was digested with Pvu ll, loaded onto a 0.8% agarose gel, electrophoresed, and blotted. Hybridization was performed with a mouse mtDNA probe.

Results

Classification of the 254 Patients Investigated by Halothane/Contracture Tests

In Table 1, it is seen that 65 patients were susceptible to MH (38 MH susceptible [MHS], 16 MHEh [halothane positive, caffeine negative], 11 MHEc [halothane negative, caffeine positive] members of known MH families) (40%), and 190 were non susceptible patients (124 MHN [halothane/caffeine negative], 51 MHN K [phenotype K], 15 MHEc patients of the other groups (EHS, ER, PIR).

TABLE 1. Classification of patients according to the European Malignant Hyperthermia Group protocol [6].

Pathology	MHS	MHEh	MHEc	MHN	MHN K	Total
MH	20	4	11	80	10	135
EHS	11	9	12	25	37	84
ER	5	3	3	11	0	22
PIR	2	0	0	7	4	13
Total	38	16	26	124	51	254

MHS, halothane and caffeine tests positive; susceptible patients: MHEh, halothane test positive, caffeine test negative, considered as susceptible; MHEc, halothane test negative, caffeine test positive; considered as positive only in MH families: MHN, halothane and caffeine tests negative; non susceptible patients: MHN K, halothane and caffeine tests negative, halothane + caffeine test positive, phenotype K, non susceptible patients; MH, members of MH families; EHS, patients who suffered from exertional heatstroke; ER, patients who suffered from exertional rhabdomyolysis; PIR, patients who suffered from different types of rhabdomyolysis.

TABLE 2. Results of histoenzymological examinations: patients for which were observed ragged red (RR) fibers or cytochrome oxidase- (COX-) negative staining of some fibers.

Pathology	MHS	MHEh	MHEc	MHN	MHN K	Total
MH	2	0	2	4	6	14
EHS	0	1	1	0	2	4
ER	1	1	0	0	1	3
PIR	1	0	0	2	1	4
Total	4	2	3	6	10	25

MHS, halothane and caffeine tests positive; susceptible patients: MHEh, halothane test positive, caffeine test negative; considered as susceptible: MHEc, halothane test negative, caffeine test positive; considered as positive only in MH families: MHN, halothane and caffeine tests negative; non susceptible patients: MHN K, halothane and caffeine tests negative, halothane + caffeine test positive, phenotype K, non susceptible patients; MH, members of MH families; EHS, patients who suffered from exertional heatstroke; ER, patients who suffered from exertional rhabdomyolysis; PIR, patients who suffered from different types of rhabdomyolysis.

Histoenzymological Results

Ragged red fibers (RR) and negative staining of COX activity were observed for 25 patients and confirmed by biochemical tests for 11 of 25; the remaining 14 patients are still under study. The distribution of the 24-patient repartition according to the results of halothane and caffeine tests is shown in Table 2.

A reduced activity (from 50% to 80%) of complex IV was observed in all 11 patients, generally associated with a reduced activity of complex I. For 1 patient, who was diagnosed as MHS in the MH group, a decrease in activity was measured for both complex III and complex IV.

In probands for whom the diagnosis of MH was raised and for whom symptoms have led us to classify the crisis as doubtful, MH was discarded by halothane/caffeine contracture tests. The only defect observed in two of these probands was a mitochondrial dysfunction.

Carnitine Palmytoyl Deficiency

Results are shown in Table 3. For 51 of 138 patients (40%), a reduced CPT activity was measured. Reduced CPT activity was considered when the activity value was less than

TABLE 3. Patients affected by a defect in carnitine palmitoyl transferase (CPT) activity.

Pathology	MHS	MHEh	MHEc	MHN	MHN K	Total
MH	2	0	1	2	2	7
EHS	1	2	8	11	10	32
ER	0	2	1	6	0	9
PIR	2	0	0	1	0	3
Total	5	4	10	20	12	51

Reduced CPT activity was considered when the activity value was less than 50% of the mean value of controls. MHS, halothane and caffeine tests positive; susceptible patients: MHEh, halothane test positive, caffeine test negative; considered as susceptible: MHEc, halothane test negative, caffeine test positive; considered as positive only in MH families: MHN, halothane and caffeine tests negative; non susceptible patients: MHN K, halothane and caffeine tests negative, halothane + caffeine test positive, phenotyhpe K, non susceptible patients; MH, members of MH families; EHS, patients who suffered from exertional heatstroke; ER, patients who suffered from exertional rhabdomyolysis; PIR, patients who suffered from different types of rhabdomyolysis.

26 nanomoles $min^{-1} mg protein^{-1}$, that is, less than 50% of the mean value of controls).

P-31 MRS

When performed, P-31 MRS showed an impairment of oxidative metabolism in good agreement with histological and biochemical results [18]. Among parameters that were measured or calculated from spectra, some of them are indicative of a defect in oxidative metabolism:

1. At rest: a reduced PCr/Pi ratio was measured, signifying a reduced content of creatine phosphate. The same effect was noted for the PCr/ATP ratio.
2. At exercise: excessive acidosis was frequently observed, and also the ATP pool was consumed. This does not occur in normal subjects.
3. During recovery: two main features appeared. First, the rate of pH_i recovery to its rest value was slow. The same trend was observed for the rate of the PCr/Pi ratio. This last ratio is the best index of the activity level of the oxidative pathway.

Family Investigations of MH-Suceptible Patients

In five families, two from the MH, one from the EHS, one from the PIR, and one from the ER group, an association of MH susceptibility and mitochondrial myopathy was observed for the MHS patient (Figs. 1–3).

In the other families of the MH group, mitochondrial myopathies were observed in relatives classified as MHEc, MHN K, or MHN. For two families, one of the MH group (Fig. 2) and one of the EHS group (Fig. 3), the father and the mother of the MHS patient were classified as MHN for one and MHN K for the other. They were not susceptible to MH but were affected by a mitochondrial defect.

mtDNA Deletions

For 1 of the 12 patients investigated, an mtDNA deletion was observed. This patient was referred after an unclear anesthesia event during eye surgery for bilateral strabismus. During and after this event, a large rise in creatine phosphokmasc (CPK) was

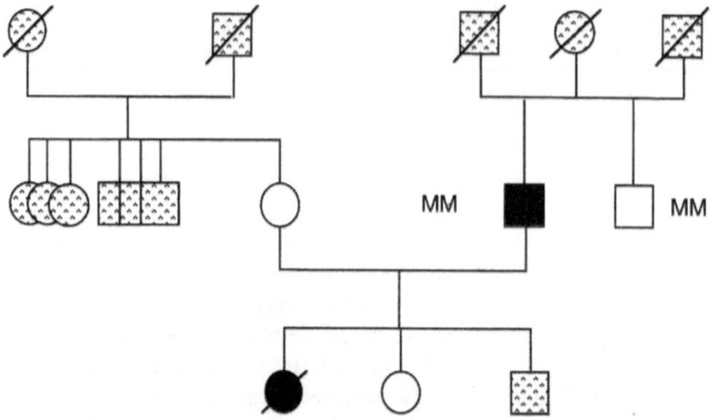

FIG. 1. Familial association of mitochondrial myopathy (*MM*) and malignant hyperthermia (MH). Note that the proband's father is malignant hyperthermia-susceptible (MHS) and affected by a mitochondrial myopathy. Women are represented by *circles* and men by *squares*

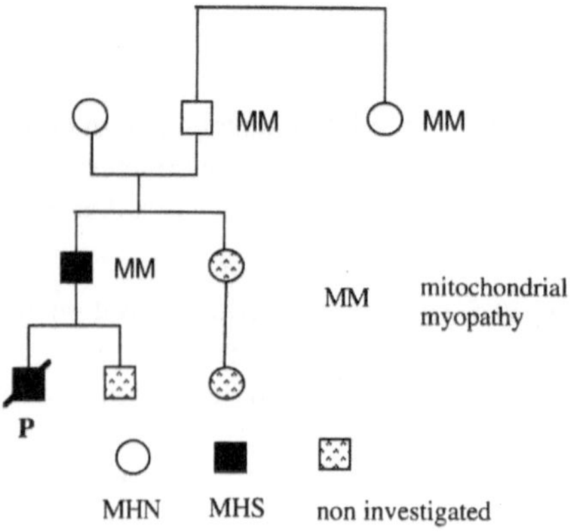

FIG. 2. Familial association of mitochondrial myopathy (*MM*) and MH. Note that the grandfather and the grandmother of the proband are both MH nonsusceptible (*MHN*). Women are represented by *circles* and men by *squares*

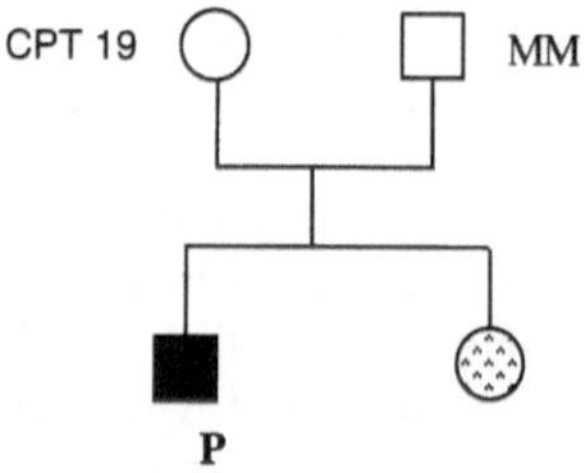

FIG. 3. Family investigation in a case of exertional heatstroke (EHS). The proband suffered from an EHS; he was recognized as MHS by halothane/contracture tests. The father and the mother were both MHN, but histological and biochemical examinations revealed defects in oxidative metabolism: reduced carnitine palmitoyl transférase (*CPT*) activity for the mother and mitochondrial myopathy (*MM*) for the father. *Circles*, women; *squares*, men; *solid* symbols, MHS: susceptible; *open* symbols, MHN: nonsusceptible

observed. This patient was classified as MHN. In fact, the appropriate diagnosis was mitochondrial myopathy with external ophthalmoplegia.

Comments

1. This report demonstrates that large investigations are useful in all these groups of patients. Such studies show that MM could be encountered in addition to the various myopathies associated with MH. In this chapter we report 25 cases in four groups of patients, the largest being the MH group.
2. The results reported here about the association of mitochondrial myopathy and MHS subjects raise three questions concerning the mode of transmission.

First: when the MH proband is recognized as MHN, if the only pathological sign observed is a defect of mitochondrial morphology or activity, this defect could be an explanation of the anesthetic event. The same is true for the MHN patients of the three other groups.

Second: when the proband is MHS, with or without mitochondrial abnormalities, and the relatives are affected by mitochondrial myopathy, the two pathologies and their transmission might be independent.

Third: when the proband is MHS, but the father and mother are nonsusceptible (MHN and MHN K), one or both parents are affected by mitochondrial defects transmitted to the proband.

To account for this observation, the first possibility is a recessive transmission of the MH trait, the mitochondrial defect being independently transmitted. The second possibility is that mitochondrial myopathy in some cases could shadow contracture tests results. This would be one of the reasons accounting for the limitation of test sensitivity. Transmission, in this point of view, should be dominant.

In contrast, mitochondrial myopathy could be an associated factor that enhances the calcium release channel defect of the proband's muscle so the MH trait can be revealed.

Conclusion

Finally, we suggest that a multidisciplinary approach is necessary in the management of MH and other related diseases. Thus, appropriate studies should be performed on muscle biopsies to evaluate the prevalence and the role of mitochondrial dysfunction in these pathologies.

Acknowledgment. The results presented in this chapter were obtained in collaboration with L. Rodet, D. Bendahan, S. Confort Gouny, and P.J. Cozzone, with the technical assistance of C. Marie dit Moisson, CRMBM, Faculté de Médecine; D. Figarella Branger and J.F. Pellissier, Laboratoire d'Anatomie et de Neuropathologie; and C. Desnuelle, with the technical assistance of M. Mayan, Secteur d'explorations fonctionnelles du tissu musculaire, CHU La Timone Adultes, Marseille; C. Desnuelle and V. Paquis, Laboratoire de Neurobiologie Cellulaire, Faculté de Médecine, Nice; R. Petrognani, O. Deslangle, and M. Aubert, Department of Anesthesia, Hôpital d'Instruction des Armées, A. Laveran, Marseille. This work was supported by grants from the Association Française Contre les Myopathies (AFM), the Assistance

Publique à Marseille (APM), the Centre National de la Recherche Scientifique (CNRS), the Direction de la Recherche du Service de Santé des Armées (DRSSA), and the European MH Association (EMHA).

References

1. Ohtani Y, Milke T, Ishitsu T, Matsuda I, Tamari H (1985) A case of malignant hyperthermia with mitochondrial dysfunction. Brain Dev 7:249
2. Kozak-Ribbens G, Figarella Branger D, Rodet L, Coulom Pontier F, Confort Gouny S, Bendahan D, Pellissier JF, Desnuelle C, Cozzone PJ (1992) Malignant hyperthermia, heat stroke and mitochondrial myopathies. In: 6th International Workshop, Hershey, PA
3. Harriman DGF (1992) The pathology of malignant hyperthermia. In: Mastaglia FL, Lord Walton of Detchant (eds) Skeletal muscle pathology. Churchill, London, pp 541–562
4. Figarella-Branger D, Kozak-Ribbens G, Rodet L, Aubert M, Borsarelli J, Cozzone PJ, Pellissier JF (1993) Pathological findings in 165 patients explored for malignant hyperthermia susceptibility. Neuromusc Disord 5–6:553–556
5. European MH Group (1984) A protocol for the investigation of malignant hyperpyrexia (MH) susceptibility. Br J Anaesthesia 56:1267–1269
6. Gronert GA, Milde JH (1981) Variations in onset of porcine malignant hyperthermia. Anesth Analg 60:499–503
7. Mac Lennan D, Phillips MS (1992) Malignant hyperthermia. Science 256:789–794
8. Ball SP, Johnson KJ (1993) The genetics of malignant hyperthermia. J Med Genet 30:89–93
9. Britt BA, Endenyi L, Eng D, Cadman DL, Man Fan H, Fung HYK (1975). Porcine malignant hyperthermia: effect of halothane on mitochondrial respiration and calcium accumulation. Anesthesiology 42:292–300
10. Ruttenbeek W, Verburg MP, Janssen AJM, Stadhouders AM, Sengers RCA (1984) In vivo induced malignant hyperthermia in pigs. II. Metabolism of skeletal muscle mitochondria. Acta Anaesthesiol Scand 28:9–13
11. Cheah KS, Cheah AM (1981) Skeletal muscle mitochondria phospholipase A2 and the interaction of mitochondria and sarcoplasmic reticulum in porcine malignant hyperthermia. Biochim Biophys Acta 638:40–49
12. Fletcher JE, Rosenberg H, Michaux K, Tripolitis L, Lizzo FH (1989) Triglycerides, not phospholipides, are the source of elevated free fatty acids in muscle from patients susceptible to malignant hyperthermia. Eur J Anaesthesiol 6:355–362
13. Gronert GA, Heffron JJA (1979) Skeletal muscle mitochondria in porcine malignant hyperthermia: respiratory activity, calcium functions, and depression by halothane. Anesth Analg 58:76–81
14. Sengers RCA, Stadhouders AM (1987) Secondary mitochondrial pathology. J Inherited Metab Dis 10(S1):98–104
15. Vladutiu GD, Hogan K, Saponara I, Tassini L, Conroy J (1993) Carnitine palmitoyl transferase deficiency in malignant hyperthermia. Muscle Nerve 16:485–491
16. Desnuelle C, Birch Machin M, Pellissier JF, Bindoff LA, Ackrell BAC, Turnbull DM (1989) Multiple defects of the respiratory chain incoding complex II in a family with myopathy and encephalopathy. Biochem Biophys Res Commun 163:695–700
17. Bendahan D, Confort Gouny S, Kozak-Ribbens G, PJ Cozzone (1993) Investigation of metabolic myopathies by P31 MRS using a standardized rest-exercise-recovery protocol: a survey of 800 explorations. MAGMA 1:91–104
18. Bendahan D, Desnuelle C, Vanuxem D, Confort Gouny S, Figarella Branger D, Pellissier JF, Kozak-Ribbens G, Pouget J, Serratrice G, Cozzone PJ (1992) ^{31}P NMR spectroscopy and ergometer exercise test as evidence for muscle oxydative performance improvement with coenzyme Q in mitochondrial myopathies. Neurology 42:1203–1208

Muscle Mitochondria and Malignant Hyperthermia

Khay S. Cheah

Introduction

Malignant hyperthermia (MH) is a genetically inherited potentially fatal disorder affecting primarily the skeletal muscle of humans, stress-susceptible pigs, and, to a lesser extent, dogs. In MH-susceptible humans and animals, it is generally accepted that an increase in the level of myoplasmic free Ca^{2+} is the initiator of the syndrome [1–5]. Various hypotheses have been suggested to account for the increase in myoplasmic Ca^{2+}, such as an excess formation of fatty acids [2,6,7], an abnormality in the excitation-contraction coupling mechanism [8], a defect in the Ca^{2+} release channel of the sarcoplasmic reticulum [9–11], free radical-mediated peroxidation of membrane lipids [12], and a lower than normal Mg^{2+} inhibition of the Ca^{2+} release channel of the sarcoplasmic reticulum [13]. However, considerable evidence also supports MH to be due to a widespread membrane defect affecting various organelles, for example, platelets, erythrocytes, lymphocytes, and mitochondria [review, 1,2,14,15], in addition to skeletal muscle.

The purpose of this article is to review pertinent findings on skeletal muscle mitochondria in MH-susceptible humans and pigs, and to discuss their probable involvement in the development of MH syndrome.

Mitochondrial Structure and Function in MH-Susceptible Individuals

In biopsy skeletal muscle samples of MH-susceptible humans [16–20] and pigs [21–24], the mitochondria showed large amplitude swelling with disrupted cristae. However, present evidence [review, 2] suggests that there is no apparent abnormality in oxidative phosphorylation [2,25–29], or in electron transport [2,25], or a deficiency in the electron transport components [25,29] in tightly-coupled skeletal muscle mitochondria isolated from MH-susceptible individuals. The large amplitude swelling observed in biopsy skeletal muscles of MH-susceptible humans and pigs was due to an increase in Ca^{2+} uptake stimulating the greater than normal phospholipase A_2 activity. This resulted in an enhanced formation of free long chain unsaturated fatty acids and lyso-derivatives of membrane phospholipids [2,6,25] which were directly responsible

Department of Agriculture, The University of Melbourne, Parkville, Victoria 3052, Australia for the abnormal mitochondrial configuration, as observed both in the in vivo [30] and in vitro [25,31] experiments of MH-susceptible humans [31] and pigs [25,30].

Skeletal muscle mitochondria isolated from MH-susceptible pigs contained significantly ($P < 0.001$) higher endogenous phospholipase A_2 activity, fatty acids, and Ca^{2+} [2,25,32] than normal. These differences were observed only in mitochondria isolated from adult and not from young MH-susceptible pigs [32]. A marked difference in response was also observed in mitochondria of biopsy skeletal muscle taken from MH-susceptible and normal pigs when subjected to a halothane plus succinylcholine challenge. Muscle mitochondria from MH-susceptible pigs during the halothane plus succinylcholine-induced MH syndrome accumulated large amounts of Ca^{2+} in their matrix compartment, which ultimately resulted in gross swelling and structural damage [30]. No such changes were, however, observed in the skeletal muscle mitochondria from normal pigs [30] under similar conditions. With isolated skeletal muscle mitochondria from MH-susceptible pigs [25] and humans [31], large amplitude swelling, Ca^{2+}-induced uncoupling, and a lowering of 9°C in the transition temperature were observed during succinate oxidation. The change in the mitochondrial configuration which was accompanied by Ca^{2+}-induced uncoupling occurred when these mitochondria took up approximately half the amount of exogenous Ca^{2+} as compared with normal mitochondria [25,31]. These data imply that skeletal muscle mitochondria from both MH-susceptible humans [31] and pigs [25] have a lower capacity for accumulation of Ca^{2+} than normal, and that halothane could also substantially reduce the Ca^{2+} accumulating ability of skeletal muscle mitochondria isolated from MH-susceptible pigs [33]. The swelling of mitochondria and Ca^{2+}-induced uncoupling were prevented by inhibitors of phospholipase A_2 activity, by bovine serum albumin, a binder of free fatty acids, and by Mg^{2+} [25,31]. The difference of 9°C in the mitochondrial transition temperature for succinate oxidation, which was observed for Ca^{2+} but not for ADP for inducing the State 3 respiration, was corrected to normal by phospholipase A_2 inhibitor and by Mg^{2+} [25]. Mitochondria from MH-susceptible pigs also showed a significantly ($P < 0.001$) faster release of Ca^{2+} than normal when treated with halothane [34], an uncoupling reagent [25], and at the onset of anaerobiosis [34].

Mitochondrial ATPase Activity

The mitochondrial ATPase activity of skeletal muscle of MH-susceptible humans and pigs has not been extensively investigated. No difference in the mitochondrial ATPase activity was detected in preparations isolated at 5 min postmortem [35], but when isolated at 30 mins postmortem, the ATPase activity of MH-susceptible pigs showed a 37% reduction as compared with a complete stability in normal mitochondria isolated at 1 h post-mortem. In both MH-susceptible humans [31] and pigs [25], the increase in the ATPase activity was attributed to a higher than normal amount of endogenous fatty acids, and this was reduced by fatty acid-free bovine serum [25,31], resulting in an improvement in the mitochondrial coupling integrity. A higher than normal oligomycin-insensitive ATPase activity in mitochondria from MH-susceptible pigs was also observed, and this was suggested to be due to the destruction of the oligomycin sensitivity by the higher than normal phospholipase A_2 activity [25,36].

The ATPase activity of EDTA-submitochondrial particles (ESMP) of cardiac muscle from MH-susceptible pigs was 34% higher than the activity in those prepared from normal pigs; this ATPase activity was restored to normal level by bovine serum albumin and was stimulated by 29% with 0.5 nmole arachidonic acid per mg protein (Cheah et al., unpublished data, 1993). The ATPase activity of ESMP from normal pigs was stimulated, but to a lower extent, by a similar concentration of arachidonic acid, and its overall activity was 43% lower than that observed for MH-susceptible pigs under the same condition. The ATPase activity of F_1 particles prepared from cardiac muscle mitochondria of MH-susceptible pigs was also less sensitive to oligomycin than normal, and 83% more oligomycin was required to decrease its activity by 50% (Cheah et al., unpublished data, 1993). Like the ATPase activity of ESMP of cardiac muscle mitochondria, the proton release of EMSP from skeletal muscle of MH-susceptible pigs was significantly ($P < 0.001$) faster, the half-time for proton release being 2.5 times faster than normal (Cheah et al., unpublished data, 1993).

Evidence for Skeletal Muscle Mitochondrial Involvement in Malignant Hyperthermia

The possible involvement of mitochondria in the development of MH was first demonstrated in dogs. Halothane induced the MH syndrome in normal dogs that had been treated with dinitrophenol [37], a classical uncoupling reagent for mitochondrial oxidative phosphorylation. This initial hypothesis of uncoupling of oxidative phosphorylation being responsible for the initiation of MH syndrome was also supported using ADP to induce the State 3 to State 4 transition of NAD+-linked glutamate plus malate oxidation in mitochondria isolated from skeletal muscle of MH-susceptible pigs [38]. The findings of an aberrant uncoupling of oxidative phosphorylation in isolated skeletal muscle mitochondria were, however, not generally supported by various investigators [2,25,27,28,30,31]. The discrepancies between different findings on oxidative phosphorylation were attributed mainly to differences in the quality of the isolated mitochondria; for example, the isolated organelles were mostly swollen and/or damaged [39], thereby contributing to the reported abnormal respiratory control index with ADP.

Differences in the Ca^{2+} transport system of skeletal muscle mitochondria were also observed by various investigators. For example, mitochondria isolated from skeletal muscle at 3 min postmortem [25], and from biopsy skeletal muscle of MH-susceptible pigs after 10 min in vivo induction of MH with halothane plus succinylcholine [22] contained a significantly higher than normal amount of endogenous Ca^{2+}. The Ca^{2+} was found to be exclusively located in the mitochondrial matrix, intracristal space, and in the outer mitochondrial compartments [22]. The process of Ca^{2+} uptake was shown to be responsible for the induction of considerable swelling of the matrix compartment and for the destruction of mitochondria [22,25]. This occurred as a consequence of Ca^{2+} taken up into the mitochondria stimulating the significantly ($P < 0.001$) higher than normal endogenous phospholipase A_2 activity in MH-susceptible pigs [2,6,25]. Skeletal muscle mitochondria isolated from MH-susceptible pigs treated in vivo with neuroleptic drugs (azaperone or spiperone) to delay the onset of halothane-induced MH, on the other hand, showed normal function with respect to oxidative phosphorylation and Ca^{2+} transport [40].

A possible involvement of mitochondria in MH syndrome was also shown by in vivo studies using the non-invasive ^{31}P-nuclear magnetic resonance (NMR) technique [41]. In MH-susceptible patients, a significantly higher than normal value for the P_i/

PCr ratio and a slower than normal post-exercise recovery rate for the value of PCr/P_i, an indicator of mitochondrial function [42], were observed.

Evidence that a basic defect in mitochondria might be responsible for MH syndrome was provided by experiments involving the interaction of isolated skeletal muscle mitochondria and sarcoplasmic reticulum from genetically selected MH-susceptible and normal British Landrace pigs. Tightly-coupled skeletal muscle mitochondria isolated from these MH-susceptible pigs inhibited Ca^{2+} uptake by sarcoplasmic reticulum isolated from both normal and MH-susceptible pigs, but mitochondria from normal pigs had no such effect on sarcoplasmic reticulum from both types of pigs [6]. Skeletal muscle mitochondria of MH-susceptible pigs contained a significantly ($P < 0.001$) higher than normal amount of endogenous fatty acids and phospholipase A_2 activity [6,25], and both of these had inhibitory effects on the Ca^{2+} transport system of sarcoplasmic reticulum [6,43].

Porcine MH is a developmental disorder, the syndrome being generally not expressed in young genetically MH-susceptible pigs [32,44]. When the porcine MH syndrome was fully expressed, this was accompanied by abnormal changes in the skeletal muscle mitochondria [32]. The skeletal muscle mitochondrial endogenous Ca^{2+}, fatty acids, and phospholipase A_2 activity, which were at normal levels prior to full expression of MH, increased significantly in parallel with the increase in MH-sensitivity to halothane.

Uncoupling of mitochondria could enhance the rate of development of MH syndrome, as elegantly shown by Belani and co-workers, using tri-iodothyronine (T_3) in MH-susceptible and normal pigs (Belani et al., unpublished data, 1994). T_3 is a hormone with established properties of causing mitochondrial uncoupling, changes in the properties of mitochondrial inner membrane, and alteration in the structural configuration [45]. Injection of T_3 into normal healthy pigs triggered hyperthyroid hyperthermia ("thyroid storm"), which mimicked malignant hyperthermia without muscle rigidity [46,47]. In the absence of any triggering agent, the rates of increase in the core temperature and end-tidal CO_2 concentration ($ETCO_2$) in hyperthyroid hyperthermia of normal pigs were considerably slower than those in the acute MH syndrome in MH-susceptible pigs [47]. With MH-susceptible pigs, the induction of thyroid storm with T_3 without a triggering agent was considerably faster than in MH-susceptible pigs receiving placebo (sodium hydroxide) (Belani et al., personal communication, 1994). Furthermore, MH-susceptible pigs pretreated with T_3 to induce hyperthyroidism immediately developed MH when challenged with succinylcholine plus halothane by showing a very rapid rise of $ETCO_2$ to 75 mm Hg, and in the esophageal temperature, to 43.5°C. The MH-susceptible pigs that received placebo, on the other hand, showed a delay of about 45 min before developing full-blown MH syndrome on exposure to succinylcholine plus halothane. The esophageal temperature of these pigs was significantly lower, and the maximal temperature was also approximately 4°C lower than in the MH-susceptible hyperthyroid pigs. The difference in the rate of development of MH syndrome between hyperthyroid and placebo MH-susceptible pigs could be explained by T_3-induced uncoupling of the skeletal muscle mitochondria, resulting in their inability to sequester Ca^{2+} released by the sarcoplasmic reticulum. The Ca^{2+} was then freely available to trigger MH instantaneously in the MH-susceptible hyperthyroid pigs. With the placebo MH-susceptible pigs, the mitochondria were able to sequester the Ca^{2+} initially released from the sarcoplasmic reticulum until Ca^{2+}-induced uncoupling, gross swelling, and destruction of the mitochondria occurred, ultimately resulting in the release of mitochondrial Ca^{2+} for triggering full-blown MH [2,25,30].

Mitochondrial myopathies have also been reported in MH-susceptible patients by histochemical examinations on biopsy skeletal muscles [48–50]. Abnormal cytochrome *c* oxidase activity [49,50] and various defects in the complexes of the mitochondrial respiratory chain system were detected [50]. Further details of mitochondrial myopathies in malignant hyperthermia were described in the presentation of Dr. G. Kozak-Ribbens at this symposium (see the chapter 1, "Mitochondrial Disorders in Malignant Hyperthermia," this volume).

Role of Mitochondria in Malignant Hyperthermia

It is generally agreed that, in both human and porcine MH syndrome, there is an apparent loss in the control of Ca^{2+} regulation in the skeletal muscle. This results in an elevation of myoplasmic Ca^{2+} level [1–4], which is then responsible for the increase in muscle rigidity and in the rate of glycolysis, as muscle contraction is dependent on the concentration of free Ca^{2+} in the myoplasm [51,52], and the enhancement in the rate of glycolysis is dependent on Ca^{2+} stimulating the activity of phosphorylase kinase [53–55] and of myofibrillar ATPase [56]. The increase in temperature rate by 1°C per 5 min to 42°C or higher [14] appears to be contributed by a combination of several processes. For example, heat production in the initial 10 min in porcine MH syndrome is shown to be attributed to an increase in aerobic metabolism [57] and could also be contributed by uncoupling of mitochondria [25,31,37,47], as well as by shortening and contraction of muscles [58]. Later, heat production is due to an increase in the formation of lactate [57,59].

In MH-susceptible pigs, the increase in the level of myoplasmic Ca^{2+} in the MH syndrome is derived from the Ca^{2+}-regulating organelles rather than from an influx of Ca^{2+} from the extracellular fluid [30], since the total Ca^{2+} in the muscle starts to decrease within 5 min of induction with halothane, accompanied simultaneously by an increase in the level of Ca^{2+} in the plasma [59].

In skeletal muscle, the sarcoplasmic reticulum is the principal regulator of Ca^{2+} level [60,61]; the mitochondria, with a large capacity for accumulating Ca^{2+} either by electron transport or by hydrolysis of ATP [62–66], function as an efficient Ca^{2+} storage system in various pathological conditions which are frequently linked with a large influx of Ca^{2+} into the cells [67]. In recent years, evidence suggests that the primary defect of MH lies within the mechanism of Ca^{2+} release of the sarcoplasmic reticulum [68–71] as a result of a mutation in the ryanodine receptor [10,70,72–74], but this appears not to be completely essential for the development of MH [7,75–77].

Evidence also supports the idea that skeletal muscle mitochondria are involved in the development of MH syndrome (see above section). The role of mitochondria in MH syndrome is strongly suggested, particularly by the findings that: (i) skeletal muscle mitochondria of MH-susceptible pigs [2,25] and humans [31] have half the normal capacity for Ca^{2+} accumulation; (ii) are more easily uncoupled by Ca^{2+} than normal [25,31]; (iii) skeletal muscle mitochondria of only MH-susceptible pigs decrease the Ca^{2+} uptake of sarcoplasmic reticulum of both normal and MH-susceptible pigs [6]; (iv) the uncoupling of mitochondria by either dinitrophenol [37] or T_3 [45,46] could induce MH; (v) the rate of development of MH is significantly faster in T_3-reated than in placebo-treated MH-susceptible pigs (Belani et al., personal communication, 1994); and (vi) skeletal muscle mitochondria of MH-susceptible pigs show considerable uptake of Ca^{2+}, resulting in large amplitude swelling within 10 min following in vivo induction of MH with halothane plus succinylcholine [22]. Furthermore, in normal mammalian muscle, release of Ca^{2+} from the sarcoplasmic reticulum by A23187 produces major muscle structural damage and myofilament degradation [78]. Electron microscopic studies show a marked swelling of the skeletal muscle mitochondria as a result of excessive uptake of the Ca^{2+} released from the sarcoplasmic reticulum. After 40 min, however, the mitochondria are shrunken following the release of the accumulated Ca^{2+} originally derived from the sarcoplasmic reticulum. This situation is analogous to that reported for MH-susceptible pigs triggered with halothane plus succinylcholine [22].

It would thus appear that the principal route for Ca^{2+} release in MH could be the cycling of Ca^{2+} from the sarcoplasmic reticulum into the mitochondria, and then from the mitochondria into the myoplasm, ultimately leading to a full-blown MH syndrome. Alternatively, the involvement of mitochondria in MH syndrome could be due to mitochondrial uncoupling by long chain fatty acids and/or Ca^{2+} following the activation of phospholipase A_2. The excess release of long chain fatty acids and/or Ca^{2+} could then induce additional release of Ca^{2+} from the sarcoplasmic reticulum, probably by the mechanism of Ca^{2+}-induced Ca^{2+} release and/or fatty acid-induced Ca^{2+} release [6,7,43]. There is no doubt that uncoupling of mitochondria can induce MH syndrome and can substantially enhance the rate of development of MH, as has been clearly shown by Belani and co-workers (unpublished data, 1994), using T_3-treated and placebo-treated MH-susceptible pigs. Under these conditions, the development of MH appears to be almost instantaneous.

However, other abnormalities exist in MH-susceptible humans and pigs, in addition to the abnormalities in the sarcoplasmic reticulum and skeletal muscle mitochondria. Peroxidation of membrane lipids by free radicals [79,80] derived from the excess formation of fatty acids generated by phospholipase A_2, inhibition of mitochondrial phospholipase A_2 activity, stabilization of membrane permeability, and inhibition of Ca^{2+} release by vitamin E in MH-susceptible pigs, as described previously by Cheah at this symposium (see the chapter "Vitamin E and Porcine Malignant Hyperthermia", this volume), and the prevention of porcine MH syndrome with N-acetycysteine [81] all suggest that a general alteration membrane integrity is the basic defect in MH.

Acknowledgment. The author wishes to express his thanks to Drs. K.G. Belani, R.J. Carr, D.S. Beebe, and P.A. Laizzo for permission to cite their unpublished data on the effect of T_3 on MH-susceptible pigs.

References

1. Britt BA (1987) In: Britt BA (ed) Malignant hyperthermia. Martinus Nijhoff, Boston, p 11
2. Cheah KS (1987) In: Britt BA (ed) Malignant hyperthermia. Martinus Nijhoff, Boston, p 79
3. Mickelson JR, Louis CF (1992) In: Puolanne E, Demeyer DI, Ruusunen DI, Ellis S (eds) Pork quality: Genetic and metabolic factors CAB International, Wallingford, p 160

4. Lopez-Padrino JR (1994) In: Ohnishi ST, Ohnishi T (eds) Malignant hyperthermia—A genetic membrane disease. CRC, Boca Raton, p 133
5. O'Brien PJ (1994) In: Ohnishi ST, Ohnishi T (eds) Malignant hyperthermia—A genetic membrane disease. CRC, Boca Raton, p 105
6. Cheah KS, Cheah AM (1981) Biochim Biophys Acta 638:40
7. Fletcher JE, Tripolitis L, Rosenberg H, Beech J (1993) Biochem Mol Biol Int 29:763
8. Okumura F, Crocker BD, Denborough MA (1980) Br J Anaesth 52:377
9. Mickelson JR, Gallant EM, Litterer LA, Johnson KM, Rempel WE, Louis CF (1988) J Biol Chem 263:9310
10. Fujii J, Otsu K, Zorzato F, De Leon S, Khanna VK, Weiler JE, O'Brien PJ, MacLennan DH (1991) Science 253:448
11. Ohnishi ST (1994) In: Ohnishi ST, Ohnishi T (eds) Malignant hyperthermai—A genetic membrane disease. CRC, Boca Raton, p 45
12. Duthie GG, Arthur JR (1993) Free Radic Biol Med 14:435
13. Lamb GD (1993) J Muscle Res Cell Motility 14:554
14. Gronert GA (1980) Anesthesiology 53:395
15. Cheah KS, Cheah AM (1984) In: Kates M, Manson LA (eds) Membrane fluidity. Biomembranes, vol. 12 Plenum, New York, p 661
16. Carpenter GC (1966) Soc Paediat Res 175:29
17. Isaacs H, Frere G, Mitchell J (1973) Br J Anaesth 45:860
18. Schiller HH, Mair WGP (1974) J Neurol Sci 21:93
19. Reske-Nielsen E, Haase J, Kelstrup J (1975) Acta Path Microbiol Scand A 83:651
20. Isaacs M (1978) In: Aldrete JM, Britt BA (eds) Second international symposium on malignant hyperthermia. Grune and Stratton, New York, p 89
21. Brucker RF, Williams CH, Popinigis J, Galvez TL, Vail WJ, Taylor CA (1973) In: Gordon RA, Britt BA, Kalow W (eds) International symposium on malignant hyperthermia. Charles C. Thomas, Springfield, p 238
22. Stadhouders AM, Vierling WAL, Verburg MP, Ruitenbeek W, Sengers RCA (1984) Acta Anaesthesiol Scand 28:14
23. Hull MT, Muller J, Albrecht WH (1978) Anesthesiology 48:223
24. Bergmann V (1979) Exp Pathol 17:243
25. Cheah KS, Cheah AM (1981) Biochim Biophys Acta 634:70
26. Cheah KS (1973) J Sci Food Agric 24:51
27. Brooks GA, Cassens RG (1973) J Anim Sci 37:688
28. Campion DR, Olson JC, Topel DC, Christian LL, Kuhlers DL (1975) J Anim Sci 41:1314
29. Ruitenbeek W, Verburg MP, Janssen AJM, Stadhouders AM, Sengers RCA (1984) Acta Anaesthesiol Scand 28:9
30. Stadhouders AM, Vierling WAL, Verburg MP, Ruitenbeek W, Sengers RCA (1984) Acta Anaesthesiol Scand 28:14
31. Cheah KS, Cheah AM, Fletcher JE, Rosenberg H (1989) Int J Biochem 21:913
32. Cheah KS, Cheah AM, Waring JC (1986) Meat Sci 17:37
33. Britt BA, Endrenyi IL, Cadman DL, Fan HM, Fung HYK (1975) Anesthesiology 42:292
34. Cheah KS, Cheah AM (1976) J Sci Food Agric 27:1137
35. Greaser ML, Cassens RG, Briskey EJ, Hoekstra WG (1969) J Food Sci 34:120
36. Kagawa J, Racker E (1966) J Biol Chem 241:2461
37. Wilson RD, Nichols Jr. RJ, Dent TE, Allen CR (1966) Anesthesiology 27:231
38. Eikelenboom G, Van den Bergh SG (1973) J Anim Sci 37:692
39. Greaser ML, Cassens RG, Briskey EJ, Hoekstra WG (1969) J Food Sci 34:125
40. Somers CJ, McLoughlin JV (1982) J Comp Pathol 92:191
41. Olgin J, Argove Z, Rosenberg H, Tuchler M, Chance B (1968) Anesthesiology 68:507
42. Argove Z, Bank WJ, Maris J, Peterson P, Chance B (1987) Neurology 37:257
43. Cheah AM (1981) Biochim Biophys Acta 648:113
44. Fay RS, Gallant EM (1990) Am J Physiol 259:R133
45. Hafner RP, Nobes CD, McGown A, Brand MD (1988) Eur J Biochem 178:511
46. Stevens JJ (1983) Anesthesiology 59:263

47. Belani KG, Carr RJ, Reardon R, Beebe DS, Komaduri V, Sweeney MF, Iaizzo PA (1994) In: VIIth international workshop on malignant hyperthermia, Hiroshima. p 50 (abstract)
48. Ohtani Y, Miike T, Ishitsu T, Matsuda I, Tamari H (1985) Brain and Development 7:249
49. Branger DF, Pellissier JF (1993) In: Aubert M, Borsarelli J, Kozak-Ribbens G, Khambata HJ (eds) Malignant hyperthermia. Normed, Bad Humberg, p 27
50. Kozak-Ribbens G, Rodet L, Branger DF, Desnuelle C, Deslangles O, Bendahan D, Gouny SC, Aubert M, Pellissier JF, Cozzone PJ (1994) In: Third international symposium on malignant hyperthermia, Hiroshima. p 38 (abstract)
51. Ebashi S, Endo M, Ohtsuki I (1969) Quart Rev Biophys 2:351
52. Weber A, Murray JM (1973) Physiol Rev 53:612
53. Ozawa E, Hosoi K, Ebashi S (1967) J Biochem (Tokyo) 61:531
54. Heilmeyer LMG, Jr, Meyer F, Hasche RH, Foscher EH (1970) J Biol Chem 245:6649
55. Brostrom CO, Hunkeler FL, Krebs EC (1971) J Biol Chem 246:1961
56. Scopes RK (1974) Biochem J 142:79
57. Hall GM, Bendall JR, Lucke JN, Lister D (1976) Br J Anaesth 48:305
58. McMahon TA (1984) In: McMahon TA (ed) Muscles, reflexes and locomotion. Princeton University Press, Princeton, p 27
59. Berman MC, Kench JE (1973) In: Gordon RA, Britt BA, Kalow W (eds) International symposium on malignant hyperthermia. Charles C. Thomas, Springfield, p 287
60. Fleischer S, Inui M (1989) Annu Rev Biophys Biochem 18:333
61. Lytton J, MacLennan DH (1992) In: Fozzard HA, Haber E, Jennings RB, Katz AM, Morgan HE (eds) The heart and cardiovascular system, 2nd edn, vol 2 Raven, New York, p 1203
62. Bygrave FL (1978) Biol Rev 53:43
63. Carafoli E, Crompton M (1978) Curr Top Membr Transp 10:151
64. Fiskum G, Lehninger A (1980) Fed Proc 39:2432
65. Carafoli E (1982) In: Martonosi AN (ed) Membranes and transport, vol. 1. Plenum, New York, p 611
66. Akerman KEO, Nicholls DG (1983) Rev Physiol Biochem Pharmacol 95:149
67. Martonosi AN (1984) Physiol Rev 64:1240
68. Donaldson SK, Gallant EM, Huetterman DA (1989) Pflugers Arch 414:15
69. Fill M, Coronado R, Mickelson JR, Vilven J, Ma J, Jacobsen BA, Louis CF (1990) Biophys J 50:471
70. MacLennan DH, Duff C, Zorzato F, Fujii J, Philips M, Korneluk RG, Frodis W, Britt B A, Worten RG (1990) Nature 343:559
71. O'Brien PJ (1990) Mol Cell Biochem 93:53
72. Gillard EF, Otsu K, Fujii J, Khanna VK, De Leon S, Derdemezi J, Britt BA, Duff CL, Worton RG, MacLennan DH (1991) Genomics 11:751
73. Hogan K, Couch F, Powers PA, Gregg RC (1992) Anesth Analg 75:441
74. MacLennan DH, Phillips MS, Zhang Y (1994) In: Ohnishi ST, Ohnishi T (eds) Malignant hyperthermia. CRC, Boca Raton, p 26
75. Rempel WE, Lu MY, Kandelgy SE, Kennedy CFH, Irvin LR, Mickelson JR, Louis FL (1993) J Anim Sci 71:395
76. Fletcher JE, Calvo PA, Rosenberg H (1993) Br J Anaesth 71:410
77. Hogan K (1994) In: Ohnishi ST, Ohnishi T (eds) Malignant hyperthermia. CRC, Boca Raton, p 293
78. Duncan CJ (1978) Experientia 34:1531
79. Duthie GG, Arthur JR (1993) Free Radic Biol Med 14:435
80. Deboer GE, Pippenger CE, Mitsumoto H, Solano R (1994) In: VII international workshop on malignant hyperthermia, Hiroshima. p 61 (abstract)
81. Peacock J, Valentine SJ, Williams KA, McPhee CP (1994) In: VII international workshop on malignant hyperthermia, Hiroshima. p 62 (abstract)

Titles were not provided by author.

Part 6. Malignant Hyperthermia Related Syndromes

Association Between Malignant Hyperthermia and Severe, Chronic Muscle Pain

BEVERLEY A. BRITT, C. ANNE MILDON, WANDA FRODIS, and ELIZABETH SCOTT

Summary. The purpose of this study is to examine the relationship between anesthetic-induced malignant hyperthermia (MH) and severe, generalized, and persistent muscle pain. We have performed the caffeine–halothane contracture test (CHCT) (3% halothane-induced contractures, caffeine-specific concentrations [CSCs], and 2 m*M* caffeine-induced contractures) on 34 healthy volunteer controls, 114 patients who have had known or suspected malignant hyperthermic (MH) reactions, and 143 patients who have had severe and persistent muscle pain that has prevented full-time work. Serum creatine kinase (CK) levels, 3% halothane contractures, and 2 m*M* caffeine contractures were substantially greater and CSCs were significantly smaller in the MH and muscle pain (MP) patients than in the control patients. There were no significant differences in these parameters between the MH and the MP patients except that CK levels were greater in the MP than in the MH patients. There seem to be at least three possible scenarios for the etiology of MP. The first is that MP patients are simply MH genotypes who have not yet had an MH reaction. They possess the MH gene, and their symptoms are triggered by environmental factors such as excessive exercise, extreme heat or cold, infection, or chemicals in the workplace. The second is that MP individuals are actually chronic fatigue immune dysfunction syndrome (CFIDS) patients. It may be that muscle changes occur in CFIDS patients that mimic MH, at least so far as the CHCT is concerned. The third possibility is that the MP patients have some third, as yet unidentified, condition that is different from both MH and CFIDS. We have shown that a cohort of chronic muscle pain patients possess an organic muscle defect as evidenced by abnormal CHCTs even though the microscopic appearance of their muscles was normal and in spite of a lack of objective physical abnormalities. This defect is probably similar to MH, but whether it is identical to MH has not been shown by this study.

Introduction

The purpose of this study was to examine the relationship between anesthetic-induced malignant hyperthermia (MH) and severe, generalized, and persistent muscle pain.

Malignant Hyperthermia Investigation Unit, Department of Anaesthesia, University of Toronto, Toronto General Hospital, Toronto, Ontario M5G 2C4, Canada

We have observed that some, but by no means all, adult malignant hyperthermic-susceptible (MHS) patients complain of incapacitating, chronic, and migratory-type muscle pains, cramps, and stiffness. These pains are often so severe as to prevent the individual from working. Some of these individuals also have repeatedly increased serum concentrations of creatine kinase (CK), even in the absence of unusual exercise or other muscle disorders. Such an association of anesthetic-induced MH with muscle pain and increased CK has also been reported by other workers [1–8].

We have performed the caffeine halothane contracture test (CHCT) on adult individuals who have not had adverse anesthetic histories, but rather have had histories of similar disabling and continuing muscle pain and often, but not always, increased CK levels. These have often begun with a flu-like illness, muscle injury, or violent and prolonged exercise in hot weather or paradoxically violent shivering in the winter time. All have had pain so severe as to prevent full-time work. The pain is usually migratory and is described as being burning, crampy, and or aching. In addition to muscle pain, often but not always with an increased CK, many also have had chronic fatigue, blinding headaches, poor sleep patterns, and episodes of excessive sweating in the absence of any external heat stimulus. These patients will, therefore, be termed MP (muscle pain).

Many of the MP patients, at least in more recent years, came to our clinic labeled as having chronic fatigue immune dysfunction syndrome (CFIDS) or fibromyalgia (FM) or both [9–11]. These are disorders that have been described under various names, including neurasthenia, since 1840. Both have been considered "organic" disorders with a psychiatric component [12–17]. However, research in recent years has shown numerous biochemical [18–22], pathophysiological [19,22–26], and immunological [19,22,25,27,28] changes in CFIDS and FM patients involving brain and other tissues in the disease process [20–22,27–29].

CFIDS patients sometimes complain of severe muscle pain with subjective muscle weakness after sustained muscular effort that improves with rest [18,19,21,23]. These muscle symptoms are confined largely to the proximal limb muscles [20,21]. As the organic nature of these diseases became established in the past decade, criteria were set to help clinicians with diagnosis [30]. These criteria can be summarized as having a typical history; having two of three physical findings; and having no other fatigue and muscle pain-producing diseases, either psychiatric, organic, or posttherapeutic, such as drug induced [30].

Because some MH patients have chronic muscle pain and sometimes also an elevated CK and because there is a cohort of patients with muscle pain and an elevated CK but no history of MH reactions, we have performed the CHCT in the latter group of patients and compared the results with the former group to see if there is any relationship between the two groups.

Methods

This study was approved by the University of Toronto Human Experimentation Committee, and all patients and controls provided an informed consent. The study began on January 5, 1989 (the date of first measurements of halothane in the gas and liquid phases) and continued until July 6, 1994.

Selection of Subjects

The patients were divided into the following three groups: (1) healthy volunteer patients (control) who had no history of an MH reaction in themselves or in their relatives (34 individuals); (2) MHS patients who had a personal history of an abnormal anesthetic reaction suspected but not proven to be MH at least 1 year before the study with or without severe chronic muscle pain or a CK elevation (reference point, 150 IU) (114 individuals); and (3) MP patients who had no history of MH reactions in themselves or in their relatives but did have severe chronic muscle pain, often migratory and worse in cold weather, and often but not always associated with generalized, persistent, and severe headaches, sleep disturbances, abnormal increases in sweating, and CK elevations in the absence of unusual exercise for 4 weeks preceding the measurements (at least three measurements, 3 months apart) (143 individuals).

All patients in group 3 were unable to work full time because of their muscle pain. None of the patients in group 3 had dark urine, muscle swelling, objective muscle weakness, or any other objective clinical sign of a muscle disease, although many subjectively thought that their muscles were weak. All had felt unwell with the foregoing symptoms for at least 4 years. None had chronic sore throats or enlarged cervical lymph nodes. All patients in this group were referred to us by neurologists, rheumatologists, psychiatrists, or physicians specializing in CFIDS. Additionally, neurological and general physical examination and laboratory investigation by the senior author were unremarkable in all category-3 patients. In particular, these patients had no evidence of chronic infection (normal history and physical examination, normal temperatures, normal white blood cell counts and differentials, and normal sedimentation rates); cancer (normal history and physical examination); autoimmune disorders such as disseminated lupus erythematosus, scleroderma, or rheumatoid arthritis (normal history and physical examination, no anemia, normal white blood cell counts, normal sedimentation rates, no rheumatoid factor, and no antinuclear antibodies); thyroid abnormalities (normal history and physical examination, normal thyroid-stimulating hormone (TSH) level, T3, T3 uptake, and T4 studies). Any patients who did not meet these criteria were excluded from the study.

Microscopic examination was performed on all excised muscle samples. Any patients whose muscle exhibited evidence of a myopathy other than MH were excluded from the study.

Restrictions

All patients in the study were required to refrain from all drugs, caffeine, cigarettes, and alcohol for 4 weeks before the study. Their exercise level was held constant (the same type of exercise and the same level of exercise) for 4 weeks before the study.

Epideomiology

The number of previous anesthetics (uneventful and associated with possible MH) was calculated for each patient in groups 1, 2, and 3. These data were analyzed on SAS (SAS Institute, Cary, NC, USA) by means of Duncan's multiple range test [31].

The sex and age of subjects were recorded. These data were analyzed on SAS by means of chi-square (χ^2) analysis, procedure general linear models, and Duncan's multiple range test [31].

Procedures

After admission to hospital, blood for the serum CK was withdrawn through a three-way stopcock in an IV line without a tourniquet being applied to the line. Immediately after withdrawal the samples were placed in the dark at 22°C. The samples were measured for CK by the Boehringer Mannheim (Mannheim, Germany) CKNAC activated method using a System 3 kit. Measurements were made on a Hitachi 737 automated analyzer (Hitachi, Tokyo, Japan). All patients had their CKs measured at least four times (at the beginning of and immediately after anesthesia, immediately after recovery, and 24 h postoperatively). The level analyzed in the study was the mean of these four values. The reference point for the CK values was 150 IU. The mean CKs were compared on SAS using procedure general linear models, Duncan's multiple range test, procedure Npar1way Wilcoxon (paired rank sum test), and χ^2 analysis [31].

The CHCT test was performed on the muscle excised from each individual as previously described and according to the North American Malignant Hyperthermia Registry [32,33]. In brief, the patients were anesthetized with Innovar, midazolam, thiopentone, and nitrous oxide. They were artificially ventilated after oral intubation. Muscle was excised from the vastus lateralis, gracilis, or rectus abdominus muscles.

The excised muscle was then divided into fascicles 2 cm × 0.2 cm. Each fascicle was tied at each end with a black silk suture. The lower end was secured via its suture to a plastic frame housing a platinum stimulating electrode that was allowed to touch the belly of the fascicle. The fascicle on its frame was then placed in a 100-ml bathing chamber filled with Krebs Ringer solution, pH 7.4, at 37°C. The upper black silk suture was tied to a force displacement transducer mounted above the chamber. Each chamber was bubbled with carbogen (95% O_2 and 5% CO_2). The muscle was stimulated via the platinum electrodes from a custom-built stimulator once every 5 s for 2 ms. The resting tension of the muscle was initially to be set at 1 g. Supramaximal twitch tension was then achieved by first stretching the muscle until no further increase in twitch height occurred, then by increasing the voltage to further increase twitch height to the maximum possible, and finally by reversing the polarity to see which polarity produced the higher twitch height.

The following tests were performed in triplicate on separate muscle fascicles after the fascicles had been in the bathing chambers for at least 15 min. To three muscle strips, incremental doses of caffeine (0.0, 0.5, 1.0, 2.0, 4.0, 8.0, and 32.0 m*M*) were added directly to the bathing chamber. The parameters measured for caffeine were the dose of caffeine required to raise the resting tension by 1.0 g (normal, ≥4.0 m*M*) (caffeine-specific concentration, CSC) and the amplitude of contracture at 2.0 m*M* (normal, <0.3 g). To three other muscle strips, 3% halothane was added to the bathing chamber via the carbogen line, and the amplitude of any resulting contracture was measured in grams from the baseline to the peak of the contracture (normal, <0.7 g).

To ensure that the concentration of halothane flowing in the gas line actually was 3% continuously, it was monitored throughout each study with a Puritan-Bennett anaesthetic agent monitor 222 (Puritan-Bennett, Wilmington, MA, USA). Additionally, the concentration of halothane in the gas phase entering each bathing chamber was checked immediately before the beginning of each study with a Rikken model 18 portable anaesthetic gas indicator (Riken, Tokyo, Japan). Finally the concentration of halothane in the liquid phase in each bathing chamber was measured intermittently throughout the study with a Shimadzu GC-8a gas chromatograph (Shimadzu, Kyoto, Japan) [34]. These three halothane measuring instruments were each calibrated against standardized concentrations of 1% and 3% halothane in carbogen obtained from Pressure Pak cylinders (Scott Medical Products, Plumsteadville, PA, USA). Concentrations between 2.7% and 3.3% were considered to be within acceptable limits; data obtained at halothane concentrations outside these limits were discarded.

The data for the CHCT were calculated by using the means of each triplicate and also by using the most abnormal response of each triplicate. However, only the latter calculations are presented in this chapter because there was no statistical difference between the methods of calculation and because the latter is the method used by the North American Malignant Hyperthermia Registry. The calculated CHCT data were then compared on SAS using procedure general linear models, Duncan's multiple range test, procedure Nparlway Wilcoxon (paired rank sum test), and χ^2 analysis [31].

For all parameters analyzed, $P < .05$ was considered to be statistically significant.

Results

Specificity [true negatives/(true negatives + false positives)] for all four parameters measured (CSC, contractures at 2 m*M* caffeine and at 3% halothane, and serum CK values). For all four, specificity was in excess of 91% (Table 1). Specificity was least for CSC (91.17%) and most for contractures at 2 m*M* caffeine and 3% halothane and for CKs (Table 1). It was not possible to do sensitivity studies because it was not ethical to deliberately challenge the MH patients with halothane and because an insufficient number of reactions sustained by the MH patients fell into category 6 on the Delphi scoring system described recently by Larach et al. [35].

The incidence of muscle pain in the three groups is shown in Table 2. Although 48 of the 114 MH patients exhibited severe and incapacitating muscle pain, only 1 control

TABLE 1. Specificity of parameters of caffeine Iialothane contracture test.

Parameter	Specificity
CSC	91.17
2 m*M* Caffeine	97.06
3% Halothane	97.06
Creatine kinase (CK)	97.06

CSC, caffeine-specific concentration (see text); 2 m*M* caffeine, amplitude of contractures in the presence of 2 m*M* caffeine; 3% halothane, amplitude of contractures in the presence of 3% halothane; specificity, true negatives/(true negatives + true positives).

TABLE 2. Frequency of pain in malignant hyperthermia (MH), control, and muscle pain patients.

Pain present	MH		Control		MP	
	Frequency	Percent	Frequency	Percent	Frequency	Percent
No	66	57.9	33	97.1	0	0.0
Yes	48	42.1	1	2.9	143	100.0

MP, muscle pain patients.

patient had muscle pain and this pain was not especially severe. Because of the method of patient selection, all the MP patients suffered from severe muscle pain.

The number of previous anesthetics (uneventful and associated with evidence of MH) is shown in Table 3. Because of the method of selection no controls and no MP patients had had previous anesthetics associated with MH while all MHS patients had had such previous MH reactions. The incidence of previous normal anesthetics was MP > control > MH ($P < .0001$). The incidence of total previous anesthetics (both uneventful anesthetics and those associated with MH reactions) was MH > MP > control ($P < .0073$).

No significant difference resulted from gender in any of the three groups of patients (Table 4). MH patients were significantly younger than MP or control patients, but the ages of the latter two groups were not significantly different between the groups (Table 5).

CSCs were significantly smaller while 2 m*M* caffeine-induced contractures, 3% halothane-induced contractures, and CKs were greater in the MH and MP patients than in the control patients (Table 6). These differences were most substantial for

TABLE 3. Previous anaesthetics with and without reactions in MH-susceptible, control and muscle pain (MP) patients.

Anesthetics	MH	Control	MP	*F*
Total number	288	46	316	5.36
Mean	2.53	1.35	2.21	0.0073
Duncan	A	B	B	
Normal	121	45	316	14.07
Mean	1.06	1.32	2.21	0.0001
Duncan	B	B	A	
MH reactions	167	0	0	194.16
Mean	1.47	0	0	0.0001
Duncan	A	B	B	

Duncan, Duncan's multiple range test.

TABLE 4. Sex of MH, MP, and control patients.

Sex	MH	Control	MP
Male	64	18	72
Female	50	16	71

$\chi^2 = .854$; $P = .653$.

TABLE 5. Statistics for age of MH, MP, and control patients.

Statistic	Age (years)		
	MH	Control	MP
Mean	34.841	45.147	42.098
Standard deviation	15.472	15.605	13.039
Nonparametric probabilities of MH/MP versus control	0.0007		0.6196
Nonparametric probability of MH versus MP		0.0001	
F value for MH, MP, and control		10.94	
Probability of *F* value		0.0001	
Duncan	B	A	A

TABLE 6. Statistics for caffeine-specific concentrations (CSC), 2 m*M* caffeine-induced contractures, 3% halothane-induced contractures, and creatine kinase values.

Statistics	CSC			2 m*M* Caffeine			3% Halothane			Creatine kinase (CK)		
	MH	Control	MP	MH	Control	MP	MH	Control	MP	MH	Control	MP
Mean	3.48	4.75	3.49	0.45	0.02	0.31	1.67	0.21	1.09	167.95	36.97	247.91
Standard deviation	1.57	1.34	1.25	1.09	0.06	0.34	2.39	0.22	1.47	427.02	28.08	596.76
Nonparametric probabilities of MH/MP versus control	0.001		0.0001	0.0001		0.0001	0.0001		0.0001	0.0001		0.0001
Nonparametric probability of MH versus MP		0.7335			0.3943			0.1153			0.0003	
F Value for MH, MP, and control		12.36			4.69			8.83			2.69	
Probability of *F* value		0.0001			0.0099			0.0002			0.0696	
Duncan	B	A	B	A	B	A	A	B	A	A/B	B	A

TABLE 7. Numbers of patients normal and abnormal for CSC, amplitude of 2 m*M* caffeine-, 3% halothane-induced contractures, and CK.

Parameter	Group	Normal		Abnormal		X^2, *P* (all)	X^2, *P* (MH/MP)
		Frequency	Percent	Frequency	Percent		
CSC	MH	50	43.9	64	56.1	22.865	1.220
	Control	31	91.2	3	8.8	0.000	0.269
	MP	53	37.1	90	62.9		
2 m*M* Caffeine	MH	49	43.0	65	57.0	39.119	0.538
	Control	33	97.1	1	2.9	0.000	0.463
	MP	55	38.5	88	61.5		
3% Halothane	MH	53	46.5	61	53.5	27.41	0.701
	Control	33	97.1	1	2.9	0.000	0.402
	MP	74	51.7	69	48.3		
CK	MH	91	79.8	23	20.2	25.906	10.739
	Control	33	97.1	1	2.9	0.000	0.001
	MP	87	60.8	56	39.2		

CSCs and were most modest for serum CK values (Table 6). There were no significant differences in the CSCs, 2 m*M* caffeine-induced contractures, or 3% halothane-induced contractures between the MH and the MP patients (Table 6) except that CK values were slightly less in MH than in MP patients (see Table 6).

The number of patients abnormal for CSC, 2 m*M* caffeine-induced contractures, 3% halothane-induced contractures, and CKs was markedly greater in the MH and MP groups than in the control group, but did not differ statistically between the MH and MP groups except that more MP patients had abnormal CKs than did MH patients according to Duncan's Multiple Range test (Table 7).

Discussion

When both previous normal anesthetics and anesthetics during which MH developed are pooled, MH patients had significantly more previous anesthetics than either controls or MP patients even though MH patients were younger than the MP or control patients. This difference probably occurred because MH reactions are much more common in children and young adults than in older individuals [32]. Also, patients in the MH group did presented to us as soon as they had had enough anesthetics to produce an MH reaction while such was not the case in the latter two groups.

The data show that the MP patients have abnormal CHCTs that are similar to those of the MH patients. The CSCs and the 3% halothane-induced contractures, especially the CSC, yield the best diagnostic discrimination between the MH and MP patients on the one hand and the normal or control patients on the other hand, although the specificity of the 2 m*M* caffeine-induced contractures is better than that of the CSCs. Poorer discriminatory results are probably obtained with 2 m*M* caffeine-induced contractures because this test is based upon only a single point on the curve.

The greater CK levels observed in MP than in MH patients probably reflect a tendency to select MH patients for the severity of their anesthetic reaction with little or no attention paid to their CKs. On the other hand, there may have been a tendency to select at least some MP patients on the basis of the degree of their serum CK elevations. It is also of note that the statistical differences between the MH and MP patients on the one hand and the controls on the other hand are quite small, which illustrates the inadequacy of using the CK as even a screening test for MH. While in a previous and much larger study [36] we did find a statistical difference between MHS and control patients, the differences were again small and there were many false negatives and false positives. Both Ellis et al. [37] and Paasuke and Brownell [38] have actually found no statistical difference in CK values between patients with positive CHCTs and those with negative CHCTs.

The data in Table 1 show that the specificity [true negative/(true negative + false positive)] of the CSC, 2 m*M* caffeine-induced contractures, 3% halothane-induced contractures, and CKs are high. These data, however, give no information about the sensitivity [true positive/(true positive + false negative)] of these tests. Such information can only be obtained by deliberately challenging human patients suspected to have MH with halothane, which is unethical, or before biopsy, by selecting patients who had very severe reactions according to some predetermined scoring system. This will be the subject of a future paper. Such a scoring system cannot be applied to the MP patients because they have not had any previous anesthetic-induced MH reactions.

Whether the MP patients are truly identical to the MH patients was not discerned by this study. Such information will have to await the outcome of ongoing genetic restriction fragment length polymorphism and sequencing studies. What has been determined, however, is that the MP patients have an organic muscle abnormality, shown by an abnormal CHCT, and that this abnormality is similar but perhaps not identical to that observed in MH patients.

There seem to be at least three possible scenarios for the etiology of the symptoms and signs exhibited by the MP patients. The first is that they are simply MH genotypes who have not yet had an MH reaction. It is already known that MH is a condition that requires, in addition to an abnormal gene, an environmental trigger (anesthetic agents in humans and pigs and stress in pigs). It could be that in some genetically predisposed humans other environmental triggers, such as excessive exercise, extreme heat or cold, infections, or environmental chemicals can also initiate abnormal events in the muscle cells. These stresses then may trigger the patients' pain, headaches, chronic fatigue, insomnia, and sweating. The second possibility is that MP individuals are actually CFIDS patients. It may be that muscle changes occur in CFIDS patients which mimic MH, at least as far as the CHCT is concerned. In fact, CFIDS may be a syndrome encompassing many different primary disorders, of which viral and autoimmune etiologies may account for some while some muscle disorder closely related to MH may account for others. The third possibility is that MP patients have some third, as yet unidentified, condition that is different from both MH and CFIDS for example abnormal calcium release from the mitochondria.

Patients presenting with muscle pain, often in combination with an increased CK, chronic fatigue, headaches, insomnia, and abnormal sweating, often look well. Their physical examinations are usually unremarkable. The tendency, therefore, is to treat these individuals as psychiatric problems and in particular to consider them as depressed [12]. Psychiatrists in particular are anxious to place such patients within their realm. Our biopsy results, however, show that many of these patients do have an organic muscle disease, as evidenced by their chronically high CKs and by their positive CHCTs. The problem is that of the cart and the horse. The treating physician considers that depression is causing the pain and fatigue and other symptoms when in reality the pain, fatigue, headaches, and sleeplessness may be causing the depression. Remove these symptoms and the depression frequently clears rapidly.

What can be done to treat these patients? No controlled studies have ever been done to assess which therapy is most beneficial. We can only speak here from our clinical experience. We have found that the most important first step is to reassure MP patients that their condition is organic and that their pain is not imaginary. This often does more to relieve their stress than any other single factor. We have noticed that therapies that relax the muscles such as heat, massage, electrical stimulation, and mechanical vibration are often helpful. On the other hand, therapies that force the muscles to actively contract are frequently harmful. Thus, such patients should be advised that strenuous exercise programs, especially programs involving prolonged periods of exercise, will only aggravate their pain and fatigue. They should try to stay fit by engaging in short periods of moderate exercise fairly frequently. A nap after each meal is beneficial.

We have observed that dantrolene sodium is helpful for some but by no means all MP patients. The side effects of dantrolene may well outweigh the benefits. If it is used, the initial dose should be very small. At biweekly intervals the dosage may then be increased until pain relief has been achieved, up to a maximum of 400 mg. Higher doses should not be used, because at higher doses a rash of giant pimples may develop on the back, and the drug will deposit in the skin, turning it to an unpleasant yellow colour. Liver function should be monitored at bimonthly intervals while the patients are on dantrolene.

Flunarizine, a selective calcium entry blocker, may relieve the headaches and improve sleep. Drugs that in our clinical experience are usually of no benefit include other calcium channel blockers, nonsteroidal antiinflammatory agents, tricyclics, and tranquilizers. The role of essential fatty acids in the management of these patients is as yet unknown because the exact relationship between MP and CFIDS has not yet been established.

Conclusions

In conclusion, we have shown that a cohort of chronic muscle pain patients possesses an organic muscle defect, as evidenced by abnormal CHCTs, even though the microscopic appearance of their muscles is normal and in spite of a lack of objective clinical physical abnormalities. This defect is probably similar to MH, but whether it is identical to MH has not been shown by this study. Such determination must await future genetic studies.

References

1. Poels PJE, Joosten EMG, Sengers RCA, Stadhouders AM, Veerkamp JH, Benders AAGM (1991) In vitro contraction test for malignant hyperthermia in patients with unexplained recurrent rhabdomyolysis. J Neurol Sci 105:67–72
2. Gronert GA, Thompson RL, Onofrio BM (1980) Human malignant hyperthermia: awake episodes and correction by dantrolene. Anesth Analg 59:377–378
3. Wingard D, Gatz EE (1977) Some observations on stress susceptible patients. In: Aldrete JA, Britt BA (eds) Second international symposium on malignant hyperthermia. Grune and Stratton, New York, pp 363–367

4. Wingard D (1977) Malignant hyperthermia—acute stress syndrome of man? In: Henschel EO (ed) Malignant hyperthermia current concepts. Appleton-Century-Crofts, New York
5. Wingard DW (1980) A stressful situation. Anesth Anal 59:1041–1042
6. Wingard DW (1974) Malignant hyperthermia: a human stress syndrome? Lancet i:408
7. Krivosec-Horber AP, Reyford H (1991) Relationship between exercise-induced myolysis and malignant hyperthermia. Br J Anaesth 67:221–228
8. Heytons L, Martin JJ, Van de Kelft E, Bossnert LL (1992) In vitro contracture tests in patients with various neuromuscular diseases. Br J Anaesth 68:72–75
9. Proceedings of a workshop on chronic fatigue syndrome, Toronto, Ontario, 28/29 September 1989. Organized by the Laboratory Centre for Disease Control, Health Protection Branch, Dept. of National Health and Welfare, Canada
10. Goldenberg D, Simms RW, Geiger A, Komaroff AL (1990) High frequency of fibromyalgia in patients with chronic fatigue seen in primary care practice. Arthritis Rheum 33:381–387
11. Goldenberg D, Komaroff AL (1989) The chronic fatigue syndrome: definition, current studies and lessons for fibromyalgia research. J Rheumatol (suppl) 19:23–27
12. Abbey SE, Garfinkel PE (1991) Neurasthenia and chronic fatigue syndrom. Am J Psychiatty 148:1638–1646
13. Ichise M, Salit JE, Abbey SE, Chung D-G, Gray B, Kirsh JC, Freedman M (1992) Assessment of regional cerebral perfusion by technetium-99 m HM-PAO SPECT in chronic fatigue syndrome. Nucl Med Commun 13(10):767–772
14. Ahles TA, Yunus MB, Masi AT (1987) Is chronic pain a variant of depressive illness? The case of primary fibromyalgia syndrome. Pain 29:105–111
15. Buchwald D, Goldenberg DL, Sullivan JL, et al (1987) The chronic active Epstein–Barr virus infection syndrome and primary fibromyalgia. Arthritis Rheum 30:1132–1136
16. David A, Wesseley S, Pelosi A (1988) Post viral fatigue: time for a new approach. Br Med J 296:696–698
17. Wesseley S (1990) Old wine in new bottles: neurasthenia and ME. Psychol Med 20:35–53
18. Arnold DL, Radde GK, Bore PJ, Styles P (1984) Excessive intracellular acidosis of skeletal muscle on exercise in a patient with a post-viral exhaustion/fatigue syndrome. Lancet 1:293
19. Behan PO, Bchan WMH, Bell EJ (1985) The post-viral fatigue syndrome—an analysis of the findings in 50 cases. J Infection 10:211–222
20. Mildon A (1991) Chronic fatigue syndrome. Can Dis Wkly Rep (suppl) 17S1E:18
21. Archard LC, Bowles NC, Behan PO, Bell EJ, Doyle D (1988) Postviral fatigue syndrome: persistence of enterovirus RNA in muscle and elevated creatine kinase. J R Soc Med 81:326–329
22. Russell IJ (1989) Neurohormonal aspects of fibromyalgia syndrome. Rheum Dis Clin North Am 15:141
23. Montague TJ, Marrie TJ, Klassen GA, Bewick DJ, Horacek BM (1989) Cardiac function at rest and with exercise in the chronic fatigue syndrome. Chest 95:779–784
24. Gow JW, Behan WMH, Clements GB, Woodall C, Riding M, Behan PO (1991) Enteroviral RNA sequences detected by polymerase chain reaction in muscle of patients with postviral fatigue syndrome. Br Med J 302:692
25. Ritz J (1989) The role of natural killer cells in immune surveillance (editorial). N Engl J Med 320:1748–1749
26. Archard LC, Cunningham L (1992) Molecular virology of muscle disease: persistent virus infection of muscle in patients with postviral fatigue syndrome. In: Hyde BM, Goldstein J, Levine P (eds) The clinical and scientific basis of myalgic encephalitis/chronic fatigue syndrome. Nightingale Research Foundation, Ottawa, Canada, P 343
27. Klimas NG, Salvato FR, Morgan R, Fletcher MA (1990)) Immunologic abnormalities in chronic fatigue syndrome. J Clin Microbiol 18:1403–1410
28. Landay AL, Jessop C, Lennette ET, Levy JA (1991) Chronic fatigue syndrome: clinical condition associated with immune activation. Lancet 338:707

29. Hickie I, Lloyd A, Wakefield D, Parker G (1990) The psychiatric status of patients with chronic fatigue syndrome. Br J Psychiatry 156:534–540
30. Holmes GP, Kaplan JE, Grantz NM, et al (1988) Chronic fatigue syndrome: a working case definition. Ann Intern Med 108:387–389
31. Cody RP, Smith JK (1991) Applied statistics and the SAS programming language. North-Holland, New York
32. Britt BA (1991) Malignant hyperthermia—a review. In: Schonbaum E, Lomax P (eds) Thermoregulation: pathology, pharmacology and therapy. Pergamon Press, New York, pp 179–292
33. Larach MG (1989) Standardization of the caffeine halothane muscle contracture test. Anesth Analg 69:511–515
34. Gronert GA, Chun K, Martucci R, Jones BR (1990) Concentration of halothane in Kreb's medium. Anesth Analg 71:305–314
35. Larach MG, Localio AR, Allen GC, Denborough MA, Ellis FR, Gronert GA, Kaplan RF, Muldoon SM, Nelson TE, Ording H, Rosenberg H, Waud B (1994) A clinical grading scale to predict malignant hyperthermia susceptibility. Anesthesiology 80:771–779
36. Britt BA, Endrenyi L, Peters PL, et al (1976) Screening of malignant hyperthermic susceptible families by CPK measurement and other clinical investigations. Can Anaesth Soc J 23:263–284
37. Ellis FR, Clarke IMC, Modgill M, Currie S, Harriman DGF (1975) Evaluation of creatine phosphokinase in screening patients for malignant hyperpyrexia. Br Med J 3:511–513
38. Paasuke RT, Brownell KW (1986) Serum creatine kinase level as a screening test for susceptibility to malignant hyperthermia. JAMA 255:769–771

Porcine Malignant Hyperthermia

Gerald A. Gronert

Porcine malignant hyperthermia (MH) differs from human MH in that it is a homozygotic expression of a single point mutation in all muscular species. This mutation probably occurred about 100–150 years ago. Pigs were bred then, as now, with the idea that selection of the most muscular offspring, followed by line breeding, would produce the most valuable marketable animal: lean, muscular, with little fat and rapid development of the muscular build. Unfortunately, not all characteristics were desirable. Although rapid development of lean body mass represented a true hybrid vigor, a rather marked disadvantage was an obvious overt sensitivity to stress. Certain stress, such as fright, shipping, coitus, and fighting, led to a process identical to MH episodes generally produced by triggering anesthetics: rigidity, tachycardia, tachypnea, acidosis, high fever, and death. This end result could occur in as few as 10 min, with a body temperature greater than 43°C.

This mutation bred true and, with selection and sharing of breeding stock, spread with time throughout the world of serious swine breeders. It was well recognized how consistent were the unique properties of these pigs, but this pattern was not recognized as a specific mutation until 1991 [1]. At that point scientists first realized the value of genetic analysis of this stress-related condition, because they had discovered that one single specific genetic alteration was the basis for the phenomenon recognized as stress susceptibility in swine.

Over the years, this phenomenon had been recognized in a variety of ways: in 1914, a report described pork that was of poor quality for making sausage because it had deteriorated so much during the slaughter process that it appeared spoiled [2]. In 1953, this type of pork was termed degenerated, and a genetic relationship was observed in the pigs that produced such pork [3]. In 1964, Briskey described this pork as pale, soft, and exudative at slaughter, because the hypermetabolism induced at slaughter was out of control until the pork was well cooled [4]; effective cooling did not occur until late in abattoir processing. In 1968, Topel et al. [5] described porcine stress syndrome, an apt and useful descriptive term.

This phenomenon began to surface in porcine anesthesia in 1966. Hall et al. [6] observed classic MH findings during halothane and succinylcholine anesthesia. Further, this type of response had also been related to exercise [5,7]. In 1970, Berman et al. [8] from South Africa described the biochemical changes of MH in a classic paper, a study that has not been improved on as a description of the overall clinical syn-

Department of Anesthesia, University of California, Davis, CA 95616, USA

drome. This anesthetic response was directly and indisputably correlated with the occurrence, after stressful death, of pale, soft, exudative pork [9].

The porcine model of MH has provided the basis for the bulk of the studies on MH and their concomitant correlation with the human MH syndrome. It has been shown that skeletal muscle is the tissue primarily responsible for the increase in whole body oxygen consumption [10]. Additionally, stress is the major factor responsible for the genesis of the MH response, whether it is manifested as the porcine stress syndrome or anesthetic-induced MH [11]. Finally, the porcine model provided the basic findings for the cure and prevention of MH, whether porcine or human, through the discovery by Harrison from South Africa of the miracle drug dantrolene [12].

In time, genetic investigations defined the chromosomal-based abnormality on chromosome 6 in pigs; an arginine to cysteine mutation conferred an altered peptide map via elimination of an arginine cleavage site on the ryanodine receptor, which is situated between the dihydropyridine receptor of the transverse tubule and the sarcoplasmic reticulum [13]. Pigs heterozygous for this mutation demonstrated functional abnormalities of their skeletal muscle, but these pigs do not feature typical explosive fulminant MH on exposure to triggers such as is observed in the homozygotic, MH-susceptible pig [14]. The ryanodine receptor is a huge homotetramer, the largest receptor in the body and four times the size of the acetylcholine receptor; the four subunits could be altered by a genetic disorder.

Other porcine studies [15] have evaluated the role of the sympathetic nervous system in MH; overall, sympathetic manifestations of MH appear to be secondary to the hypermetabolic state. Calcium antagonists are not therapeutic in MH as they tend to alter primarily surface membrane calcium transients rather than those of the sarcoplasmic reticulum [16]. The porcine skeletal muscle is sensitive to stress, particularly that involving depolarization and volatile anesthetic agents, because of the undue sensitivity of the ryanodine receptor and the lowered mechanical threshold of affected skeletal muscle [17,18]. Intracellular free unbound ionized calcium is increased during MH because of the malfunction of the ryanodine receptor. However, it is important to emphasize that this does not confer triggering sensitivity to ions or drugs that alter calcium fluxes from extracellular fluid [19].

Increased plasma concentrations of calcium to 13 mEq/l, digoxin to 9 ng/ml, or arterial P_{CO_2} to 150 mmHg do not trigger the susceptible pig [19]. These studies were conducted in vivo, the first two on cardiopulmonary bypass and the third during controlled mechanical ventilation. However, increases in plasma potassium to 7 mEq/l depolarize skeletal muscle sufficiently to trigger typical abnormal MH responses [19]. The MH abnormality appears to result in an increased resting intracellular ionized calcium concentration [20] because of the altered homeostasis. This change is in part related to the balance between calcium and magnesium [21].

This brief summary has highlighted some of the features of porcine MH.

References

1. Fujii J, Otsu K, Zorzato F, et al (1991) Identification of a mutation in porcine ryanodine receptor associated with malignant hyperthermia. Science 253:448–451
2. Herter M, Wilsdorf G (1914) Die Bedeutung des Schweines für die Fleischversorgung. Arbeiten der Deutscher Landwirtschaft-Gesellschaft, Berlin, Heft 270
3. Ludvigsen J (1953) Muscular degeneration in hogs (preliminary report). In: Proceedings, 15th international veterinary congress, Stockholm, 1953, vol VI

4. Briskey EJ (1964) Etiological status and associated studies of pale, soft, exudative porcine musculature. Adv Food Res 13:89–178
5. Topel DG, Bicknell EJ, Preston KS, et al (1968) Porcine stress syndrome. Mod Vet Pract 49:40–41; 59–60
6. Hall LW, Woolf N, Bradley JWP, Jolly DW (1966) Unusual reaction to suxamethonium chloride. Br Med J 2:1305
7. Sybesma W, Eikelenboom G (1969) Malignant hyperthermia syndrome in pigs. Neth J Vet Sci 2:155–160
8. Berman MC, Harrison GG, Bull AB, Kench JE (1970) Changes underlying halothane-induced malignant hyperpyrexia in Landrace pigs. Nature 225:653–655
9. Nelson TE, Jones EW, Henrickson RL, Falk SN, et al (1974) Porcine malignant hyperthermia: observations on the occurrence of pale, soft, exudative musculature among susceptible pigs. Am J Vet Res 35:347–350
10. Gronert GA, Heffron JJA, Milde JH, Theye RA (1977) Porcine malignant hyperthermia: role of skeletal muscle in increased oxygen consumption. Can Anaesth Soc J 24:103–109
11. Lucke JN, Hall GM, Lister D (1979) Malignant hyperthermia in the pig and the role of stress. Ann NY Acad Sci 317:326–337
12. Harrison GG (1975) Control of the malignant hyperpyrexic syndrome in MHS swine by dantrolene sodium. Br J Anaesth 47:62–65
13. Mickelson JR, Knudson CM, Kennedy CFH, et al (1992) Structural and functional correlates of a mutation in the malignant hyperthermia-susceptible pig ryanodine receptor. FEBS Lett 301:49–52
14. Gallant EM (1992) Lentz LR. Excitation-contraction coupling in pigs heterozygous for malignant hyperthermia. Am J Physiol 262:C422–C426
15. Gronert GA, Milde JH, Theye RA (1977) Role of sympathetic activity in porcine malignant hyperthermia. Anesthesiology 47:411–415
16. Gallant EM, Foldes FF, Rempel WE, Gronert GA (1985) Verapamil is not a therapeutic adjunct to dantrolene in porcine malignant hyperthermia. Anesth Analg 64:601–606
17. Gallant EM, Fletcher TF, Goettl VM, Rempel WE (1986) Porcine malignant hyperthermia: cell injury enhances halothane sensitivity of biopsies. Muscle Nerve 9:174–194
18. Gallant EM, Godt RE, Gronert GA (1980) Mechanical properties of normal and malignant hyperthermia susceptible porcine muscle: effects of halothane and other drugs. J Pharmacol Exp Ther 213:91–96
19. Gronert GA, Ahern CP, Milde JH, White RD (1986) Effect of CO_2, calcium, digoxin, and potassium on cardiac and skeletal muscle metabolism in malignant hyperthermia susceptible swine. Anesthesiology 64:24–28
20. Lopez JR, Allen PD, Alamo L, et al (1988) Myoplasmic free [Ca^{2+}] during a malignant hyperthermia episode in swine. Muscle Nerve 11:82–88
21. Lamb GD (1993) Ca^{2+} inactivation, Mg^{2+} inhibition and malignant hyperthermia. J Muscle Res Cell Motil 14:554–556

Vitamin E and Porcine Malignant Hyperthermia

KHAY S. CHEAH

Summary. In malignant hyperthermia (MH), an increase in the level of myoplasmic free Ca^{2+} is generally accepted to be the initiator of the syndrome, but the primary cause responsible for this increase has yet to be fully characterized. In MH-susceptible pigs, which are widely used as a model for understanding human MH syndrome, an excess formation of fatty acids [1,2], an abnormality in the excitation–contraction coupling mechanism [3], a defect in the Ca^{2+} release channel of the sarcoplasmic reticulum [4,5], a defect in membrane integrity [6,7], free radical-mediated peroxidation of membrane lipids [8], and a lower than normal Mg^{2+} inhibition of the Ca^{2+} release channel of the sarcoplasmic reticulum [9] have all been postulated to be responsible for the development of porcine MH. Vitamin E supports the idea that porcine MH syndrome is associated with an alteration in membrane permeability induced by fatty acids. Dietary supplementation of vitamin E in the form of DL-alpha-tocopheryl acetate reduces the increased leakage of pyruvate kinase and creatine kinase, and normalizes the membrane permeability of skeletal muscle and erythrocytes, the rate of glycolysis, and the amount of Ca^{2+} released in skeletal muscle of MH-susceptible pigs. Vitamin E also reduces the enhanced level of endogenous fatty acids and phospholipase A_2 activity of skeletal muscle mitochondria of MH-susceptible pigs to normal, in addition to preventing the formation of meat quality deficiencies associated with the halothane gene. The principal beneficial effect of vitamin E results from its ability to prevent the destabilization of membrane permeability by fatty acids, and probably also from inhibiting phospholipase A_2 activity, thereby preventing the increase in the level of myoplasmic Ca^{2+}.

Introduction

Malignant hyperthermia (MH), a genetically inherited and potentially fatal disorder, affects primarily the skeletal muscle of humans and pigs. In apparently healthy humans [10,11], MH-susceptible pigs [12–14], and dogs [15], MH can easily be induced with halothane, a fluorinated hydrocarbon anesthetic. In MH-susceptible pigs, the syndrome can also be triggered by physiological or environmental stress such as excitement and changes in temperature [16,17]. The characteristic features of MH

Agriculture Department, University of Melbourne, Parkville, Victoria 3052, Australia syndrome show striking similarities in humans and pigs [11,14], and MH-susceptible pigs are therefore frequently used as experimental models for understanding human MH.

An increase in the level of myoplasmic Ca^{2+} is generally accepted to be the initiator of MH syndrome [2,7,18,19], but the primary cause for the Ca^{2+} increase has yet to be fully characterized. In MH-susceptible pigs, an excess formation of fatty acids [1,2], a defect in the Ca^{2+} release channel of the sarcoplasmic reticulum [4,5], a defect in membrane integrity [6,7], free radical-mediated peroxidation of membrane lipids [8], and inhibition of the Ca^{2+} release channel of the sarcoplasmic reticulum by Mg^{2+} that is lower than normal [9] have all been postulated to be responsible for the increase in myoplasmic free Ca^{2+}. However, considerable evidence also suggests that MH appears to be a widespread membrane defect not only in skeletal muscle but also in membranes of other organelles, for example, in platelets, erythrocytes, and lymphocytes [6,7].

A single-point mutation of arginine to cysteine in the gene for the ryanodine receptor (i.e., Ca^{2+} release channel) of the sarcoplasmic reticulum [20] has been reported to correlate with MH in five major breeds of lean, heavily muscled pigs. This has provided strong support for the sarcoplasmic reticulum as primarily responsible for the increase in the level of myoplasmic Ca^{2+} in porcine MH syndrome. However, unlike pigs, only a small proportion of MH-susceptible humans exhibit mutation of the ryanodine receptor [21]; thus, the DNA test based on the mutation of the ryanodine receptor could only be successfully applied to identify MH-susceptible pigs [22,23].

The purpose of this chapter is to discuss the beneficial effects of dietary supplementation of vitamin E in porcine MH syndrome. It reviews pertinent findings on vitamin E that support the association of porcine MH with an alteration in membrane integrity, and it also discusses current data illustrating the importance of fatty acids in inducing the increase in the level of myoplasmic Ca^{2+} in MH syndrome.

Vitamin E

The term vitamin E applies to a family of structurally related compounds known as tocopherols, which consist of a main hydroxylated chromanol ring linked to an isoprenoid side chain. Naturally occurring vitamin E is composed of a mixture of tocopherol isomers (alpha, beta, gamma, and delta), which differ mainly by the number and position of the methyl group substitution to the chromanol ring. The biological activity differs among the tocopherols; the rank of vitamin E activity is alpha > beta > gamma, delta [24]. Only the D-alpha-tocopherol occurs substantially in nature, and the DL-alpha-tocopherol is the synthetic form of vitamin E most frequently used in research.

Located almost exclusively in cellular membranes, vitamin E, is a lipid-soluble antioxidant [25] and an inhibitor of peroxidation of membrane lipids [25–27], and acts as a membrane stabilizer through its ability to form complexes with free fatty acids [28]. Thus, vitamin E is commonly regarded as an efficient free radical scavenger, protecting membranes from damage through lipid peroxidation [25,26,29].

Effect of Vitamin E on Porcine Stress Syndrome

Duthie and co-workers, among the earlier investigators on the effect of vitamin E, concluded that MH-susceptible British Landrace (BL) pigs showed evidence of vitamin E deficiency by exhibiting signs of free radical-mediated damage to cell membranes. These pigs showed significantly higher than normal amounts of pyruvate kinase (PK) and of creatine kinase (CK) in the plasma, increased formation of pentane in m. longissimus dorsi (LD), and increased malonaldehyde (MAD) in erythrocytes when incubated with H_2O_2 [30]. In these studies, membrane integrity was monitored by plasma PK and CK, and lipid peroxidation by pentane and MAD formation. All these parameters were reduced to normal by a daily dietary supplementation of 235 IU vitamin E/kg diet for 5–6 weeks, in spite of no differences being detected in the vitamin E content in plasma and tissues between MH-susceptible and normal BL pigs [30,31]. This implies that MH-susceptible pigs probably require more than the normal amount of vitamin E to maintain their membrane integrity so as to counteract the effect of stress and development of MH. Indeed, transportation stress-induced increase of plasma CK and MAD production in MH-susceptible pigs were significantly reduced by a daily supplementation of 235 IU vitamin E/kg diet, and halothane-induced increased plasma PK and CK plus MAD formation in MH-susceptible pigs were also significantly reduced by a daily supplementation of 250 IU vitamin E/kg diet in conjunction with 500 mg vitamin C [30]. An antioxidant abnormality was also suggested in liver microsomes of MH-susceptible BL pigs [32] on the basis of higher ($P < .05$) erythrocyle sedimentation rate (ESR) signal heights and of greater production of thiobarbituric acid reactive substances (TBARS). However, no difference in either vitamin E content or major polyunsaturated fatty acids in microsomes was detected between MH-susceptible and normal BL pigs.

Effect of Vitamin E on Erythrocyte Fragility, Mitochondrial Phospholipase A_2 Activity, and Ca^{2+} Released in MH-Susceptible Pietrain Pigs

Table 1 summarizes the beneficial effect of a daily supplementation of 200 mg vitamin E/kg diet for 130 days on erythrocyte fragility, LD mitochondrial phospholipase A_2 (PLA_2) activity, and Ca^{2+} released in LD muscle of MH-susceptible Pietrain (nn) pigs as compared with those of untreated Pietrain (nn), and of MH-susceptible (nn) and normal (NN) BL pigs. In untreated BL pigs, both erythrocyte fragility and LD mitochondrial PLA_2 activity of MH-susceptible pigs were significantly ($P < .001$) higher than normal, but were very similar to those of untreated MH-susceptible Pietrain pigs. Following vitamin E supplementation to MH-susceptible Pietrains, these two parameters were reduced to levels not significantly different from those of untreated normal BL pigs. The amount of Ca^{2+} released by MH-susceptible BL and Pietrain pigs was also very similar. After vitamin E supplementation, Ca^{2+} released in LD muscle of MH-susceptible Pietrain pigs also was reduced ($P < .001$) to a level not different from that of normal BL pigs.

The amount of endogenous fatty acids in LD mitochondria of MH-susceptible Pietrain pigs ($n = 3$) was also significantly reduced ($P < .01$) by vitamin E treatment from 197 ± 12 nmol/mg protein to 144 ± 5 nmol/mg protein, that is, to a level very similar to that of normal BL pigs (Fig. 1). These results showed that daily dietary supplementation of vitamin E to MH-susceptible pigs could stabilize the membrane integrity of erythrocytes when subjected to an osmotic shock in 0.7% NaCl, and could also inhibit mitochondrial PLA_2 activity and Ca^{2+} released in LD muscle.

Inhibition of PLA_2 activity by vitamin E had previously been reported for rat platelets [33]. Rats fed with 1000 ppm vitamin E diet showed a diminished PLA_2

TABLE 1. Effect of daily dietary supplementation with vitamin E on erythrocyte fragility, Ca^{2+} released in LD muscle, and LD mitochondrial phospholipase A_2 activity of malignant hyperthermia-susceptible (MH-susceptible) Pietrain pigs.

Pigs	Vitamin E supplementation (mg/kg diet)	Erythrocyte fragility (%)	Phospholipase A_2 activity (nmol fatty acid/hr per mg protein)	Ca^{2+} released (μg/g LD)
British Landrace (nn)	0	1.46 ± 0.10 ($n = 7$)[a]	9.6 ± 0.6 ($n = 4$)[a]	10.4 ± 0.3 ($n = 6$)[a]
British Landrace (nn)	0	0.35 ± 0.06 ($n = 9$)[b]	4.0 ± 0.3 ($n = 4$)[b]	0.7 ± 0.1 ($n = 4$)[b]
Pietrain (nn)	0	1.46 ± 0.17 ($n = 5$)[a]	9.9 ± 0.9 ($n = 3$)[a]	9.9 ± 1.1 ($n = 4$)[a]
Pietrain (nn)	200	0.52 ± 0.07 ($n = 4$)[b]	3.0 ± 0.3 ($n = 3$)[b]	0.9 ± 0.2 ($n = 4$)[b]

LD, Longissimus dorsi muscle; SE, standard error of the mean.
[a] The results are expressed as means ± SE with the number of pigs in parentheses.
[b] Within columns, means having different superscripts are significantly different to at least $P < 0.01$.

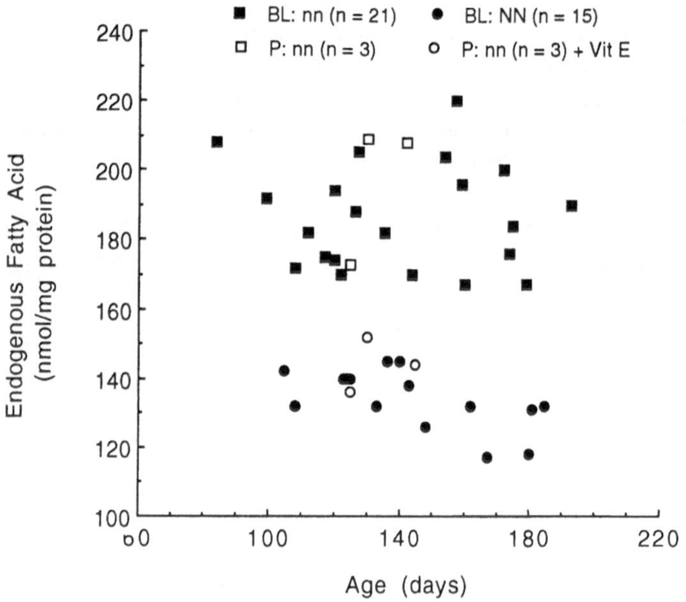

FIG. 1. Effect of vitamin E on longissimus dorsi (LD) mitochondrial fatty acids in malignant hyperthermia- (MH-) susceptible and normal pigs. *BL*, British Landrace (*solid symbols*); *P*, Pietrain (*open symbols*)

activity compared to those fed on a vitamin E-free diet. Inhibition of PLA_2 activity by vitamin E was also demonstrated with partially purified rat platelet PLA_2 [33].

Effect of Vitamin E on Membrane Permeability, Rate of Glycolysis, and Ca^{2+} Released in LD Muscle of MH-Susceptible Pigs

This effect was investigated by treating MH-susceptible Landrace × Large White (L × LW) pigs with a daily dietary supplementation of 1 g vitamin E/kg diet for 46 days. Membrane permeability (a), rate of glycolysis (b), and Ca^{2+} released (c) were deter-

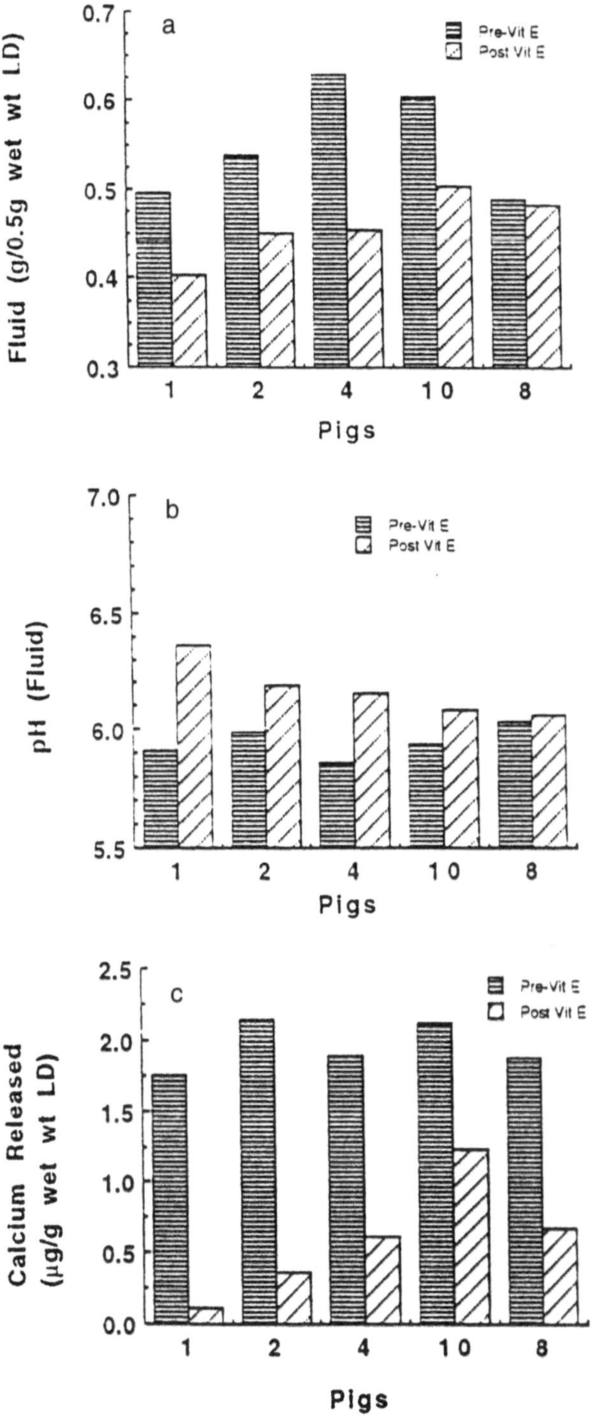

FIG. 2a–c. Effect of vitamin E on MH-susceptible pigs. **a** Membrane permeability. **b** Rate of glycolysis. **c** Ca^{2+} released

mined from biopsy LD samples before and after vitamin E supplementation (Fig. 2). All three parameters were substantially reduced by vitamin E to levels observed for normal pigs. Membrane permeability, measured as Fluid and expressed in grams per 0.5 g wet weight LD, was reduced ($P < .02$) from 0.55 ± 0.03 to 0.46 ± 0.02; rate of glycolysis, measured as pH (Fluid), from 5.95 ± 0.03 to 6.17 ± 0.05 ($P < .01$) and Ca^{2+} released from 1.95 ± 0.08 to 0.60 ± 0.19 ($P < .001$).

Vitamin E Content and Meat Quality

The content of vitamin E in skeletal muscles was increased by 460% after a daily dietary supplementation of 500 mg vitamin E/kg diet for a period of 46 days [34], with the red muscle m. masseter taking up 51% more vitamin E than LD, a predominantly white muscle. The increase of vitamin E in LD was accompanied by the prevention of pale, soft, and exudative meat formation.

Discussion

The principal effect of vitamin E in porcine MH appears to be its ability to inhibit fatty acid-induced membrane destabilization, thereby preventing an increase in the level of myoplasmic Ca^{2+} and the ultimate formation of poor meat quality associated with the halothane gene. This is achieved by inhibition of PLA_2 activity [33] (Table 1 and Fig. 1) and lipid peroxidation [30].

MH-susceptible pigs, in a way, resemble animals deficient in vitamin E with respect to loss of membrane integrity, which had been postulated to be attributed to free radical-mediated peroxidation of membrane lipids [8]. However, MH-susceptible pigs are not vitamin E deficient even though these pigs appear to have a defect in antioxidant defense mechanisms [8] that could be partly corrected biochemically by a daily dietary supplementation of 235 IU vitamin E/kg diet [31].

The fatty acid hypothesis [1,2] involving mitochondrial PLA_2 [1,2,6] is consistent with the hypothesis of an abnormality in the antioxidant defense system [8]. The enhanced PLA_2 activity of MH-susceptible pigs could generate a rapid release of peroxidizable substrates (i.e., fatty acids and lipid hydroperoxides) to overwhelm the antioxidant defense mechanisms, thereby resulting in MH-susceptible pigs showing characteristic features of animals deficient in vitamin E.

The fatty acid hypothesis for porcine MH is also consistent with that proposed for human MH [35,36], which exhibited fatty acid-induced uncoupling of skeletal muscle mitochondria in the presence of exogenous Ca^{2+} [2,35] and a marked reduction in the threshold of halothane-induced Ca^{2+} release in heavy sarcoplasmic reticulum by oleic acid at 37°C but not at 25°C [36]. This has led to current suggestions that fatty acids act as initiators [1,2,6,35] or modulators [36] for MH syndrome.

The uptake of vitamin E following dietary supplementation varies in different tissues [37–39]. Pigs supplemented with 405 mg vitamin E/kg diet for 7 weeks showed a threefold higher content of vitamin E in myocardium than in skeletal muscle, with the oxidative fibers showing the highest capacity for vitamin E uptake [38]. Red muscles have a greater capacity for vitamin E uptake than the predominantly white LD muscle [34,38]. This is most likely to result from the higher number of mitochondria in red muscles because isolated mitochondria were shown to have higher storage for vitamin E than microsomal and myofibrillar proteins [39]. This, in turn, could account for the significant ($P < .003$) inhibition of LD mitochondrial PLA_2

activity in MH-susceptible Pietrain pigs following supplementation with vitamin E (see Table 1).

The reduction of fatty acid formation in LD mitochondria of MH-susceptible Pietrains treated with vitamin E (Fig. 2) implies that vitamin E stabilizes mitochondrial membranes by inhibiting PLA_2 activity (Table 1) [33] and by forming complexes with the saturated and unsaturated fatty acids [28] in the mitochondrial membranes. Mitochondria contain PLA_2 [1,2], and the substantial reduction of fatty acid formation and of PLA_2 activity in LD mitochondria of MH-susceptible pigs by vitamin E supplementation to levels similar to those of normal (NN) pigs supports our original hypothesis that fatty acids generated by an enhanced PLA_2 activity [1,2,6,35] are intimately involved in the development of the MH syndrome.

Erythrocyte fragility observed in MH-susceptible pigs, which is greater than normal (see Table 1) is also most likely caused by enhanced PLA_2 activity [40] that has been shown to be associated with erythrocyte fragility [40–42] and erythrocyte membrane abnormality [43]. The reduction in erythrocyte fragility after vitamin E supplementation in our current experiments (Table 1) could thus be explained by an inhibition of PLA_2 activity by vitamin E, as was also previously reported for rat platelet PLA_2 [33].

Conclusion

Evidence supports the concept that porcine MH syndrome is attributed to an alteration in membrane integrity induced by an enhanced formation of fatty acids. This can be prevented by dietary supplementation of vitamin E, which stabilizes the membrane permeability by inhibiting PLA_2 activity and peroxidation of membrane lipids, thereby preventing the increase in the level of myoplasmic Ca^{2+} and the ultimate development of poor meat quality in MH-susceptible pigs. Vitamin E may perhaps be useful in the preoperative treatment of MH.

Acknowledgments. The research conducted on Landrace × Large White pigs was supported by grants (UM 53P and UM 59P) by the Australian Pig Research and Development Corporation, and those on British Landrace and Pietrain pigs were conducted at the Agricultural Food Research Council, Food Research Institute, Langford, Bristol (U.K.).

References

1. Cheah KS, Cheah AM (1981) Skeletal muscle mitochondrial phospholipase A_2 and the interaction of mitochondria and sarcoplasmic reticulum in porcine malignant hyperthermia. Biochim Biophy Acta 638:40–49
2. Cheah KS (1987) Mitochondria and malignant hyperthermia. In: Britt BA (ed) Malignant hyperthermia. Nijhoff, Boston, pp 79–102
3. Okumura F, Crocker BD, Denborough MA (1980) The site of the muscle cell abnormality in swine susceptible to malignant hyperpyrexia. Br J Anaesth 52:377–383
4. Mickelson JR, Gallant EM, Litterer LA, Johnson KM, Rempel WE, Louis CF (1988) Abnormal sarcoplasmic reticulum ryanodine receptor in malignant hyperthermia. J Biol Chem 263:9310–9315
5. Ohnishi ST (1994) How was the abnormal calcium release channel from the sarcoplasmic reticulum identfied? In: Ohnishi ST, Ohnishi T (eds) Malignant hyperthermia—a genetic membrane disease. CRC Press, Boca Raton, pp 45–66

6. Cheah KS, Cheah AM (1985) Malignant hyperthermia: molecular defects in membrane permeability. Experientia (Basel) 41:656–661
7. Britt BA (1987) Aetiology and pathophysiology of malignant hyperthermia In: Britt BA (ed) Malignant hyperthermia. Nijhoff, Boston, pp 11–42
8. Duthie GG, Arthur JR (1993) Free radicals and calcium homeostasis: relevance to malignant hyperthermia. Free Radical Biol & Med 14:435–442
9. Lamb GD (1993) Ca^{2+} inactivation, Mg^{2+} inhibition and malignant hyperthermia. J Muscle Res Cell Motil 14:554–556
10. Britt BA, Kalow W (1970) Malignant hyperthermia: a statistical review. Can Anaesth Soc J 17:293–315
11. Gronert GA (1980) Malignant hyperthermia. Anesthesiology 53:395–423
12. Mitchell G, Heffron JJA (1982) Porcine stress syndromes. In: Chichester CO (ed) Advances in food research. Academic Press, New York, pp 167–230
13. Cheah KS, Cheah AM (1984) Membrane permeability in porcine malignant hyperthermia. In: Kates M, Manson LA (eds) Membrane fluidity. Biomembranes, vol 12. Plenum, New York, pp 661–687
14. Harrison GG (1987) Porcine malignant hyperthermia—the saga of the "hot" pig. In: Britt BA (ed) Malignant hyperthermia. Nijhoff, Boston, pp 103–136
15. O'Brien PJ (1994) Canine malignant hyperthermia/canine stress syndrome. In: Ohnishi ST, Ohnishi T (eds) Malignant hyperthermia—a genetic membrane disease. CRC Press, Boca Raton, pp 105–116
16. Allen WM, Hebert CN, Smith LP (1974) Deaths during and after transportation of pigs in Great Britian. Vet Rec 94:212–215
17. Williams CH, Houchins C, Shanklin MD (1975) Pigs susceptible to energy metabolism in the fulminant hyperthermia stress syndrome. Br Med J 3:411–413
18. Lopez-Padrino JR (1994) Free calcium concentration in skeletal muscle of malignant hyperthermia susceptible subjects: effects of ryanodine. In: Ohnishi ST, Ohnishi T (eds) Malignant hyperthermia—a genetic membrane disease. CRC Press, Boca Raton, pp 133–150
19. Mickelson JR, Louis CF (1992) Calcium regulation in porcine skeletal muscle-review. In: Puolanne E, Demeyer DI, Ruusunen M, Ellis S (eds) Pork quality: genetic and metabolic factors. CAB International, Wallingford, UK, pp 160–184
20. Fujii J, Otsu K, Zorzato F, De Leon S, Khanna VK, Weiler JE, O'Brien PJ, MacLennan DH (1991) Identification of a mutation in the porcine ryanodine receptor that is associated with malignant hyperthermia. Science 253:448–451
21. Hogan K (1994) Molecular genetic diagnosis of human malignant hyperthermia. In: Ohnishi ST, Ohnishi T (eds) Malignant hyperthermia—a genetic membrane disease. CRC Press, Boca Raton, pp 293–317
22. Otsu K, Philips MS, Khanna VK, De Leon S, MacLennan DH (1992) Refinement of diagnostic assays for a probable causal mutation for porcine and human malignant hyperthermia. Genomics 13:835–837
23. Hughes IP, Moran C, Nicholas FW (1992) PCR genotyping of the ryanodine receptor gene for a putative casual mutation for malignant hyperthermia in Australian pigs. J Anim Breed Genet 109:465–476
24. Kasparek S (1980) Chemistry of tocopherols and tocotrienols. In: Machlin LJ (ed) Vitamin E—a comprehensive treatise. Dekker, New York, pp 7–65
25. Burton GW, Cheeseman KH, Doba D, Ingold KU, Slater TF (1983) Vitamin E as an antioxidant in vitro and in vivo. In: Porter R, Whelan J (eds) Ciba Found Symp Biol Vitamin E 101:4–18
26. McCay PB (1985) Vitamin E: interactions with free radicals and ascorbate. Annu Rev Nutr 5:323–340
27. Sokol RJ (1988) Vitamin E deficiency and neurological disease. Annu Rev Nutr 8:351–373
28. Erin NE, Spirin MM, Tabidze LV, Kagan VE (1984) Formation of α-tocopherol complexes with fatty acids. A hypothetical mechanism of stabilization of biomembranes by vitamin E. Biochim Biophys Acta 774:96–102

29. McCay PB, King MM (1980) Vitamin E: its role as a biological free radical scavenger and its relationship to the microsomal mixed function oxidase system. In: Machlin LJ (ed) Vitamin E—a comprehensive treatise. Dekker, New York, pp 289–317
30. Duthie GG, Arthur JR (1989) The antioxidant abnormality in the stress susceptible pig: the effects of vitamin E supplementation. Ann NY Acad Sci 570:322–234
31. Duthie GG, Arthur JR, Nicol F, Walker M (1989) Increased indices of lipid peroxidation in stress susceptible pigs and effects of vitamin E. Res Vet Sci 46:226–230
32. Duthie GG, McPhail DB, Arthur JR, Goodman BA, Morrice PC (1990) Spin trapping of free radicals and lipid peroxidation in microsomal preparations from malignant hyperthermia susceptible pigs. Free Radical Res Commun 8:93–99
33. Douglas CE, Chan AC, Choy PC (1986) Vitamin E inhibits platelet phospholipase A_2. Biochim Biophys Acta 876:639–645
34. Cheah KS, Cheah AM, Krausgrill DI (1995) Effect of dietary supplementation of vitamin E on pig meat quality. Meat Sci 39:255–264
35. Cheah KS, Cheah AM, Fletcher JE, Rosenberg H (1989) Skeletal muscle mitochondrial respiration of malignant hyperthermia-susceptible patients. Calcium-induced uncoupling and free fatty acids. Int J Biochem 21:913–920
36. Fletcher JE, Tripolitis L, Rosenberg H, Beech J (1993) Malignant hyperthermia: halothane- and calcium-induced release in skeletal muscle. Biochem Mol Biol Int 29:763–772
37. Behrens WA, Madere R (1987) Mechanisms of absorption, transport and tissue uptake of RRR-α-tocopherol and *d*-γ-tocopherol in the white rat. J Nutr 117:1562–1569
38. Jensen M, Essen-Gustavsson B, Hakkarainen J (1988) The effect of a diet with a high or low content of vitamin E on different skeletal muscles and myocardium in pigs. J Vet Med A35:487–497
39. Asghar A, Gray JI, Booren AM, Gomaa EA, Abouzied MM, Miller ER (1991) Effects of supranutritional dietary vitamin E levels on subcellular deposition of α-tocopherol in the muscle and on pork quality. J Sci Food Agric 57:31–41
40. Iyer SL, Katyare SS, Howland JL (1976) Elevated erythrocyte phospholipase A associated with Duchenne and myotonic muscular dystrophy. Neurosci Lett 2:103–106
41. Lloyd SJ, Nunn MG (1978) Osmotic fragility of erythrocytes in Duchenne muscular dystrophy. Br Med J 11:252–259
42. Van Deenen LLM (1981) Topology and dynamics of phospholipids in membranes. FEBS Lett 123:3–15
43. Kalofoutis A, Jullien G, Spanos V (1977) Erythrocyte phospholipids in Duchenne muscular dystrophy. Clin Chim Acta 74:85–90

Malignant Hyperthermia in Dogs

Thomas E. Nelson[1]

Our knowledge and understanding of the pharmacogenetic disease malignant hyperthermia (MH) has been significantly advanced through investigations of the porcine animal model. The distinction between "trigger" and "safe" anesthetics for MH-susceptible (MHS) human patients was originally based on the results of testing these drugs in the MHS pig [1,2]. The efficacy of dantrolene sodium for treating and preventing MH was discovered by Harrison [3] via experiments with MH pigs. Subsequent to its discovery in pigs, MHS was reported to occur in other species including the dog [4], cat [5], and horse [6]. Because MH in man represents a spectrum of susceptibility [7] and is genetically heterogenous [8], thorough investigations of MH among different animal species may provide important genetic and epigenetic information that will further advance our understanding of MH in man. In this chapter, I attempt to document the current state of knowledge regarding MH in dogs, focusing on the interesting genetic and phenotypic differences between MH in dogs and in pigs and how understanding the basis for these differences may have an impact on our understanding of MH in man.

Clinical Presentation of MH in Dogs

Malignant hyperthermia in dogs was first reported by Short in 1973 [4]; subsequently, other clinical case reports were published [9–13]. The occurrence of MH susceptibility in a breeding colony of dogs was reported by O'Brien et al. [12], but a detailed, systematic study of the clinical syndrome was not performed. A colony of MH-susceptible dogs was established in my laboratory (from MHS male and female progenitors that originated from the University of Saskatchewan [Saskatoon, Canada]), and a systematic evaluation of the syndrome was performed using a standard halothane/succinylcholine challenge protocol developed previously for investigation of the MH pig model [7]. A total of 102 offspring from MHS × MH-nonsusceptible (MHN) crosses were exposed to the MH anesthetic challenge protocol, and the clinical responses, including heartrate, blood pressure, arterial blood gases and lactate, temperature, carbon dioxide production, and skeletal muscle rigidity assessments, were measured at 5- to 10-min intervals during exposure to anesthesia.

[1]Department of Anesthesia, Bowman Gray School of Medicine, Wake Forest University, Winston-Salem, NC 25157-1009, USA

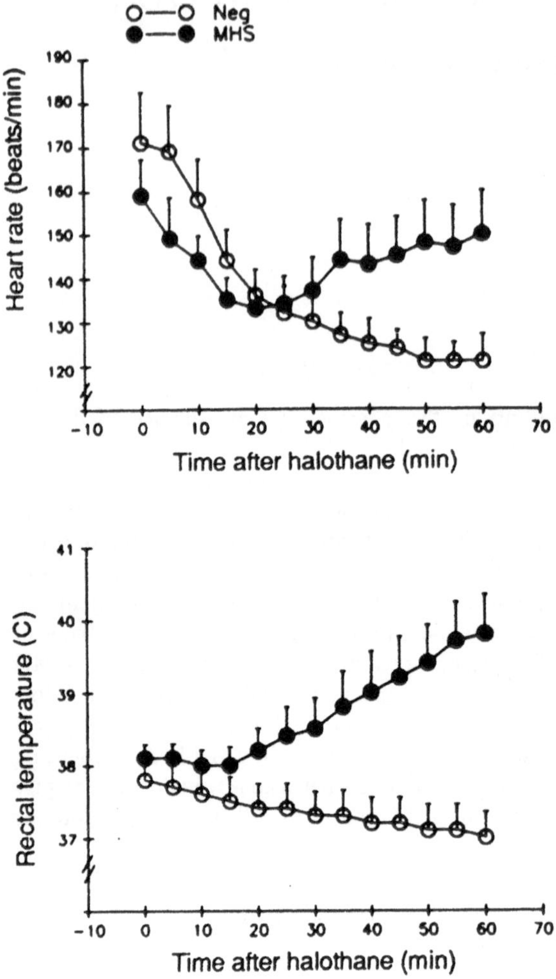

FIG. 1. Changes in heart rate, systolic blood pressure, rectal temperature, and end-tidal CO_2 excretion during halothane-induced malignant hyperthermia (MH) in dogs. Values are means ± SE for 10 malignant hyperthermia-susceptible (*MHS*) and 10 MH-nonsusceptible MHN dogs. (Reproduced with permission from [14]

Of the 102 offspring tested, 47 developed clinical signs of a hypermetabolic response. These signs included marked increases in end-tidal CO_2 excretion, rectal temperature, and heartrate [14] (Fig. 1). The maximum rates of CO_2 production were 7.8 times the basal rates before halothane anesthesia among 10 MHS dogs. In this study, when the MH hypermetabolic state produced life-threatening conditions (i.e., tachyarrhythmias, acidosis, rapidly rising temperature) the triggering anesthetics were discontinued and therapeutic measures were initiated. Consequently, the MH syndrome was never allowed to progress to a degree that resulted in mortality or morbidity among the animals.

In MHS pigs, the MH syndrome is associated with severe skeletal muscle rigidity and lactic acidosis, but neither of these conditions was observed among the MHS dogs tested. Whether the absence of rigidity and lactic acidosis is a consequence of therapeutic intervention or a species difference in the manifestation of the syndrome is a question that needs to be addressed. If the syndrome were allowed to progress without therapeutic intervention, the MHS dogs might then develop skeletal muscle rigidity

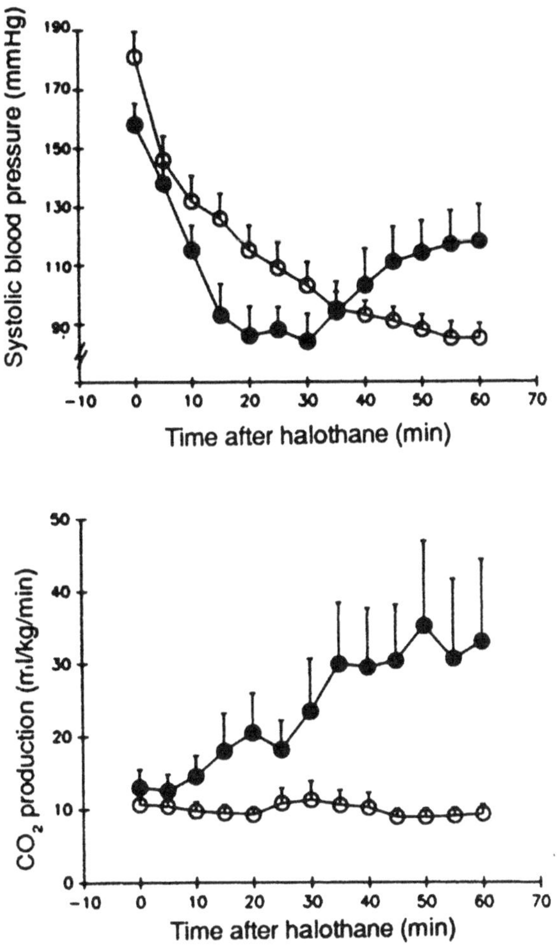

FIG. 1. *continued*

and lactic acidosis. However, if the progression of the MH syndrome in dogs is compared to that in pigs on the basis of increased CO_2 production and increased body temperature, it becomes clear that at comparable increases in metabolic rate and body temperature the pigs develop skeletal muscle rigidity and lactic acidosis while the MHS dogs do not [14].

On this basis, the difference between clinical expression of the MH syndrome between susceptible dogs and pigs is more likely a species difference or a different mutation predisposing to the disease. It is of interest to note that the clinical manifestations of the MH syndrome in man are also variable and that nonrigid variants have been described [15]. Differences in clinical expression of the MH syndrome are a product of the genotype and the environment [16,17], and it is often difficult to separate these influential factors. As it is known that MH is genetically heterogenous in man, it is very important to determine if the differences observed between MHS dogs and pigs are genetically or epigenetically based.

Trigger Anesthetic Agents

The published clinical reports of anesthetic agent-induced MH in dogs indicate that halothane and succinylcholine can trigger the syndrome [9–13]. The occurrence of an MH-like hypermetabolic state several hours after surgical anesthesia with apparent "non-trigger" agents has been reported [11]. O'Brien et al. [12] reported that halothane but not methoxyflurane triggered MH in susceptible animals. These animals were however anesthetized in a student surgical laboratory environment completely devoid of scientifically controlled experimental conditions. We have shown that environmental conditions, especially body temperature, can have a marked effect on the development of the MH syndrome in susceptible pigs [17].

It is possible that lack of adequate anesthetic exposure (i.e., concentration × time) or a drop in body temperature could have prevented the MHS dogs in O'Brien's report [12] from developing MH when anesthetized with methoxyflurane. Under carefully controlled experimental conditions in which body temperature was not allowed to decrease after anesthesia and the concentration of methoxyflurane was maintained at 2.3 × M.A.C. (minimum alveolar concentration), five of five MHS dogs developed severe signs of MH. We illustrate this by data from three of these animals by comparing the responses to halothane and methoxyflurane in these same animals (Fig. 2). If a fall in body temperature and suboptimal concentrations of methoxyflurane protect against the development of MH in dogs, this may explain why this dominantly inherited syndrome is rarely observed in routine veterinary medicine clinical practice.

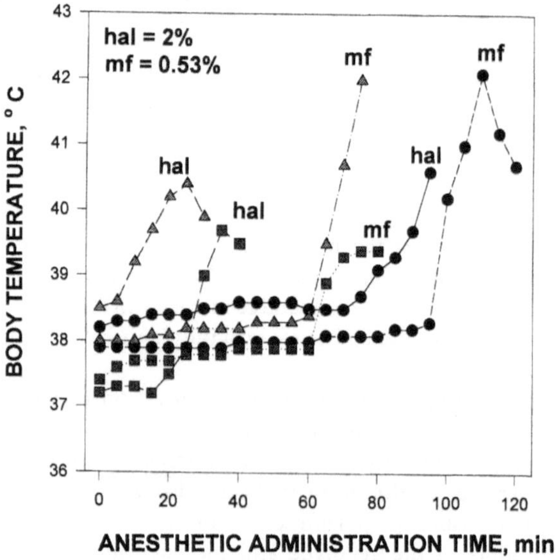

FIG. 2. Comparison of halothane versus methoxyflurane induction of malignant hyperthermia in three MHS animals. Each dog was first exposed to halothane, 2.3 × M.A.C. (minimum alveolar concentration) and subsequently to 2.3 × M.A.C. in a separate procedure. Each symbol represents a single animal. *hal*, Halothane; *mf*, methoxyflurane

Pharmacological Contracture Responses of Biopsied MHS Dog Skeletal Muscle

Skeletal muscle biopsied from MHS humans and pigs is distinguished from normal muscle by abnormal contracture response to various pharmacological agents [7]. It has been clearly established that such biopsied skeletal muscle has abnormal contracture responses when exposed to halothane, caffeine, or ryanodine [16]. When gracilis muscle biopsied from MHS dogs is exposed to these drugs, qualitatively similar abnormal contracture responses are observed [16]. These drugs are known to act at the sarcoplasmic reticulum (SR) membrane. The drugs produce a rate of calcium release from the SR stores that exceeds the rate and amount of calcium sequestration mechanisms with a net effect of increased myoplasmic [Ca^{2+}]. When myoplasmic [Ca^{2+}] increases above the concentrations found under resting cell conditions (i.e., $100 \times 10^{-9} M$) but below the critical concentration to initiate contracture, the cell responds with an increase in metabolism [18].

When myoplasmic [Ca^{2+}] increases to levels sufficient to activate the contractile elements, then sustained contracture will occur. In skeletal muscle biopsied from MHS humans, pigs and dogs, much lower concentrations of caffeine or ryanodine will produce contracture in the affected muscle as compared to normal muscle from each species [16]. Although halothane 3% (vol/vol) bubbled through the muscle bath produces either very little or no contracture in normal muscle, contractures ranging from 1 to 6g tension can be observed in muscle from MHS humans, dogs, and pigs [16]. This finding suggests that the pathophysiological events that produce the MH syndrome share a common etiological pathway among these three species. Because the contracture response of skeletal muscle is initiated by increases in myoplasmic [Ca^{2+}], this probably represents a common link of abnormality in the muscle of each affected species.

The contracture responses of skeletal muscle to ryanodine deserve further discussion. Ryanodine is a plant alkaloid that binds with high affinity and specificity to the ryanodine receptor protein. When administered to animals, it produces skeletal muscle rigidity and is lethal; when applied to skeletal muscle in a contracture chamber, it slowly produces a contracture [19]. Whether the slow onset of ryanodine-induced contracture is a function of the rate of binding to the receptor or is caused by binding to the receptor, or both, is not known. It is well known that binding of ryanodine to the RYR1 receptor is increased when the open-state probability of the channel is increased [20]. Thus, the greater sensitivity of MH skeletal muscle to the contracture-producing effects of ryanodine may simply be because the RYR1 calcium channel has a higher open-state probability in MHS muscle. Assuming this to be the case, then the fact that MHS human, pig, and dog muscle are abnormally sensitive to the contracture-producing effects of ryanodine suggests that MH in each of these species is expressed either directly or indirectly in the RYR1 protein molecule. The end result is an increase in the open-state probability of this calcium release channel.

Inheritance

Each dog represented in this chapter has been phenotyped for MH susceptibility by two independent methods. An anesthetic challenge protocol determines if an individual animal will develop signs of MH when exposed to halothane and succinylcholine. The difference between responders who develop the MH hypermetabolic syndrome and the nonresponders is unequivocal, and no spectrum of susceptibility has been observed among MHS dogs. Results of the anesthetic challenge MH phenotyping are paralleled by a second method of phenotyping by pharmacological contracture testing of biopsied skeletal muscle. On the basis of these phenotyping outcomes, it appears that MH is inherited as an autosomal dominant trait in this MH dog colony and that the MHS animals we have investigated are heterozygous for the MH trait (Fig. 3). Our primary breeder, an MHS mixed-breed male, has developed the anesthetic-induced MH syndrome, and gracilis muscle biopsied from this dog has developed abnormal contractures when exposed to halothane, caffeine, and ryanodine. When this animal was bred to randomly selected female dogs with normal MH phenotypes, approximately 50% of the offspring (male and female) had MHS-positive phenotypes (Fig. 3). This is strong evidence for an autosomal dominant form of MHS inheritance in the dog, but not proof.

The cysteine to arginine mutation in the RYR1 protein in MHS pig skeletal muscle is closely associated with, if not causal for, susceptibility to MH in pigs [21] and in some human families [22]. At this time we know that MHS dogs do not have the *RYR1* pig mutation, but we do not know if a different mutation exists in RYR1 or some other protein in MH dogs (Dr. Kirk Hogan 1992, personal communication).

Sarcoplasmic Reticulum Function in MHS Dog Skeletal Muscle

The skeletal muscle SR membrane system is most important for the regulation of myoplasmic [Ca^{2+}] during the contraction–relaxation cycle. Unlike cardiac muscle, skeletal muscle is not dependent on extracellular calcium for contractility but utilizes calcium stored within the SR terminal cisternae. In response to the muscle action potential, calcium is released from the terminal cisternae through a Ca^{2+}-specific channel. The skeletal muscle ryanodine receptor protein (RYR1) is considered to be the primary pathway for calcium release from the terminal cisternae [23]. The SR membrane also functions to remove calcium from the myoplasm to initiate the relaxation phase. This is accomplished by the 100-kDa calcium pump protein, which utilizes adenosine triphosphate (ATP) to pump calcium into the SR lumen. It has been shown that the rate of calcium uptake is normal in SR membrane vesicles isolated from human and pig MHS skeletal muscle [18]. In addition, it was shown that halothane activates the calcium pump protein in normal [24] and in MHS [25] muscle SR, an effect opposite to that expected to cause MH.

We have measured the rate of calcium uptake by SR from MHS and MHN dog skeletal muscle and found no evidence for a functional defect in this protein (Table 1). In a light SR fraction (LSR) that is devoid of calcium release channels there is no difference in calcium uptake rate between LSR from MHS and MHN muscle (Table 1). In a heavy SR (HSR) fraction, however, Ca^{2+} uptake rate is slower in MHS membranes; this is more likely caused by Ca^{2+} leak from a channel blocked by ruthenium red than a defective Ca^{2+} pump. Preliminary investigations into the calcium release channel from MHS dog skeletal muscle suggest that this protein may be functionally altered.

We have used two different methods for measuring calcium release from isolated SR membrane vesicles. One method involves actively loading calcium into the SR

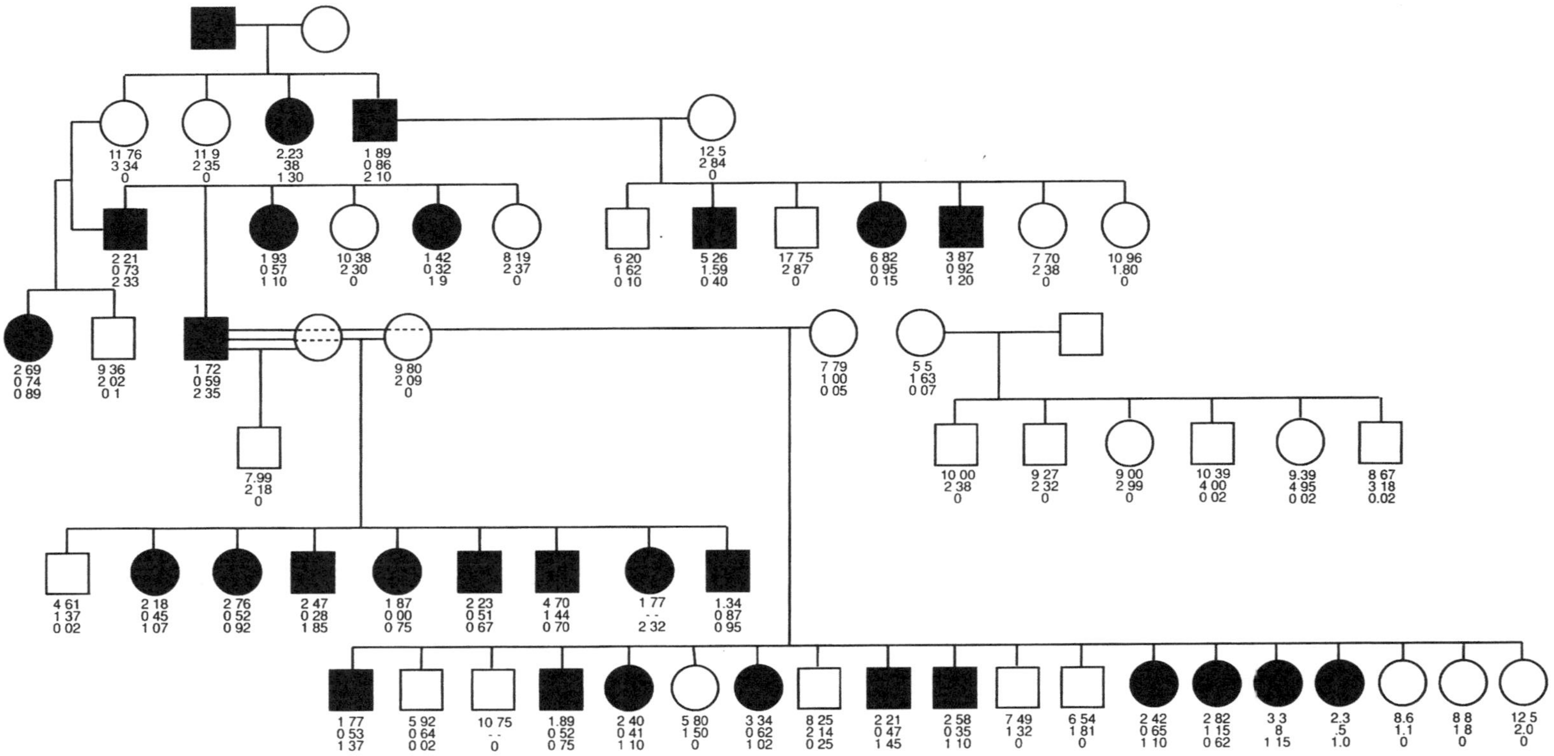

FIG. 3. Canine MH pedigree of 48 offspring from MHS crossed to MHN dogs; 28 are MHS, supporting an autosomal dominant mode of inheritance. *Filled symbols*, MHS; *open symbols*, MHN; *squares*, males; *circles*, females. Values under each symbol are from in vitro contractive tests and represent, in order, caffeine-specific concentration to produce 1 g isometric contracture without (CSC) and with (HCSC) 1% halothane; last value is contracture response to halothane 3%

TABLE 1. Calcium uptake rates compared to calcium preload threshold for Ca^{2+}-induced Ca^{2+} release (CICR) in skeletal muscle sarcoplasmic reticulum from malignant hyperthermia-susceptible (MHS) and malignant hyperthermia-nonsusceptible (MHN) dogs.

	MHS dogs	MHN dogs
Calcium uptake rate (nmol/mg min^{-1}):		
Without ruthenium red		
LSR	12.30 ± 2.8	12.95 ± 1.6
HSR	7.0 ± 1.3*	13.0 ± 1.6
With ruthenium red		
LSR	19.22 ± 2.3	16.64 ± 1.8
HSR	18.82 ± 1.6*	24.94 ± 2
Calcium preload threshold (nmol Ca^{2+}/mg):		
With pyrophosphate	950 ± 110* (n = 5)	1380 ± 98 (n = 5)
Without pyrophosphate	50.6 (n = 4)	52.5 (n = 5)

LSR, light sarcoplasmic reticulum fraction; HSR, heavy sarcoplasmic reticulum fraction.
* MHS values significantly different from corresponding MHN values ($P < 0.05$).

lumen with ATP until further addition of calcium outside the SR lumen causes calcium-induced calcium release (CICR) [26]. This CICR will not occur until a critical amount of calcium is loaded into the SR lumen [26]. It was shown previously that the critical calcium threshold for CICR is lower in SR from MHS pig muscle [27]. When this method was applied to SR from MHS dog muscle, no difference was found for the critical calcium threshold when compared to SR from MHN dogs (see Table 1).

Another method for measuring calcium release from isolated SR membrane vesicles utilizes the permeant anion pyrophosphate, which acts to lower the intralumenal [Ca^{2+}] by formation of insoluble calcium pyrophosphate (CaPPi) complex. In this case, a much higher concentration of calcium loading is necessary before CICR can occur. When this method was tested in SR from MHS pig muscle, a lower calcium preload for CICR was observed [28]. This method also detected a lower calcium preload threshold for SR membrane vesicles from MHS dog muscle (see Table 1). Although it is not possible at this time to explain why these two methods give different results for the membranes from MHS dog muscle, the results with calcium loading in the presence of pyrophosphate suggest that an abnormality in calcium release is present.

Potential Importance of MH in Dogs

It is now clearly established that a single mutation does not explain MH in man but that different mutations in RYR1 or other proteins can be causal for the disease [29]. What is interesting to note is that even with this apparent genetic heterogeneity, phenotypic expression of the trait by abnormal pharmacological contracture response in affected skeletal muscle is consistent within and across species. The common denominator to all MH-predisposing mutations appears to be an abnormal, sustained increase in myoplasmic calcium when the muscle cell is exposed to certain pharmacological agents.

If MH in dogs were identical in all respects to MH in pigs, then its value as an animal model would be limited. The fact that MH in dogs has a different clinical presentation than in pigs, is autosomally dominantly inherited versus recessive in pigs, and does not have the same mutation makes it most attractive in solving the mysteries of heterogeneously predisposed MH in man. The absence of skeletal muscle rigidity during MH in this dog model can bring new impetus to the long-standing discussion about a nonrigid variant for MH in man. The development of a severe hypermetabolic state without lactic acidosis in MH dogs suggests that differences in skeletal muscle metabolism among species can determine the metabolic end product of this hypermetabolic state.

We propose that lactic acidosis does not occur in MH dogs because dog skeletal muscle has greater aerobic metabolic capacity than does pig skeletal muscle. Thus, the conversion of glucose to lactate is accompanied by a release of 47 kcal, whereas the conversion of glucose to carbon dioxide and water releases 686 kcal of heat. If the pig skeletal muscle produces a large amount of lactate while the dog muscle breaks down glucose to carbon dioxide and water, then the dog will produce 14 times more heat for each glucose molecule than will the pig. Stated another way, to release the same amount of metabolic heat from glucose, the pig must utilize about 10 times more substrate than does the dog. If these simplistic theories are correct, then a comparison of MH dog versus MH pig metabolism indicates that similar differences may apply to MHS humans. Consequently, it may not be appropriate to expect that the clinical aspects of MH will be the same from one human patient to another.

The MH pig animal model has been extremely valuable in augmenting our understanding of the disease in man and has contributed to a reduction in morbidity and mortality from the disease. However, MH in dogs may be more relevant to MH in man because (1) it is inherited as an autosomal dominant trait, as it is in most if not all humans; (2) the nonrigid form of MH in dogs may represent a similar presentation of MH in some human families; and (3) the mutation predisposing to MH in dogs, while not yet discovered, may represent yet another of several mutations occurring in man. More in-depth investigation of MH in dogs is necessary before many of its potential benefits to understanding MH in man can be realized.

Acknowledgment: This work was supported by NIH Grant GM23875.

References

1. Hall LW, Trim CM, Woolf N (1972) Further studies of porcine malignant hyperthermia. Br Med J 2:145–148
2. Harrison GG, Saunders SJ, Biebuyck JF, Hickman R, Dent DM, Weaver V, Terblanche J (1969) Anaesthetic-induced malignant hyperpyrexia and a method for its prediction. Br J Anaesth 41:844–854
3. Harrison GG (1975) Control of the malignant hyperpyretic syndrome in MHS swine by dantrolene sodium. Br J Anaesth 47:62–65
4. Short CE, Paddleford RR (1973) Malignant hyperthermia in the dog. Anesthesiology 39:462–463
5. De Jong RH, Hoarner JE, Amory DW (1974) Malignant hyperthermia in the cat. Anesthesiology 41:608–609
6. McClure JJ (1975) Malignant hyperthermia in the horse. Minn Vet 15:11–12
7. Nelson TE, Flewellen EH, Gloyna DF (1983) Spectrum of susceptibility of malignant hyperthermia: diagnostic dilemma. Anesth Analg 62:545–552
8. Levitt RC, Nouri N, Jedlicka AE, McKusick VA, Marks AR, Shuttack JG, Fletcher JE, Rosenberg H, Meyers DA (1991) Evidence for genetic heterogeneity in malignant hyperthermia. Genomics 11:543–547
9. Bagshaw RJ, Cox RH, Knight DH, et al (1978) Malignant hyperthermia in a greyhound.

10. Kirmayer AH, Klide AM, Purvance JE (1984) Malignant hyperthermia in a dog: case report and review of the syndrome. J Am Vet Med Assoc 185:978–982
11. Leary SL, Anderson LC, Manning PJ, et al (1983) Recurrent malignant hyperthermia in a greyhound. J Am Vet Med Assoc 182:521–522
12. O'Brien PJ, Cribb PH, White RJ, et al (1983) Canine malignant hyperthermia: diagnosis of susceptibility in a breeding colony. Can Vet J 24:172–177
13. Rand JS, O'Brien P (1987) Exercise-induced malignant hyperthermia in an English springer spaniel. J Am Vet Med Assoc 190:1013–1014
14. Nelson TE (1991) Malignant hyperthermia in dogs. J Am Vet Med Assoc 198:989–994
15. Britt BA, Kalow W (1970) Malignant hyperthermia: a statistical review. Can Anaesth Soc J 17:293–315
16. Nelson TE (1993) Pharmacogenetics and malignant hyperthermia. Anaesth Pharmacol Rev 1:362–376
17. Nelson TE (1990) Porcine malignant hyperthermia: critical temperatures for in vivo and in vitro responses. Anesthesiology 73:449–454
18. Nelson TE (1987) Skeletal muscle sarcoplasmic reticulum in the malignant hyperthermia syndrome. In: Britt BA (ed) Malignant hyperthermia. Nijhoff, Boston, pp 43–78
19. Jenden DJ, Fairhurst AS (1969) The pharmacology of ryanodine. Pharmacol Rev 21:1–24
20. Chu AM, Diaz-Munoz M, Hawkes MJ, Brush K, Hamilton SL (1990) Ryanodine as a probe for the functional state of the skeletal muscle sarcoplasmic reticulum calcium release channel. Mol Pharmacol 37:735–741
21. Fujii J, Otsu K, Zarzato S, de Leon VK, Khanna JE, Weiler JE, O'Brien PJ, MacLennan DH (1991) Identification of a mutation in porcine ryanodine receptor associated with malignant hyperthermia. Science 253:448–451
22. MacLennan DH, Duff C, Zorzato F, Fujii J, Phillips M, Korneluk RG, Frodis W, Britt BA, Worton RG (1990) Ryanodine receptor gene is a candidate for predisposition to malignant hyperthermia. Nature 343:559–561
23. McPherson PS, Campbell KP (1993) The ryanodine receptor/calcium release channel. J Biol Chem 268:13675–13678
24. Nelson TE, Sweo T (1988) Ca^{2+} uptake and Ca^{2+} release by skeletal muscle sarcoplasmic reticulum: differing sensitivity to inhalational anesthetics. Anesthesiology 69:571–577
25. Louis CF, Zualkernan K, Roghair T, Mickelson JR (1992) The effects of volatile anesthetics on calcium regulation by malignant hyperthermia-susceptible sarcoplasmic reticulum. Anesthesiology 77:114–125
26. Nelson TE, Nelson KE (1990) Intra- and extrahumenal sarcoplasmic reticulum regulatory sites for Ca^{2+}-induced Ca^{2+} release. FEBS Lett 263:292–294
27. Nelson TE (1983) Abnormality in calcium release from skeletal sarcoplasmic reticulum of pigs susceptible to malignant hyperthermia. J Clin Invest 72:862–870
28. Nelson TE, Levin M, Volpe P (1991) Evidence for intraluminal Ca^{2+} regulatory site defect in sarcoplasmic reticulum from malignant hyperthermia pig muscle. J Pharmacol Exp Ther 256:645–649
29. MacLennan DH, Phillips MS (1992) Malignant hyperthermia. Science 256:789–794

Heatstroke

Koh Shingu[1] and Kenjiro Mori[2]

Heat illness is a disorder caused by a high-temperature environment alone or combined with physical exercise. Heat illness has been classified into eight conditions by the Climate Physiology Committee of the Medical Research Council of Great Britain [1]: (1) heatstroke, (2) sun stroke, (3) heat cramp, (4) heat exhaustion and collapse, (5) sunburn, (6) diseases of sweat gland—anhidrosis, (7) asthenic reaction, and (8) others. Of these eight, three conditions are of dominant importance in relation to malignant hyperthermia: heat cramps, heat exhaustion, and heatstroke. The following is a brief summary of these three conditions, especially heatstroke [2–4].

Heat Cramp

Heat cramp is an acute disorder of the skeletal muscle characterized by brief, intermittent, and painful cramp in skeletal muscles that have been subjected to intensive muscular work. It may occur in acclimatized persons, even in a normal environmental temperature. Its precise mechanism has not been elucidated. Predisposing factors include profuse perspiration, a large water intake, and failure to replace sodium loss. It usually disappears spontaneously within several 10 minutes. Prompt relief occurs following replacement of the sodium deficiency. Based on these findings, an acute sodium deficiency has been postulated as its cause. Hyponatremia prevents a sufficient number of sodium ions from entering the cell, might interfere with Na^{+}/Ca^{2+}-exchange adenosine triphosphatase (ATPase), and induces an accumulation of calcium ion in the muscle cell [2].

Heat Exhaustion

Heat exhaustion is a common disorder resulting from intense muscular work in high environmental temperature and high humidity. This disorder results from profound loss of either body water or sodium chloride.

1. Water-depletion type (hypertonic dehydration): symptoms of this disorder consist of intense thirst, fatigue, weakness, discomfort, anxiety, and impaired judgment. Hperventilation with paresthesia, tetany, agitation, hysteria, muscular incoordina-

[1] Department of Anesthesiology, Kansai Medical University, Osaka, 570 Japan
[2] Department of Anesthesia, Kyoto University Hospital, Kyoto, 606 Japan

tion, and psychosis are also manifested in some patients. Body temperature is invariably elevated.

2. Salt-depletion type: in contrast to heat cramp, heat exhaustion caused by salt depletion is accompanied by systemic symptoms, such as profound weakness, fatigue, severe frontal headache, giddiness, anorexia, nausea, vomiting, and diarrhea. Heat cramp often occurs in acclimatized persons, but heat exhaustion from salt depletion tends to occur in unacclimatized persons. Body temperature is usually normal or subnormal. Treatment consists of supplying salt (normal saline).

Heatstroke

Heatstroke is a severe systemic disorder characterized by high body temperature (more than 41°C), stupor or coma, and anhidrosis. Heatstroke is classified into two entities: classic heatstroke and exertion-induced heatstroke [2,5].

Classic Heatstroke

Classic heatstroke occurs during heat waves, in epidemic form, in advanced age, and in patients with chronic diseases such as congestive heart failure, diabetes mellitus, or alcoholism that interfere with heat loss. The major cause is failure in acclimatization and thermal regulation resulting from anhidrosis secondary to heart failure or dysfunction of the sweat glands. Loss of consciousness is an early sign. The muscles are always flaccid. Tachycardia and hypotension are common.

Exertion-Induced Heatstroke

Exertion-induced heatstroke was first reported as a single independent disease entity by Knochel [2,6]. This disorder has aroused interest since the time of World War II. It occurs during exercise in young healthy men, with sudden onset of mental confusion followed by coma and hyperthermia [7]. Hanson and Zimmerman [7] postulated that development of this disorder occurs in young men who are not familiar with prolonged exercise and are not aware of the early signs of overexertion. This disorder does not occur in marathon athletes, who are aware of their own exercise limit. When heat production exceeds the maximum capacity of the heat loss mechanism, the core temperature is elevated, leading to heatstroke. Hyperthermia is sustained and is resistant to conventional cooling and infusion of cooled fluids. Contracture of various degrees is noted in the skeletal muscles [8].

Multiple organ failure is noted in the central nervous system (CNS), hemostasis, cardiovascular system, lungs, kidneys, gastrointestinal tract, skeletal muscles, and endocrine organs; water and electrolyte imbalance is also noted. Organ failure results from tissue damage caused by high temperature, metabolic acidosis, cellular hypoxia, or a combination of these factors. The CNS dysfunction may be caused by edema or congestion of the brain, and intracranial hemorrhage is found in some cases postmortem. Cerebral hypoxia from brain hypoperfusion, cellular damage caused by the high body temperature (>41°C), and hypoglycemia may be possible causes. The most prominent and common cardiovascular sign is an extreme tachycardia of more than 160 beats/min, which is frequently associated with rhythm disturbances. Overall vasodilation and hypotension are prominent. In moderate cases, a hyperdynamic state enhances heat dissipation. However, a hypodynamic response becomes apparent in severe cases, and left heart failure ensues. In almost all cases, proteinuria is seen with granular casts and red cells in the urine specimen.

Acute renal failure is a common complication in severe cases. Myoglobinuria, disseminated intravascular coagulation, and hypoperfusion may be additional causes of renal failure as well as heat damage to the kidneys. Damage to the skeletal muscles is noted, and high muscle enzyme levels in plasma and myoglobinuria are often present. Gastrointestinal symptoms such as diarrhea and vomiting are common. Liver damage as indicated by jaundice and elevated hepatic enzymes is also common. Dehydration and hypoperfusion to the skin diminish perspiration, but sweating persists in more than 50% of cases. Heat and decreased pH damage capillary endothelium, release tissue kinases, induce dysfunction of platelets, and affect clotting factors. All these factors relate to a disturbed hemostasis, and intravascular coagulation is induced.

The pathophysiological features in exertion-induced heatstroke are very smilar to those of anesthesia-induced malignant hyperthermia [3,8–10]. Wingard and Gatz [11] pointed out that malignant hyperthermia-susceptible patients are at risk from other stress such as trauma, emotions, and severe physical exercise, and they named this syndrome the "human stress syndrome" [11]. Several case reports have described exertion-induced heatstroke in malignant hyperthermia-susceptible patients and their families [3,12]. However, it is still controversial whether the exertion-induced heatstroke-susceptible individual is susceptible to anesthesia-induced malignant hyperthermia. Further study is required before a conclusion is reached.

Treatment

The correlation of this disease with anesthesia-induced malignant hyperthermia and the possible favorable effects of dantrolene was first suggested by Oshima et al. [10]. This review was followed by several cases of trials of dantrolene in exertion-induced heatstroke in Japan [13–15], Canada [16], and Australia [17], and the general agreement has been that muscle contracture and hyperthermia were corrected by dantrolene. Although no controlled studies of dantrolene sodium for heatstroke affecting mortality have been carried out, dantrolene sodium should be administered for exertion-induced heatstroke. Other symptomatic treatment is also required, such as cooling, oxygenation, hydration with chilled fluids, diuresis, treatment of cardiac dysrhythmias, and correction of metabolic acidosis and electrolytic balance.

References

1. Weiner JS, Horne GO (1958) A classification of heat illness. Br Med J 1:1533–1535
2. Knochel JP (1974) Environmental heat illness. Arch Intern Med 133:841–863
3. Jardon OM (1982) Physiologic stress, heat stroke, malignant hyperthermia. Mil Med 147:8–14
4. Shibolet S, Lancaster MC, Danon Y (1976) Heat stroke. Aviat Space Environ Med 47:280–301
5. Petersdorf RG (1994) Hypothermia and hyperthermia. In: Isselbacher KJ, Braunwald E, Wilson JD, Martin JB, Fauci AS, Kasper DL (eds) Harrison's principles of internal medicine, 13th edn. McGraw-Hill, New York, pp 2473–2479

6. Knochel JP, William M, Beisel WR, Herndon EG, Gerard ES, Barry KG (1976) The renal, cardiovascular, hematologic and serum electrolyte abnormalities of heat stroke. Am J Med 30:299–309
7. Hanson PG, Zimmerman SW (1979) Exertional heat stroke in novice runners. JAMA 242:154–157
8 de la Radiguet BP, Bastaie P, Poujol C (1978) L'hypothermie maligne d'effort ou. Nouv Presse Med 27:2381–2385
9. Gronert GA (1980) Malignant hyperthermia. Anesthesiology 53:395–423
10. Oshima E, Shingu K, Tamai S, Arakawa Y, Mori K (1980) Heat stroke, a review. In: Morio M (ed) Proceedings of the IIIrd Japanese symposium on malignant hyperthermia. Hiroshima J Anesth 16:103–106
11. Wingard DW, Gatz EE (1978) Some observations on stress-susceptible patients. In: Aldrete JA, Britt BA (eds) Second international symposium on malignant hyperthermia. Grune and Stratton, New York, pp 363–372
12. Ikegaki S, Kiyaozaki K, Hachiya H, Kuze S, Ito Y (1983) Death of a boy due to malignant hyperthermic syndrome following strenous sport with malignant hyperthermia in his family tree. In: Proceedings of the VIth Japanese symposium on malignant hyperthermia. Hiroshima J Anesth 19:1–7
13. Tatekawa S, Nishi S, Li J, Okuda T, Yukioka H, Asada A, Fujimori M (1983) A case report of a serious heat stroke after a rugby game. In: Proceedings of the VIth Japanese symposium on malignant hyperthermia. Hiroshima J Anesth 19:9–14
14. Sakabe H, Ono A, Kaji A, Satani M, Nishimura K, Fujimori M (1985) A case of exertion-induced heat stroke. In: Proceedings of the VIIIth Japanese symposium on malignant hyperthermia. Hiroshima J Anesth 21:49–54
15. Fujiwara Y, Uchimoto R, Izaki A, Nomura S, Inomata I, Matayoshi Y (1985) A case of exertion-induced heat stroke associated with severe liver damage. In: Proceedings of the VIIIth Japanese symposium on malignant hyperthermia. Hiroshima J Anesth 21:99–105
16. Lydiatt JS, Hill GE (1981) Treatment of heat stroke with dantrolene. JAMA 246:41–42
17. Denborough MA (1982) Heat stroke and malignant hyperpyrexia. Med J Aust 1:204–205

Intracellular Calcium Mechanism in the Pathogenesis of Neuroleptic Malignant Syndrome

SHIGETO YAMAWAKI and TERUO HAYASHI

Introduction

Neuroleptic malignant syndrome (NMS), a potentially lethal adverse effect of neuroleptics, characterized by hyperthermia, severe autonomic dysregulation, muscular rigidity, altered consciousness, and laboratory evidence of rhabdomyolysis, has gained widespread recognition in recent years. Various conceptual pathogenetic models of this disorder have been developed, such as a dopamine (DA) receptor blockade hypothesis [1–3], a relative noradrenaline(NA)/DA excess hypothesis [4,5], a serotonin 5-hydroxtryptamine (5-HT)/DA imbalance hypothesis [6], a glutamate/DA imbalance hypothesis [7], a gama amino butyric acid (GABA) deficiency hypothesis [8], and an opiate hypothesis [9]; but the pathogenetic mechanism underlying NMS remains unclear.

Dantrolene, a well-known muscular relaxant, is used to treat malignant hyperthermia caused by anesthetic agents, a disease similar to NMS. There have also been many reports of successful treatment of NMS with dantroene [10–13]. The reported therapeutic mechanism of dantrolene in malignant hyperthermia is its inhibitory effect on the abnormal Ca^{2+} release from sarcoplasmic reticulum in skeletal muscle [14]; however, this is obscure in NMS. The authors report three cases of NMS treated with dantrolene, suggesting a central effect of dantrolene and involvement of a central serotoninergic hypermetabolic state. It is hypothesized that a neuroleptic- and/or antidepressant-induced imbalance between serotonin and dopamine in the thermoregulation center (hypothalamus) produces hyperthermia in NMS [6]. However, the hypermetabolic state of serotonin and other neurotransmitters, for example NA and glutamate, may be involved in the pathophysiology of NMS. The authors therefore focused on the intracellular Ca^{2+} mechanism, because of its important role in both neurotransmitter release and muscle contraction.

The present study illustrates and discusses the role of intracellular Ca^{2+} in the therapeutic mechanism of dantrolene in NMS, and in the pathophysiology of NMS.

Department of Psychiatry and Neurosciences, Hiroshima University School of Medicine, 1-2-3 Kasumi, Minami-ku, Hiroshima, 734 Japan

Effect of Dantrolene on Intracellular Ca^{2+}-related Serotonin Release from the hypothalamus (Yamawaki et al. [15])

Slices of hypothalamus (3 mm × 0.3 mm) obtained from male Wistar rats (150–250 g) were incubated for 60 min at 37°C, in physiological solution containing 0.1 M [^{3}H]5-HT creatinine sulphate, under gentle bubbling with 95% O_2/5% CO_2 gas. At the end of incubation, the slices were transferred to superfusion chambers and superfuged with a physiological solution with or without 1.3 mM $CaCl_2$. In Ca^{2+}-free solution, 1.0 μM tetrasodium ethylene guanine tetraacetic acid (EGTA) was added to remove extracellular Ca^{2+}. The slices were stimulated twice with 30 mM KCl, 10 μM fenfluramine, 10 μM ruthenium red, or 10 μM veratrine for 4 min each at 66 min, (stimulation period 1, S1) and 90 min (stimulation period 2, S2) after the start of superfusion. The superfusate was collected at 2-min intervals. The control was S1, and dantrolene was added to the perfusion medium 10 or 20 min before the S2 and remained present throughout the rest of the superfusion. To compare dantrolene-induced changes of stimulated tritium release, the quotient S2/S1 was calculated.

High K^+-evoked Release of [^{3}H]5-HT. High K^+-evoked [^{3}H]5-HT release was Ca^{2+}-dependent since no release occurred in the absence of 1.3 mM Ca^{2+}. As shown in Fig. 1, at the concentrations higher than 10^{-6} M, dantrolene significantly decreased high K^+-evoked [^{3}H]5-HT release from hypothalamus slices in a dose-dependent manner.

Fenfluramine- and Ruthenium Red-induced Release of [^{3}H]5-HT. Significant tritium release from [^{3}H]5-HT-loaded slices of hypothalamus occurred in response to 10 μM fenfluramine, and 10 μM ruthenium red, which were extracellular Ca^{2+}-independent, since release occurred in the absence of 1.3 mM Ca^{2+}. Dantrolene, at a concentration of 10^{-6} M, did not affect the fenfluramine- or ruthenium red-induced [^{3}H]5-HT release. (Data not shown.)

Veratrine-induced [^{3}H]5-HT Release. Since veratrine increases not only membrane permeability to Na^+ [16] but also Ca^{2+} mobilization from intracellular Ca^{2+} store sites, availability of free Ca^{2+} is increased in the presynaptic nerve terminals, leading to the release of neurotransmitter [17]. Veratrine (10 μM) significantly increased tritium release from [^{3}H]5-HT-loaded slices of hypothalamus in the absence of 1.3 mM Ca^{2+}.

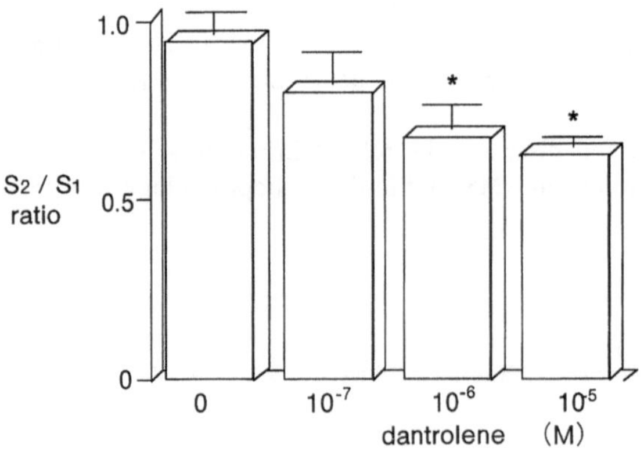

FIG. 1. Effect of dantrolene on high K^+-evoked 5-hydroxytryptamine (5-HT) release. Each point represents the mean ± SEM of six to eight experiments. *, $P < 0.05$

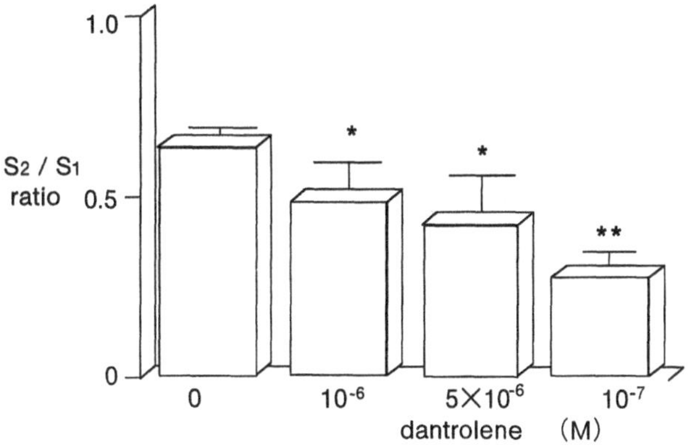

FIG. 2. Effect of dantrolene on veratrine-induced [^{3}H] 5-hydroxytryptamine (5-HT) release. Each point represents the mean ± SEM of six to eight experiments. *, $P < 0.05$, **, $P < 0.01$

As shown in Fig. 2, this veratrine-induced [^{3}H]5-HT release was significantly decreased by 10^{-6}M, 5×10^{-6}M, and 10^{-5}M dantrolene in a dose-dependent manner.

Effect of Dantrolene on Hyperthermia in an Animal Model of NMS (Kato and Yamawaki [18])

A stainless-steel cannula, consisting of a guide tube with a snug fitting trocar, was implanted into the preoptic anterior hypothalamus (PO/AH) using the stereotaxic atlas and coordinates of Paxinos and Watson [19]. Veratrine or saline was injected through the inserted cannula, which was connected to a 10 μM Hamilton microsyringe with Teflon tubing. Rectal temperature was monitored with a digital thermometer. Animals were killed by microwave application, monoamines (DA and 5-HT), and their metabolites 3,4-dihydroxyphenylacetic acid (DOPAC), homovanillic acid (HVA), and 5-hydroxy indole acetic acid (5-HIAA) in the hypothalamus including the thalamus, were measured by high-performance liquid chromatography (HPLC), with an electro-chemical detector.

Changes in Body Temperature Induced by Microinjection of Veratrine into the PO/AH in Rats Pretreated with Haloperidol. One ml of sterile saline of containing 0 μg, 5 μg, 50 μg, or 100 μg veratrine was microinjected into the PO/AH in rats, to which haloperidol (1 mg/kg) had been given intraperitoneally (i.p.) 30 min before injection of veratrine. Rectal temperature was monitored at 15 min intervals for 60 min after veratrine injection. Microinjection of more than 50 μM veratrine into the PO/AH significantly increased rectal temperature with abnormal behavior, such as severe tremor and convulsion (Fig. 3). Since these findings are very similar to the symptoms of NMS, rats given haloperidol i. p. and then treated with veratrine microinjected into the PO/AH are considered to be an appropriate animal model of NMS.

Changes in the Contents of Metabolites of DA and 5-HT in the Hypothalamus Containing Thalamus. As shown in Table 1, the intraperitoneal application of haloperidol (1 mg/kg) significantly increased the content of two major metabolites, DOPAC ($P < 0.01$) and HVA ($P < 0.05$), but produced no changes in 5-HIAA, a metabolite of 5-HT. With microinjection of veratrine (50 μg) into the PO/AH in rats pre-treated with haloperidol, the levels of DOPAC and 5-HIAA were significantly higher than those induced by haloperidol alone ($P < 0.05$).

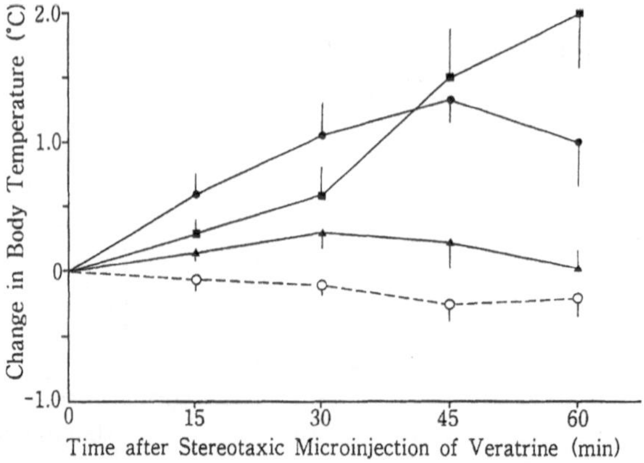

FIG. 3. Body temperature responses to different doses of veratrine injected into preoptic anterior hypothalamus (PO/AH) in rats pretreated with haloperidol. Pretreatment with haloperidol did not significantly affect base-line body temperature (38.22 ± 0.05, n = 22). Each point represents the mean ± SEM of eight determinations presented as degree of temperature change from base-line temperature. *Open circles*, haloperidol (1 mg/kg) + saline (1 μl); *closed triangles*, haloperidol (1 mg/kg) + veratrine (5 μg in 1 μl of saline); *closed circles*, haloperidol (1 mg/kg) + veratrine (50 μg in 1 μl of saline); *closed squares*, haloperidol (1 mg/kg) + veratrine (100 μg in 1 μl of saline)

TABLE 1. Changes in content of metabolites of DA and 5-HT in the region of the thalamus and hypothalamus after stereotaxic microinjection of veratrine (VRT).

Treatment (n)	Metabolites (nmol/g)				
	DOPAC	HVA	DA	5-HIAA	5-HT
SAL + SAL (6)	0.74 ± 0.06	0.21 ± 0.05	3.30 ± 0.44	3.35 ± 0.40	6.89 ± 0.13
HPD + SAL (6)	1.22 ± 0.12[a]	0.66 ± 0.13[a]	2.83 ± 0.12	3.96 ± 0.26	7.20 ± 0.15
HPD + VRT (6)	1.95 ± 0.26[b]	1.22 ± 0.29[b]	3.19 ± 0.26	5.47 ± 0.40[b]	6.97 ± 0.20

[a] $P < 0.01$, significantly higher compared to corresponding values of rats intraperitoneally pretreated with saline (SAL).
[b] $P < 0.05$, significantly higher compared to corresponding values of microinjection of SAL into PO/AH in rats pretreated with haloperidol (HPD). All values were means ± SEM of six determinations. DA, dopamine; 5-HT, 5-hydroxytyptamine (serotonin); VRT, veratrine; DOPAC, 3,4-dihydroxyphenylacetic acid; HVA, homovanillic acid; 5-HIAA, 5-hydroxy indole acetic acid.

Effect of Dantrolene on Hyperthermia Induced by Microinjection of Veratrine (50 μg) into the PO/AH in Rats Pretreated with Haloperidol (1 mg/kg, i.p.). Intraperitoneal injection of 20 mg/kg and 40 mg/kg of dantrolene significantly inhibited hyperthermia induced by microinjection of veratrine (50 μg) into the PO/AH in rats pretreated with haloperidol (1 mg/kg, i.p.), at $P < 0.01$ and $P < 0.001$, respectively (Fig. 4).

Effect of Dantrolene on Compound 48/80-induced Ca^{2+} Increase in Neuro-2A Cells (Hayashi et al. [20])

Cultured Neuro-2A mouse neuroblastoma cells were washed with Krebs Ringer Hepes buffer (145 mM NaCl, 5 mM KCl, 0.5 mM Na_2HPO_4, 10 mM Glucose, 10 mM HEPES, 1 mM $MgSO_4$, 1 mM $CaCl_2$; pH 7.4) and loaded with fura-2-AM (5 μM) at 25°C for 30 min. Fura-2-loaded cells were placed on the stage of a fluorescence microscope/

video-camera system consisting of a Nikon TDM-EF2 microscope (Nikon, Tokyo, Japan) with a Fluor 40 ×, charge-coupled-device camera and color video processor. The fluorescence intensity of fura-2 was measured with excitation wavelengths of 340 nm and 380 nm, and an emission wavelength of 510 nm. The output of fluorescent images was digitized by a color image processor Argus 50 (Hamamatsu, Japan), and calculated to the 340/380 nm fluorescence intensity ratio.

Effect of Compound 48/80 on Intracellular Ca^{2+} Concentration ($[Ca^{2+}]_i$) Levels in Neuro 2A Cells. Compound 48/80 is one of several well-known membrane perturbants which change the property and structure of the bilayer membrane [21]. The addition of Compound 48/80 to Neuro-2A cells for 1 min elicited a rapid concentration-dependent increase in $[Ca^{2+}]_i$. The maximum response of $[Ca^{2+}]_i$ mobilization was obtained at a concentration of Compound 48/80 more than 0.33 mg/ml (Fig. 5).

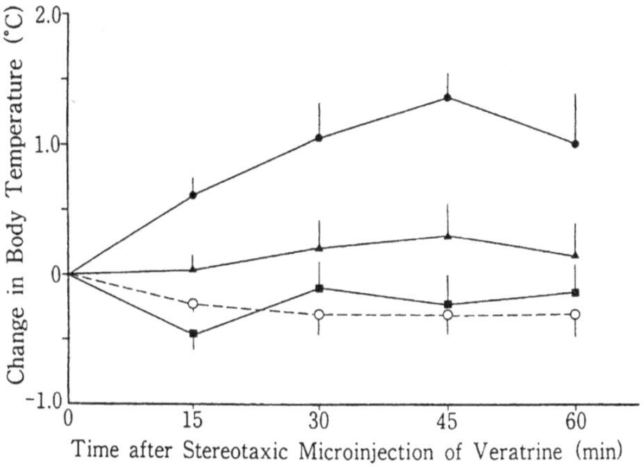

FIG. 4. Effect of dantrolene on hyperthermia induced by microinjection of veratrine into PO/AH in rats pretreated with haloperidol. Pretreatment with haloperidol and dantrolene did not significantly affect base-line body temperature (38.22 ± 0.09, $n = 16$). Each point represents the mean ± SEM of eight determinations presented as degree of temperature change from base-line temperature. *Open circles*, haloperidol (1 mg/kg) + dantrolene (40 mg/kg) + Saline (1 μl); *closed circles*, haloperidol (1 mg/kg) + saline (1 μl) + veratrine (50 μg in 1 μl of saline); *closed triangles*, haloperidol (1 mg/kg) + dantrolene (20 mg/kg) + veratrine (50 μg in 1 μl of saline); *closed squares*, haloperidol (1 mg/kg) + dantrolene (40 mg/kg) + veratrine (50 μg in 1 μl of saline)

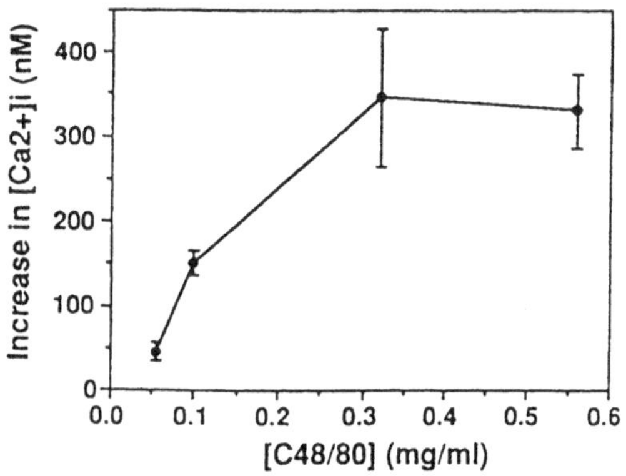

FIG. 5. Effect of Compound 48/80 on increase in intracellular calcium concentration (*$[Ca^{2+}]i$*), expressed as *nM* $[Ca^{2+}]_i$. Each point represents the mean ± SEM of five to eight separate experiments

Effect of Dantrolene on Compound 48/80-induced Increase in $[Ca^{2+}]_i$ Levels in Neuro-2A Cells. In the presence of 10 μM dantrolene, the mobilization of $[Ca^{2+}]_i$ induced by Compound 48/80 (0.56 mg/ml) was decreased to 44% of the first response (Fig. 6a). As shown in Fig. 6b, the effect was concentration-dependent, and a significant decrease was observed at more than 0.33 μM dantrolene (IC^{50} was 5 μM).

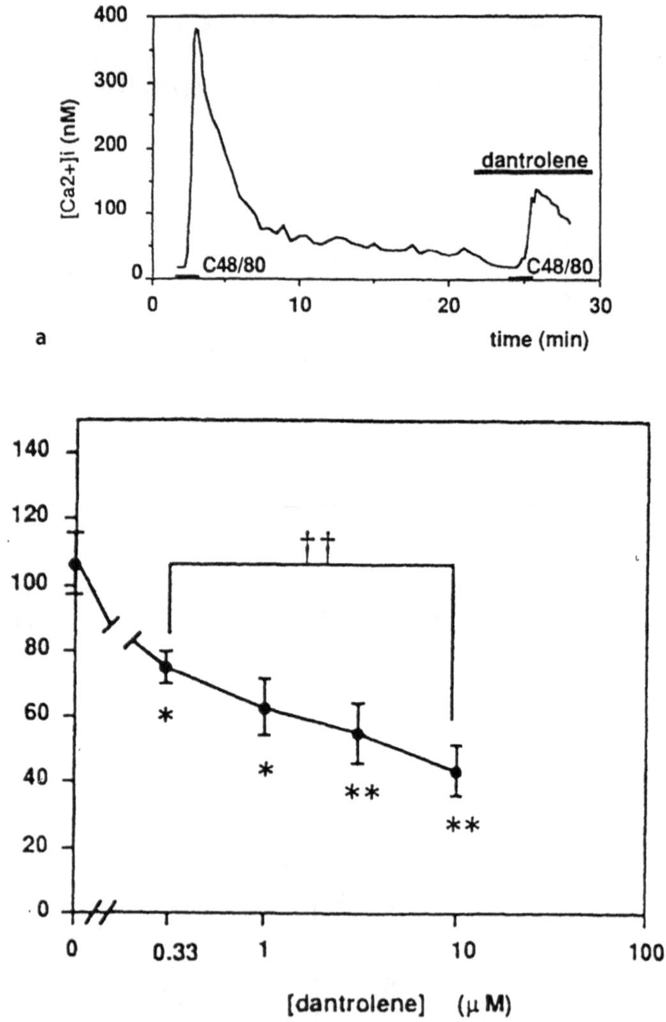

FIG. 6. Effect of dantrolene on Compound 48/80-induced intracellular calcium mobilization in Neuro-2A cells, expressed as *nM* of $[Ca^{2+}]_i$. **a** Dantrolene (10 μ*M*) was applied 2 min before the second stimulation. **b** Data are shown as the percentage of the response to the first stimulation (in the absence of dantrolene). Each point represents the mean ± SEM of five to eight separate experiments. ††, $P < 0.01$ comparing 0.33 and 10 μ*M* dantrolene; **, $P < 0.01$ compared with no dantrolene (Kruskal-Wallis test)

Discussion

Several lines of evidence provide support for the involvement of DA receptor blockade in the pathogenesis of NMS [1–3], but the view that NMS results from DA receptor blockade may be simplistic, as it fails to account for the rare occurrence of NMS and the various autonomic symptoms. Obviously, other cofactors are necessary to trigger an episode of NMS. Increased sympathetic activity was first demonstrated by Feibel and Schiffer [22], who found increased plasma and urinary levels of NA and adrenaline in a case of NMS. Some have reported a relative NA/DA excess state in NMS [4,5]. In the hypothalamic thermoregulatory center, DA is known to be related to a fall in body temperature, while 5-HT causes temperature to rise [23]. Yamawaki [6] reported a case of NMS where the increased CSF levels of 5-HIAA were normalized by treatment with dantrolene, and proposed a 5-HT/DA imbalance hypothesis of hyperthermia in NMS. This hypothesis is supported by the fact that the 5-HT hypermetabolic state, so-called serotonin syndrome, induced by 5-HT-enhancing drugs consists of symptoms similar to NMS, such as hyperthermia, tremor, diaphoresis, myoclonus, etc. [24]. However, some investigators reported that urinary 5-HIAA was normal [22], and CSF 5-HIAA was normal [25] or decreased [26] in NMS patients. Weller and Kornhuber [7] have also hypothesized an imbalance mechanism between dopamine and glutamate neurons in NMS. Considering the above reports, it is suggested that the balance between DA and other neurotransmitters, such as NA, 5-HT and glutamate, may be disturbed by neuroleptics in the onset of NMS. The mechanism underlying the hypermetabolic state of these neurotransmitters is unknown.

Dantrolene, a muscular relaxant, was originally used for treatment of malignant hyperthermia caused by anesthetic agents, and was also reported to be effective for NMS [10–13]. The pathogenetic mechanism of malignant hyperthermia is known to be an abnormal increase in the release of Ca^{2+} from the sarcoplasmic reticulum in skeletal muscle [27]. Dantrolene is reported to inhibit this abnormally increased Ca^{2+} release and then relax muscular rigidity [14]. Similar effects of dantrolene in skeletal muscle have been discussed in the therapeutic mechanism of NMS by several investigators [12,28]; however, a central effect of dantrolene has also been suggested [6,23].

As in the case of muscle contraction, Ca^{2+} plays an important role in neurotransmitter release. High K^+-evoked neurotransmitter release, including serotonin, is known to be dependent on extracellular Ca^{2+} [29]. In the present study, Ca^{2+}-dependency of [^{3}H]5-HT release evoked by high K^+ was confirmed, since no [^{3}H]5-HT release by either stimulation occurred in the absence of 1.3 mM Ca^{2+}. Dantrolene at concentrations over 10^{-6}M significantly decreased high K^+-evoked [^{3}H]5-HT release from slices of rat hypothalamus. Fenfluramine is reported to induce 5-HT release independent of intracellular or extracellular Ca^{2+}. In the absence of extracellular Ca^{2+}, veratrine- and ruthenium-red induced [^{3}H]5-HT release from the slices of rat hypothalamus. The [^{3}H]5-HT release induced by either fenfluramine or ruthenium red was not affected by 10^{-6}M dantrolene (Table 1), while dantrolene significantly inhibited the veratrine-induced [^{3}H]5-HT release in a dose-dependent manner. Veratrine was used to produce depolarization by opening sodium channels [16,30], and was reported to mobilize Ca^{2+} from the intracellular Ca^{2+} store site in rat cerebral cortex [17]. The [^{3}H]5-HT release induced by veratrine is thought to be mediated by this Ca^{2+} mobilization from intracellular Ca^{2+} store site. The inhibitory effect of dantrolene on veratrine-induced [^{3}H]5-HT release from rat hypothalamus may also be mediated by this intracellular Ca^{2+} mechanism. This mechanism may occur in the skeletal muscle of malignant hyperthermia and may also mediate the release of other neurotransmitters such as NA. These results suggest that abnormal release of neurotransmitters such as 5-HT and NA, mediated by abnormal $[Ca^{2+}]_i$ increase in the hypothalamus, may be involved in the pathogenesis of NMS.

References

1. Henderson VW, Wooten GF (1981) Neuroleptic malignant syndrome: a pathogenetic role for dopamine receptor blockade? Neurology (NY) 31:132–137
2. Toru M, Matsuda O, Makiguchi K, Sugano K (1981) Neuroleptic malignant syndrome-like state following a withdrawal of antiparkinsonian drugs. J Nerv Ment Dis 169:324–327
3. Burke RE, Fahn S, Mayeux R, Weinberg H, Louis K, Willner JH (1981) Neuroleptic malignant syndrome caused by dopamine-depleting drugs in a patient with Huntington disease. Neurology (NY) 31:1022–1026
4. Schibuk M, Schachter D (1986) A role for catecholamines in the pathogenesis of neuroleptic malignant syndrome. Can J Psychiatry 31:66–69
5. Gurrera RJ, Romero JA (1992) Sympathoadrenomedullary activity in the neuroleptic malignant syndrome. Biol Psychiatry 32:334–343
6. Yamawaki S (1986) A consideration on the pathophysiology of "syndrome malin"—three cases of successful treatment with dantrolene. Jpn J Psychiatr Treatment 1:413–422
7. Weller M, Kornhuber J (1992) A rationale for NMDA receptor antagonist therapy of the neuroleptic malignant syndrome. Med Hypotheses 38:329–333
8. Lew T, Tollefson G (1983) Chlorpromazine-induced neuroleptic malignant syndrome and its response to diazepam. Biol Psychiatry 18:1441–1446
9. Sandyk R (1985) Neuroleptic malignant syndrome and the opioid system. Med Hypotheses 17:133–138
10. Coons DJ, Hillman FJ, Marshall RW (1982) Treatment of neuroleptic malignant syndrome with dantrolene sodium: a case report. Am J Psychiatry 139:944–945
11. Goekoop JG, Carbaat PAT (1982) Treatment of neuroleptic malignant syndrome with dantrolene. Lancet ii:9–50
12. Goulon M, Rohan-Chabot P, Elkharrat K, Gajdos P, Bismuth C, Conso F (1983) Beneficial effects of dantrolene in the treatment of neuroleptic malignant syndrome: A report of two cases. Neurology 33:516–518
13. Yamawaki S, Yanagawa K, Morio M, Mori K (1986) Possible central effect of dantrolene sodium in neuroleptic malignant syndrome. J Clin Psychopharmacol 6:378–379
14. Desmedt JE, Hainaut K (1977) Inhibition of the intracellular release of calcium by dantrolene in barnacle giant muscle fibers. J Physiol (Lond) 265:565–585
15. Yamawaki S, Kato T, Yano E (1990) Studies on pathogenesis of neuroleptic malignant syndrome: Effect of dantrolene on serotonin release in rat hypothalamus. Hiroshima J Anesthes 26:45–52
16. Benforado JM (1968) The veratrum alkaloids. In: Root WS, Hofman FG (eds) Physiological pharmacology vol. 4, Part D, Academic, New York, pp 331–398
17. Schoffelmeer ANM, Mulder AH (1983) [^{3}H] noradrenaline release from brain slices induced by an increase in the intracellular sodium concentration: Role of intracellular calcium stores. J Neurochem 40:615–621
18. Kato T, Yamawaki S (1989) A pharmacological study of veratrine-induced hyperthermia in the rat: A model of neuroleptic malignant syndrome. Hiroshima J Med Sci 38:173–181
19. Paxinos G, Watson C (1982) The rat brain in stereotaxic coordinates, 2nd edn. Academic, New York

20. Hayashi T, Kagaya A, Motohashi N, Yamawaki S (1994) The possible mechanism of dantrolene stabilization of cultured neuroblastoma cell plasma membranes. J Neurochem 63:1849–1854
21. Jain MK, Rogers J (1989) Substrate specificity for interfacial catalysis by phospholipase A2 in the scooting mode. Biochem Biophys Acta 1003:91–97
22. Feibel JH, Schiffer RB (1981) Sympathoadrenomedullary hyperactivity in the neuroleptic malignant syndrome: a case report. Am J Psychiatry 138:1115–1116
23. Yamawaki S, Lai H, Horita A (1983) Dopaminergic and serotoninergic mechanisms of thermoregulation: mediation of thermal effect of apomorphine and dopamine. J Pharmacol Exp Ther 227:383–388
24. Sternbach H (1991) The serotonin syndrome. Am J Psychiatry 148:705–713
25. Ansseau M, Reynolds CF III, Kupfer DJ, et al (1986) Central dopaminergic and noradrenergic receptor blockade in a patient with neuroleptic malignant syndrome. J Clin Psychiatry 47:320–321
26. Nisijima K, Ishiguro T (1990) Neuroleptic malignant syndrome: a study of CSF monoamine metabolism. Biol Psychiatry 27:280–288
27. Endo M, Tanaka M, Ogawa Y (1970) Calcium-induced release of calcium from the sarcoplasmic reticulum of skinned skeletal muscle fibers. Nature 228:34–36
28. Granato JE, Stern BJ, Ringel A, Karim AH, Krumholz A, Coyle J, Adler S (1983) Neuroleptic malignant syndrome: Successful treatment with dantrolene and bromocriptine. Ann Neurol 14:89–90
29. Gothert M (1980) Serotonin-receptor-mediated modulation of Ca^{2+}-dependent 5-hydroxytryptamine release from neurones of the rat brain cortex. Naunyn Schmiedebergs Arch Pharmacol 310:93–96
30. Catterall WA (1984) The molecular basis of neuronal excitability. Science 233:653–661

Controversies in the Pathogenesis of Neuroleptic Malignant Syndrome

P.J. Adnet, N. Bello, H. Reyford, and R.M. Krivosic-Horber

Introduction

The neuroleptic malignant syndrome (NMS) is a relatively rare, but probably underrecognized, potentially fatal complication of the use of neuroleptic drugs. This syndrome was first described in the French medical literature with the introduction of neuroleptics in 1960, where it was referred to as "akinetic hypertonic syndrome." Over the last decade, almost a 1000 cases of NMS have been reported, but many features of this syndrome remain controversial. Indeed, a grading scale of specific signs and symptoms for the diagnosis of NMS and a spectrum of clinical severity are two issues that await resolution. Many diagnostic criteria have been proposed, but no single set of criteria has been adopted for general use. Hence, different presentations of this disorder could explain some contradictory findings associated with NMS: prospective studies have provided disparate estimates of the frequency of NMS, ranging from 0.07% to 2.20% among patients receiving neuroleptic agents; risk factors for NMS vary in different patient populations; the association between NMS and other potentially fatal syndromes such as malignant hyperthermia is unclear.

Theories

Two major, though not necessarily competing, theories to explain neuroleptic malignant syndrome (NMS) are a neuroleptic-induced alteration of central neuroregulatory mechanisms and an abnormal reaction of predisposed skeletal muscle. This latter alternative hypothesis is based on similarities between NMS and malignant hyperthermia (MH) and suggests that neuroleptic medications induce abnormal calcium availability in muscle cells of susceptible individuals, and thereby trigger muscle rigidity, rhabdomyolysis, and hyperthermia. Alternatively, another hypothesis could be that is some circumstances, neuroleptics may exert a direct toxic effect on normal skeletal muscle.

Malignant Hyperthermia Investigation Unit, Centre Hospitalier Universitaire, Lille Hôpital B, 59037 Lille Cédex, France

Central Dopamine Receptor Blockade

Dopamine plays a role in central thermoregulation in mammals. A dopamine injection into the preoptic-anterior hypothalamus causes a reduction in core temperature [1]. Since neuroleptic drugs block dopamine receptor sites, the hyperthermia associated with NMS may result from a blockade of the hypothalamic dopamine site. This was first suggested by Henderson and Wooten [2] reporting a patient with Parkinson's disease and chronic psychosis who developed NMS when dopaminergic agonists were withdrawn, whereas haloperidol was continued. Burke [3] also observed NMS in a patient with Huntington's chorea taking methyltyrosine, a cathecholamine synthesis inhibitor, and tetrabenazine which depletes central nervous system (CNS) catecholamines. This suggests that NMS is caused by dopamine depletion or blockade leading to abnormal central thermoregulation. The dopamine blockade theory is also supported by the report of a case in which NMS developed when L-dopa/carbidopa and amantadine were abruptly discontinued in a patient with Parkinson's disease who had never taken neuroleptics [4]. On the other hand some dopamine-function-enhancing drugs, such as bromocriptine [5,6] or amantadine [7] have shown efficacy in treating NMS.

The blockade of dopamine receptors in the hypothalamus is thought to lead to impaired heat dissipation. In addition, a blockade of dopamine receptors in the corpus striatum is thought to cause muscular rigidity, thus generating heat. Hence, the excess heat production in association with a decrease in heat dissipation produces hyperthermia, which is one of the main signs of the syndrome. The peripheral anticholinergic effects of neuroleptics, which reduce sweating, most probably do not play a major role in hyperthermia associated with NMS since most NMS patients (70%) are diaphoretic. However, it is unlikely that the blockade of dopamine receptors in the hypothalamus and corpus striatum could completely explain all the signs of NMS. Indeed, hypothalamic thermoregulation involves noradrenergic, serotoninergic, cholinergic, and central dopaminergic pathways [8]. Many neuroleptics may have additional selective effects on peptides co-transmitting with dopamine in the striatum and other parts of the brain.

Primary Skeletal Muscle Defect Similar to Malignant Hyperthermia

The hypothesis regarding a common pathophysiology of NMS and MH has been suggested [9–11]. This hypothesis is based mainly on three points: (1) NMS and MH have clinical features in common, including hyperthermia, rigidity, elevated creatine kinase level, and a mortality rate for both NMS and MH ranging from 10%–30%; (2) sodium dantrolene, a peripheral muscle relaxant, has been used successfully in both syndromes; (3) abnormal results have been found in in vitro contractility tests in patients with either of these two disorders. These in vitro halothane–caffeine tests are at present the most reliable diagnostic measure for patients susceptible to MH [12,13]. The tests determine the sensitivity of muscle fibers to halothane or caffeine added to the bathing solution. Muscle fibers from patients susceptible to MH have a lower contracture threshold for these drugs than do those from normal patients. Hence, in order to evaluate a possible association between NMS and MH, several investigators have used the halothane–caffeine tests on skeletal muscle fibers removed from patients with documented NMS episodes. However, conflicting results have been reported regarding the prevalence of MH susceptibility among NMS patients [14–16]. Three main series and some sporadic case reports are now published. The first was by Caroff et al. [14], who found five out of seven NMS patients were MH susceptible on the basis of the 3% halothane response. The second, by Araki et al. [15], reported abnormal contracture in six NMS patients similar to MH using a caffeine skinned fiber technique. The third was by our laboratory [16,17] and found only one NMS patient as MH equivocal, and 13 NMS patients as MH non-susceptible.

One possible explanation for these discrepancies is that patients diagnosed as having NMS may represent a heterogeneous group with great variability in clinical presentation, response to treatment, and possibly, response to test drugs. In addition, some of the variations in the in vitro results from different centers may be attributed to variations in laboratory procedures. The tests used for MH diagnosis varied in protocol: caffeine alone [15], halothane and caffeine on separate muscle bundles [11,17], halothane associated with cumulative concentrations of caffeine [10], and caffeine on skinned muscle fibers [15]. The sensitivity of this latter method may be inadequate as the technique itself excludes any detection of a possible defect in the sarcolemma [18,19]. Differences were also observed in the in vitro contracture test criteria for diagnosis of MH. Caroff et al. [14] used as the threshold the development of a contracture greater than 0.5 g in response to 1% halothane or greater than 0.7 g in response to 3% halothane. The European MH Group [12] required, for an abnormal response to halothane, a contracture greater than or equal to 0.2 g in response to 2% or less halothane. Protocol and diagnostic criteria from other sporadic case reports were not fully explained.

Delay between the adverse reaction to the neuroleptic drug and the biopsy was either not mentioned [15], one month [14], or 2–3 weeks in our previous studies [16,17]. Caroff et al. [14] found no correlation between the maximum halothane response and the time between normalization of creatine kinase (CK) and the biopsy. However, positive contracture test results in NMS patients may be falsely positive and reflect coincidental changes in muscle resulting from NMS. For example, Gallant et al. [20] reported that muscle fiber injury during biopsy and in vitro testing procedures enhances halothane sensitivity and could contribute to false-positive contracture test results. Denborough et al. [11] reported positive results in two patients who recovered from nondrug-related episodes of rhabdomyolysis. Adnet et al. [19] found that if cut fiber specimens depolarize with time after biopsy, these preparations become less sensitive to caffeine, leading to caution in the interpretation of the in vitro caffeine contracture test. This suggests that nonspecific muscle damage secondary to metabolic factors, drug exposure, abnormalities in central nervous system activity, or testing procedures may have contributed to abnormal responses observed in vitro in muscle obtained from survivors of NMS episodes. The lower caffeine thresholds of muscle bundles from patients with malignant hyperthermia are related to higher caffeine sensitivities in both type I (slow-twitch, slow-oxidative) and type II (fast-twitch, fast-glycolytic) fibers [21]. This has been demonstrated in chemically skinned muscle fibers dissected from vastus lateralis muscle from patients susceptible to malignant hyperthermia. The technique has been successfully applied to separate normal human fibers into two classes based upon their relative sensitivities to strontium (Sr^{2+}). Using a skinned fiber procedure, Araki and al. [15] found a higher caffeine sensitivity in both type I and type II fibers from patients with NMS compared to normal muscle and thus concluded that NMS patients may be at risk for MH. However, the study did not explore the possible differences in caffeine sensitivities between NMS- and MH-susceptible patients, and the caffeine skinned fiber test was not validated with the concomittant use of the in vitro contracture tests now standardized in two major procedures (European MH protocol [12], and North American MH protocol [13]). Since our previous reports, 18 additional patients have been investigated according to the European MH protocol. Muscle specimens from 12 of the 18 patients were also tested using the North American MH protocol and the caffeine skinned muscle fiber test as before [15,21]. None of these patients were MH susceptible, but one patient was equivocal according to the European MH protocol. No significant difference was found in the caffeine sensitivity of NMS and normal patients. MH-susceptible fibers, however, had a significantly lower caffeine threshold than both NMS and normal fibers. A statistical analysis for inferences based upon negative results can be calculated [22]. The increase of our patient population from 14 to 32 patients decreases the possible rate of MH among NMS patients from 31% to less than 10% with 95% confidence. Our study alone, therefore, does not justify abandoning precautions against MH in NMS patients. According to the formula described by Graysel [22], a total number of 59 consecutive NMS patients should be tested negative with similar in vitro protocol to statistically exclude any cross-reactivity between MH and NMS. Until such results are available, we suggest that all patients with clinical NMS should be tested for MH susceptibility before being considered at risk for MH during anesthesia.

Direct Toxic Effect on Skeletal Muscle Induced by Neuroleptics

Muscle contracture has been produced in vitro by neuroleptic agents such as chlorpromazine [23]. The drug is reported to influence calcium transport across the sarcoplasmic reticulum and has been studied in various experimental muscle preparations. In a study on skinned fibers, chlorpromazine influenced both the contractile system and the function of the sarcoplasmic reticulum [24]. However, in a previous report by the anothers, the drug induced contracture to the same extent in both NMS and normal muscle [16]. Likewise, Caroff et al. [14] also found no significant differences in the muscle response to another neuroleptic drug (fluphenazine) between NMS, malignant hyperthermia susceptible, and control patients.

To further explore a possible direct action of neuroleptic drugs on skeletal muscle, we analyzed the in vitro muscle contracture response to four neuroleptic agents implicated in NMS episodes. No increase in sensitivity to neuroleptic drugs was found in muscle from NMS patients as compared to muscle from normal patients [25]. These negative findings may indicate that the in vitro contracture response to neuroleptic agents does not correlate with clinical evidence of NMS. However, this may also show that neuroleptics are potentially active on skeletal muscle. Thus, it is possible that in some circumstances such as exhaustion, psychomotor agitation, or dehydration, which are not explored by the in vitro method, neuroleptics may exert a potent toxicity on skeletal muscle leading to contracture and rhabdomyolysis. Therefore, the neuroleptic contracture tests may be inadequate to reproduce in vivo conditions. It is important to consider the many factors present in vivo such as neuroleptic metabolites, central dopaminergic blockade, and risk factors for NMS in developing more precise pharmacological models to explore possible interactions between neuroleptics and skeletal muscle function. Of particular interest is the fact that further administration of the original agent may not result in a recurrence of NMS. This suggests that neuroleptics may be a necessary but not exclusive cause of the syndrome [26,27].

References

1. Cox B, Kerwin R, Lee TE (1978) Dopamine receptors in the central thermoregulatory pathways of the rat. J Physiol (Lond) 282:471–483
2. Henderson VW, Wooten GF (1981) Neuroleptic malignant syndrome: a pathogenetic role for dopamine receptor blockade? Neurology 31:132–137
3. Burke RE, Fahn S, Mayeux R (1981) Neuroleptic malignant syndrome caused by dopamine depleting drugs in a patient with Huntington's chorea. Neurology 31:1022–1026
4. Toru M, Matsuda O, Makaguchi K (1981) Neuroleptic malignant syndrome-like state following a withdrawal of antiparkinsonian drug. J Nerv Ment Dis 169:324–327
5. Bond WS (1984) Detection and management of the neuroleptic malignant syndrome. Clin Pharmacol 3:302–307
6. Figa-Talamanca L, Gualandi C, Dimeo L, Dibattista G, Neri G, Lorusso F (1985) Hyperthermia after discontinuance of levodopa and bromocriptine therapy: impaired dopamine receptors a possible cause. Neurology 35:258–261
7. McCarron MM, Boettger ML, Peck JI (1982) A case of neuroleptic malignant syndrome successfully treated with amantadine. J Clin Psychiatry 43:381–382
8. Blingh J, Cottle WH, Maskrey M (1971) Influence of ambient temperature on the thermoregulatory responses to 5 hydroxytryptamine, noradrenalin and acetylcholine injected into the lateral cerebral ventricles of sheep, goats and rabbits. J Physiol 212:377–392
9. Delacour JL, Daoudal P, Chapoutot JL, Rocq B (1981) Traitement du syndrome malin des neuroleptiques par le dantrolène. Nouvelle Presse Médicale, 10:3572–3573
10. Tollefson G (1982) A case of neuroleptic malignant syndrome: in vitro muscle comparison with malignant hyperthermia. J Clin Psychopharmacol 2:266–270
11. Denborough MA, Collins SP, Hopkinson KC (1985) Rhabdomyolysis and malignant hyperpyrexia. Br Med J 1878:ii
12. The European Malignant Hyperpyrexia Group (1984) A protocol for the investigation of malignant hyperpyrexia susceptibility. Br J Anaesth 56:1267–1269
13. North American Malignant Hyperthermia Group (1989) Standardization of the caffeine halothane muscle contracture test. Anesth Analg 69:511–515
14. Caroff SN, Rosenberg H, Fletcher JE, Heiman-Patterson TD, Mann SC (1987) Malignant hyperthermia susceptibility in neuroleptic malignant syndrome. Anesthesiology 67:20–25
15. Araki M, Takagi A, Higuchi I, Sugita H (1988) Neuroleptic malignant syndrome: caffeine contracture of single muscle fibers and muscle pathology. Neurology 38:297–301
16. Adnet PJ, Krivosic-Horber RM, Adamantidis MM, Haudecoeur G, Adnet-Bonte CA, Saulnier F, Dupuis BA (1989) The association between the neuroleptic malignant syndrome and malignant hyperthermia. Acta Anaesthesiol Scand 33:676–680
17. Krivosic-Horber RM, Adnet P, Guevart E, Theunynck D, Lestavel P (1987) Neuroleptic malignant syndrome and malignant hyperthermia. Br J Anaesth 59:1554–1556
18. Adnet PJ, Krivosic-Horber RM, Adamantidis MM, Reyford H, Cordonnier C, Haudecoeur G (1991a) Effects of calcium-free solution, calcium antagonists, and the calcium agonist BAY K 8644 on mechanical responses of skeletal muscle from patients susceptible to malignant hyperthermia. Anesthesiology 75:413–419
19. Adnet PJ, Krivosic-Horber RM, Adamantidis MM, Haudecoeur G, Reyford HG, Dupuis BA (1991b) Is resting membrane potential a possible indicator of viability of muscle bundles used in the in vitro caffeine contracture test? Anesth Analg 74:105–111
20. Gallant EM, Fletcher TF, Goettl VM, Rempel WE (1986) Porcine malignant hyperthermia: cell injury enhances halothane sensitivity of biopsies. Muscle Nerve 9:174–184
21. Adnet PJ, Bromberg NL, Haudecoeur G (1993) Fiber-type caffeine sensitivities in skinned muscle fibers from humans susceptible to malignant hyperthermia. Anesthesiology 78:168–177

22. Grayzel J (1989) A statistic for inferences based upon negative results. Anesthesiology 71:320–321
23. Kelkar VV, Jindal MN (1974) Chlorpromazine-induced contracture of frog rectus abdominis muscle. Pharmacology 12:32–38
24. Tagaki A (1981) Chlorpromazine and skeletal muscle: A study of skinned single fibers of the guinea pig. Exp Neurol 734:477–486
25. Reyford HG, Cordonnier C, Adnet PJ, Krivosic-Horber R, Bonte CA (1990) The in vitro exposure of muscle strips from patients with neuroleptic malignant syndrome (NMS) cannot be correlated with the clinical features. J Neurol Sci 98:(suppl 6.6): 527
26. Caroff SN (1980) The neuroleptic malignant syndrome. J Clin Psychiatry 41:79–83
27. Levenson JL (1985) Neuroleptic malignant syndrome. Am J Psychiatry 142:1137–1145

Neuromuscular Disorders and Malignant Hyperthermia

Frank Lehmann-Horn

Summary. With numerous neuromuscular disorders the use of volatile anesthetics or depolarizing muscle relaxants is associated with complications mimicking an episode of malignant hyperthermia (MH). These events may resemble the early symptoms of MH to a degree that MH-specific treatment is initiated before the real signs of MH occur. Yet, most of these complications are clinically and pathogenetically different from a true MH crisis. True MH crises are seen (a) in MH-susceptible individuals, (b) in patients with central core disease, and probably (c) in patients with King-Denborough syndrome. Crises similar but clinically and pathogenetically different from MH include: (a) the myotonic reaction which, in contrast to MH, can be induced or aggravated by depolarizing relaxants in the absence of volatile anesthetics, as seen for instance in myotonia congenita, hyperkalemic periodic paralysis, paramyotonia congenita, myotonia fluctuans and permanens, and in myotonic dystrophy; (b) rhabdomyolysis induced by suxamethonium and followed by elevation of extracellular potassium in hypokalemic periodic paralysis; and (c) myolysis of skeletal or cardiac muscle induced by suxamethonium and perhaps volatile anesthetics in Duchenne and Becker muscular dystrophy. This report suggests that different anesthetics and therapeutical drugs should be employed in patients with these myopathies.

Introduction

Anesthesia-related complications similar to malignant hyperthermia (MH) have been reported for patients with numerous neuromuscular disorders such as central core disease (CCD) and King-Denborough syndrome (KDS), myotonias and periodic paralyses, and Duchenne and Becker muscular dystrophies [1]. Therefore, patients having one of these diseases have been generally considered at high risk for MH. However, the pathophysiology of these disorders is different from that of MH and also the anesthesia-related muscle fiber alterations are different. In particular is it incorrect to consider these patients and their family members as having the genetic trait for MH. The following groupings are the consequence of clinical findings, in vitro contracture test results, and recent developments of molecular genetics (Table 1).

Department of Applied Physiology, University of Ulm, D-89069 Ulm, Germany

Table 1. Anesthesia-related complications in neuromuscular diseases.

Reaction and disease	Location	Gene	IVCT
"True" malignant hyperthermia			
MHS1	19q12–13.2	RYR1	+
MHS2	17q13?	?	+
MHS3	7q11.23–21.1	CACNL2A	+
MHS4	3q13.1	?	+
Central core disease (CCD)	19q12–13.2	RYR1	+
CC in cardiomyopathy patients??	14q1	MYH7	?
King-Denborough syndrome	?	?	+
Myotonic reaction			
Muscle Na^+ channel diseases	17q13.1–3	SCN4A	−(+)
Muscle Cl^- channel diseases	7q35	ClC-1	−(+)
Dystrophic muscle hypersensitivity			
Myotonic dystrophy	19q13	DM	−(+)
Muscular dystrophy (DMD, BMD)	Xp21.2	Duchenne	−(+)

MHS, malignant hyperthermia susceptibility types; CC, central cores; DMD, Duchenne muscular dystrophy; BMD, Becker muscular dystrophy; IVCT, in vitro contracture test.

"True" MH Events in Neuromuscular Disorders

Central Core Disease

In this disease, skeletal muscle hypermetabolism induced by volatile anesthetics and depolarizing relaxants can be life threatening [2–4]. CCD is a rare, dominantly inherited, congenital myopathy with high clinical variability. Most CCD patients suffer from proximal weakness with improvement rather than progression in adulthood. Usually they have difficulties in rising from the sitting position and in climbing stairs. A minority of the patients are severely affected, e.g., never learn to walk. Some family members may not reveal any clinical symptoms or signs although the characteristic histologic features such as the degenerations known as central cores can be found in the subclinical state. Such cores, consisting of damaged myofibrils and lacking mitochondria, may be abundant in all type-1 fibers of all skeletal muscles. For all patients with clinical or subclinical CCD the in vitro contracture test was positive.

Following genetic data on MH [5,6], CCD was linked to RYR1 or to the corresponding region on chromosome 19q [7–10], and several point mutations within RYR1 were discovered [11,12]. The substitutions (Arg-163-Cys; Ile-403-Met; Arg-2434-His) can lead to subclinical or clinical CCD in addition to MH susceptibility, and in some MH-susceptible family members even central cores can be absent [13]. Unexpectedly, central cores were also detected in muscle biopsies from family members of patients having hypertrophic cardiomyopathy caused by mutations in the ventricular heavy-chain β-myosin. The corresponding gene, MYH7, is located on chromosome 14q [14]. Central cores are not specific for CCD, and the association with MH susceptibility has not yet been clarified. Nevertheless, CCD could be heterogeneous both for genetics and for MH susceptibility [9,15–18].

King-Denborough Syndrome

Skeletal muscle hypermetabolism may also be life threatening in KDS. Typical MH crises and even anesthesia-related death have been reported [19–21]. The rare congenital progressive myopathy is characterized by short stature, scoliosis, pectus deformity, ptosis, low-set ears, anti-Mongolian slanted eyes, and cryptorchism [19,20] and thus resembles the more frequent Noonan syndrome. All KDS patients who have been subjected to the in vitro contracture test were MH susceptible [21]. Unfortunately, linkage data are lacking.

Anesthetic Complications Due to Aggravated Myotonic Reactions

Myotonic disorders are sometimes connected with the risk of anesthesia-related events, but they are not, normally susceptible to MH. Deaths due to muscle hypermetabolism have not been reported.

Nondystrophic Myotonias and Periodic Paralyses

These groups can be divided into the dominantly inherited muscle sodium channel disorders (hyperkalemic periodic paralysis, paramyotonia congenita, myotonia fluctuans and permanens) and the muscle chloride channelopathies (autosomal dominant and recessive myotonia congenita) [22].

Muscle Sodium Channel Diseases

The clinical variability within this group of diseases is considerable. Some patients having the mildest form, i.e., myotonia fluctuans, may not be aware of their skeletal muscle dysfunction, although a severe myotonic reaction can be induced by depolarizing agents such as potassium or suxamethonium [23,24]. In general, patients with sodium channel myotonia or hyperkalemic periodic paralysis are predisposed to anesthesia-related events [25]. Typically, the complication is characterized by a suxamethonium-induced masseter spasm or generalized muscle spasms. Muscle stiffness may impede intubation and mechanical ventilation causing severe hypoxia. Hyperthermia, but no rapid increase in body temperature, may occur. For anesthesia performed without administration of suxamethonium, complications have so far not been reported. In vitro contracture tests performed according to the European protocol yielded equivocal or negative results [23,26] whereas tests performed according to the North American protocol were positive in some patients [24,25].

The defective gene, SCN4A, encodes the α subunit of the adult skeletal muscle sodium channel and is located on chromosome 17q13.1–13.3. Sixteen point mutations have been detected in SCN4A, causing either hyperkalemic periodic paralysis, paramyotonia congenita, myotonia fluctuans, or myotonia permanens [22]. Based on linkage data revealing rather low LOD scores in each of several North American MH families, SCN4A was suggested as a candidate gene for MH susceptibility (MHS) [27]. A role for SCN4A in MH susceptibility has not been confirmed and the data are in disagreement with the fact that, for all European families that were not linked to RYR1, linkage with chromosome 17q13 was also excluded [28].

In contrast to MHS, the diagnosis of myotonia or periodic paralysis can be made using noninvasive techniques, i.e., clinical examination, electromyography, and molecular genetics. Clinical signs are slowed muscle relaxation, percussion myotonia, and electrical myotonia. The latter may also be present in the subclinical state. In contrast to the long-lasting contractures of MHS muscle bundles, myotonic muscle fibers reveal only transient elevations of the force base line when exposed to caffeine or halothane [26]. Also in contrast to typical MH responses, myotonic muscle fibers can undergo a contracture after exposure to a single does of suxamethonium. Such responses have not been observed in muscle bundles obtained from typical MHS patients [29].

Muscle Chloride Channel Diseases: Myotonia Congenita

Both the autosomal dominant and recessive forms of myotonia congenita (Thomsen- and Becker-type, respectively) are linked to the same gene coding for the chloride channel of adult human muscle [30]. Eighteen mutations in this gene have been shown to cause myotonia congenita [22]. The reason for the myotonic overexcitability of the surface membrane is a reduced chloride conductance of the muscle fiber membrane due to loss-of-function or change-of-function mutations in the channel protein.

This myotonia is also intensified by all depolarizing relaxants and anesthetics that act either directly on the sarcolemma or via the endplate [31]. In vitro, contractions could be induced by a single dose of suxamethonium [26].

Voltage-Gated Calcium Channel Disease of Muscle

Anesthesia-related complications similar to MH have also been reported for hypokalemic periodic paralysis (HypoPP) [32].

A systematic genome analysis linked HypoPP to chromosome 1q31–32 [33], a region containing the gene encoding that $\alpha 1$ subunit of the L-type calcium channel of skeletal muscle. This subunit is part of the voltage-gated dihydropyridine (DHP) receptor/calcium channel complex that is located in the transverse tubular system and is involved in voltage-dependent calcium release from the sarcoplasmic reticulum.

Point mutations have been detected in two S4 regions believed to act as voltage sensors [34,35]. The mutations enhance inactivation of the L-type calcium current [36] and might reduce calcium release by inactivating sodium channels as well as by a direct effect on its voltage control.

Several in vitro contracture tests revealed equivocal results with a predominance of pathological halothane contractures [26].

Myotonic Dystrophy

Myotonic dystrophy, the most common muscle disease in adulthood, is a progressive muscular dystrophy and a multisystemic disorder. Although life-threatening hyperthermias have not been reported (in contrast to myotonic reactions and barbiturate-induced apneas), the occurrence of positive or equivocal in vitro contracture test results could more seriously raise the question whether these patients possess the genetic trait for MH [26]. However, the dystrophia myotonica (DM) gene has been mapped to chromosome 19q and the distance to RYR1 is in the order of 25 cM which makes genetic linkage of the two conditions very unlikely [37].

In addition to events due to suxamethonium-induced myotonic reactions, anesthesia-related problems may arise from the fact that some muscle groups are dystrophic, as discussed in the following section.

Anesthetic Complications Due to Suxamethonium-Induced Myolysis

Duchenne or Becker Muscle Dystrophy

Three adverse events can occur during general anesthesia in patients with clinical or subclinical Duchenne (DMD) or Becker (BMD) muscular dystrophy: (1) episodes that are clinically similar to MH, with acidosis, elevated temperature, muscle rigidity, and hyperkalemia; (2) acute rhabdomyolysis; and (3) sudden and unexpected cardiac arrest due to the heart muscle being hypersensitive to suxamethonium, with difficult resuscitation [38,39].

In general, the resting myoplasmic calcium concentration is elevated in progressive muscular dystrophies such as DMD and BMD [40], probably by an abnormally high influx of calcium from the extracellular space into the myoplasm due to a genetically determined defect in the cytoskeleton of the surface membrane [41]. In contrast to MH, CCD, and KDS, muscle biopsies of DMD patients gave MH-negative results [38].

Conclusion

The past few years have brought a veritable explosion in our knowledge of the molecular basis of muscular disorders. Not only has information as to the specific genetic defects been plentiful, but also new insights into the pathomechanism have been gained. Such information has importance for the anesthesiologist because anesthetic complications are commonly associated with patients known to be suffering from muscular disorders or those in the preclinical stages. Knowledge of the pathogenesis of these diseases and the inherent anesthesia-related complications demand further differentiation from malignant hyperthermia regarding safe anesthesia for patients with muscular diseases and the appropriate therapy in case of complications [42].

References

1. Wedel D (1992) Malignant hyperthermia and neuromuscular disease. Neuromuscul Disord 2:157–164
2. Denborough MA, Dennett X, Anderson R (1973) Central core disease and malignant hyperpyrexia. Br Med J 1:272–273
3. Frank JP, Harati Y, Butler I, Nelson TE, Scott CJ (1990) Central core disease and malignant hyperthermia syndrome. Ann Neurol 7:11–17
4. Shuaib A, Paasuke RT, Brownell AWK (1987) Central core disease: a reappraisal of its clinical features and new management recommendations. Medicine 66:389–396
5. McCarthy TV, Healy JMS, Lehane M, Heffron JJA, Deufel T, Lehmann-Horn F, Farrall M, Johnson K (1990) Localization of the malignant hyperthermia susceptibility locus to human chromosome 19q11.2–13.2. Nature 343:562–563
6. MacLennan DH, Duff C, Zorzato F, Fujii J, Phillips M, Korneluk R, Frodis W, Britt BA, Worton RG (1990) Ryanodine receptor gene is a candidate for predisposition to malignant hyperthermia. Nature 343:559–561

7. Haan EA, Freemantle CJ, McCure JA, Friend KL, Mulley JC (1990) Assignment of the gene for central core disease to chromosome 19. Hum Genet 86:187–190
8. Kausch K, Lehmann-Horn F, Hartung EJ, Janka M, Wieringa B, Grimm T, Müller CR (1991) Evidence for linkage of the central core disease locus to the proximal long arm of human chromosome 19. Genomics 10:765–769
9. Mulley JC, Kozman HM, Phillips HA, Gedeon AK, McCure JA, Iles DE, Gregg RG, Hogan K, Couch FJ, MacLennan DH, Haan EA (1993) Refined genetic localization for central core disease. Am J Hum Genet 52:398–405
10. Schwemmle S, Wolff K, Palmucci LM, Grimm T, Lehmann-Horn F, Hübner CH, Hauser E, Iles DE, MacLennan DH, Müller CR (1993) Multipoint mapping of the central core disease locus. Genomics 17:205–207
11. Quane KA, Healy JMS, Keating KE, Manning BM, Couch FJ, Palmucci LM, Doriguzzi C, Fagerlund TH, Berg K, Ording H, Bendixen D, Mortier W, Linz U, Müller CR, McCarthy TV (1993) Mutations in the ryanodine receptor gene in central core disease and malignant hyperthermia. Nature Genetics 5:51–55
12. Zhang Y, Chen HS, Khanna VK, De Leon S, Phillips MS, Schappert K, Britt BA, Brownell KW, MacLennan D (1993) A mutation in the human ryanodine receptor gene associated with central core disease. Nature Genetics 5:46–49
13. Islander G, Henriksson K-G, Ranklev-Twetman E (1995) Malignant hyperthermia susceptibility without central core disease (CCD) in a family where CCD is diagnosed. Neuromuscul Disord 5:125–127
14. Fananapazir L, Dalakas MC, Cyran F, Cohn G, Epstein ND (1993) Missense mutations in the β-myosin heavy-chain gene cause central core disease in hypertrophic cardiomyopathy. Proc Nat Acad Sci USA 90:3993–3997
15. Deufel T, Golla A, Iles D, Meindl A, Meitinger T, Schindelhauer D, DeVries A, Pongratz D, MacLennan D, Johnson KJ, Lehmann-Horn F (1992) Evidence for genetic heterogeneity of malignant hyperthermia susceptibility. Am J Hum Genet 50:1151–1161
16. Fagerlund T, Islander G, Ranklev E, Harbitz I, Hauge G, Møkleby E, Berg K (1992) Genetic recombination between malignant hyperthermia and calcium release channel in skeletal muscle. Clin Genet 41:270–272
17. Iles D, Lehmann-Horn F, Deufel T, Scherer SW, Tsui L-C, Olde Weghuis D, Suijkerbuijk RF, Heytens L, Mikala G, Schwartz A, Ellis FR, Stewart AD, Wieringa B (1994) Localization of the gene encoding the α2/δ-subunits of the L-type voltage-dependent calcium channel to chromosome 7q and segregation of flanking markers in malignant hyperthermia susceptible families. Hum Mol Genet 3:969–975
18. Sudbrak R, Procaccio V, Klausnitzer M, Curran JL, Monsieurs K, Van Broeckhoven C, Ellis R, Heytens L, Hartung EJ, Kozak-Ribbens G, Heilinger D, Weissenbach J, Lehmann-Horn F, Mueller CR, Deufel T, Stewart AD, Lunardi J (1995) Mapping of a further malignant hyperthermia susceptibility (MHS) locus to chromosome 3q13.1. Am J Hum Genet 56:684–691
19. King JO, Denborough MA (1973) Anaesthetic-induced malignant hyperpyrexia in children. J Pediatrics 83:37–40
20. McPherson EW, Taylor CA Jr (1981) The King syndrome: malignant hyperthermia, myopathy, and multiple anomalies. Am J Med Genet 8:159–165
21. Heiman-Patterson PT, Rosenberg HR, Binning CPS, Tahmoush AJ (1986) King-Denborough syndrome: Contracture testing and literature review. Pediatr Neurol 2:175–177
22. Hoffman EP, Lehmann-Horn F, Rüdel R (1995) Over-active or inactivated: ion channels in muscle diseases. Cell 80:681–686
23. Ricker R, Moxley RT, Heine R, Lehmann-Horn F (1994) Myotonia fluctuans, a third type of muscle sodium channel disease. Arch Neurol 51:1095–1102
24. Vita GM, Olckers A, Jedlicka AE, George AL, Heiman-Patterson T, Rosenberg H, Fletcher JE, Levitt RC (1995) Masseter muscle rigidity associated with glycine^{1306}-to-alanine mutation in adult muscle sodium channel α-subunit gene. Anesthesiology 82:1097–1103

25. Heiman-Patterson PT, Martino C, Rosenberg HR, Fletcher J, Tahmoush A (1988) Malignant hyperthermia in myotonia congenita. Neurology 38:810–812
26. Lehmann-Horn F, Iaizzo PA (1990) Are myotonias and periodic paralyses associated with susceptibility to malignant hyperthermia? Br J Anaesth 65:692–697
27. Levitt RC, Olckers A, Meyers S, Fletcher JE, Rosenberg H, Isaacs H, Meyers DA (1992) Evidence for the localization of a malignant hyperthermia susceptibility locus (MHS2) to human chromosome 17q. Genomics 14:562–566
28. Sudbrak R, Golla A, Hogan K, Powers P, Gregg R, Du Chesne I, Lehmann-Horn F, Deufel T (1993) Exclusion of malignant hyperthermia susceptibility (MHS) from a putative MHS2 locus on chromosome 17q and of the α1, β1, and γ subunits of the dihydropyridine receptor calcium channel as candidates for the molecular defect. Hum Mol Gen 2:857–862
29. Ording H, Skovgaard LT (1987) In vitro diagnosis of susceptibility to malignant hyperthermia: evaluation of tests with halothane-caffeine, potassium chloride, suxamethonium and caffeine-suxamethonium. Acta Anaesthesiol Scand 31:462–465
30. Koch MC, Steinmeyer K, Lorenz C, Ricker K, Wolf F, Otto M, Zoll B, Lehmann-Horn F, Grzeschik KH, Jentsch TJ (1992) The skeletal muscle chloride channel in dominant and recessive human myotonia. Science 257:797–800
31. Rüdel R, Lehmann-Horn F (1985) Membrane changes in cells from myotonia patients. Physiol Rev 65:310–356
32. Lambert C, Blanloeil Y, Krivosic Horber R, Bérard L, Reyford H, Pinaud M (1994) Malignant hyperthermia in a patient with hypokalemic periodic paralysis. Anesth Analg 79:1012–1014
33. Fontaine B, Vale Santos JM, Jurkat-Rott K, Reboul J, Plassart E, Rime CS, Elbaz A, Heine R, Guimaraes J, Weissenbach J, Baumann N, Fardeau M, Lehmann-Horn F (1994) Mapping of the hypokalaemic periodic paralysis (HypoPP) locus to chromosome 1q31–32 in three European families. Nature Genetics 6:267–272
34. Jurkat-Rott K, Lehmann-Horn F, Elbaz A, Heine R, Gregg RG, Hogan K, Powers P, Lapie P, Vale-Santos JE, Weissenbach J, Fontaine B (1994) A calcium channel mutation causing hypokalemic periodic paralysis. Hum Mol Gen 3:1415–1419
35. Ptacek L, Tawil R, Griggs RC, Engel A, Layzer RB, Kwiecinski H, McManis PG, Santiago L, Moore M, Fouad G, Bradley P, Leppert MF (1994) Dihydropyridine receptor mutations cause hypokalemic periodic paralysis. Cell 77:863–868
36. Sipos I, Jurkat-Rott K, Harasztosi CS, Fontaine B, Kovacs L, Melzer W, Lehmann-Horn F (1995) Skeletal muscle DHP receptor mutations alter calcium currents in human hypokalaemic periodic paralysis myotubes. J Physiol 483:299–306
37. MacKenzie AE, Korneluk RG, Zorzato F, Fujii J, Phillips M, Iles D, Wieringa B, Leblond S, Bailly J, Willard HF, Duff C, Worton RG, MacLennan DH (1990) The human ryanodine receptor gene: Its mapping to 19q13.1. Placement in a chromosome 19 linkage group, and exclusion as the gene causing myotonic dystrophy. Am J Hum Genet 46:1082–1089
38. Gronert GA, Fowler W, Cardinet GH, Griz A Jr, Ellis WG, Schwartz MZ (1992) Absence of malignant hyperthermia contractures in Becker–Duchenne dystrophy at age 2. Muscle Nerve 15:52–56
39. Schulte-Sasse U (1993) Experiences and lessons from the MH hotline in Germany: problems associated with succinylcholine. MHAUS Communicator 11:3–44
40. Bodensteiner JB, Engel AG (1978) Intracellular calcium accumulation in Duchenne dystrophy and other myopathies: A study of 567 000 fibers in 114 biopsies. Neurology 28:439–446
41. Koenig M, Beggs AH, Moyer M, Scherpf S, Heindrich K, Bettecken TG, Meng G, Müller CR, Lindlöf M, Kaariainen H, de la Chapelle A, Kiuru A, Savontaus M-L, Gilgenkrantz H, Récan D, Chelly J, Kaplan J-C, Covone AE, Archidiacono N, Romeo G, Liechti-Gallati S, Schneider V, Braga S, Moser H, Darras BT, Murphy P, Francke U, Chen JD, Morgan G, Denton M, Greenberg CR, Wrogemann K, Blonden LAJ, van Paassen HMB, van Ommen GJB, Kunkel LM (1989) The molecular basis for Duchenne versus Becker

muscular dystrophy: correlation of severity with type of deletion. Am J Hum Genet 45:498–506
42. Iaizzo P, Lehmann-Horn F (1995) Anesthetic complications in muscle disorders. Anesthesiology 82:1093–1096

Part 7. Therapy of Malignant Hyperthermia

Updated Therapy of Malignant Hyperthermia

Uwe Schulte-Sasse

Anesthesia mortality is remarkably low [1], and thus mass disasters can be discounted. Therefore, it is our task today to analyze rare adverse outcomes [2] so that we can identify some basic, still existing causes for morbidity and mortality and thus be able to overcome thses dangers and approach the infinitely distant goal of zero mortality in our speciality. Regarding safety in anesthesia, the words of an American anesthetist should be quoted [3]:

> We don't need more powerful narcotics! We don't need more sophisticated formulas for the kinetic modeling! We don't need more studies to substantiate the studies that verified the studies that supported the studies that proved something in 1969! What we desperately need is information about how to better use the data and devices we already have!

The last phrase is particularly fitting for the present-day insufficient diagnosis and treatment of malignant hyperthermia (MH).

Once the symptoms are recognized and the presumptive diagnosis of MH is established, avoiding any loss of time, treatment according to a preconceived plan is of utmost importance. Therapy, particularly dissolving the dantrolene, is time consuming, so more than one person must be on the spot to support the anesthetist in charge. MH must be treated at the place of its occurrence: moving the patient to an intensive care unit (ICU) or even to another hospital has resulted in death in more than one case.

Measures To Be Taken Immediately

1. End administration of triggering agents.
2. Increase respiratory minute volume by a factor of 3 using 100% oxygen.
3. Administer dantrolene in a rapid infusion until definite signs of therapeutic success become visible; 2.5 mg/kg is an *average* therapeutic dose, BUT considerably higher doses have been needed for a successful outcome [4].

A case of rapid development of the syndrome in a 31-year-old well-built male should be presented. In 1978 we were called to an emergency in another hospital.

German Hotline for Malignant Hyperthermia Emergencies, Department of Anaesthesia and Critical Care Medicine, Heilbronn Community Hospital of the University of Heidelberg, 74078 Heilbronn, Germany

When we first saw the patient, he was covered with snow; nevertheless, a temperature probe read 41.7°C. The intubated patient ventilated 31 l/min spontaneously, in spite of 400 mg pethidine and 0.6 mg fentanyl! His heartrate was 172 beats/min. During the rapid infusion of dantrolene, all these parameters were definitely returning toward normal: after 25 min, pulse rate was 132, temperature 40.7°C, arterial P_{CO_2} 43 mmHg, pH 7.5, and a base excess of +5.6 mEq/l indicated a somewhat vigorous administration of bicarbonate. At that moment the patient had received 3 mg/kg of dantrolene, which continued to be infused rapidly, and 35 min after start of therapy his pulse rate was 84/min and temperature 39.5°C. The fast change of parameters toward normal confirmed the diagnosis and the effectiveness of the therapy. Administration of dantrolene was continued, now at a slower pace. Repeated blood gas analyses and close clinical observation confirmed the correct course of therapy. At 120 min after the start of dantrolene, 10 mg/kg had been infused and a maintenance dose was begun.

It must be stressed again and again that the patient will be saved only under the condition that the rapid administration of dantrolene is not slowed before definite signs of therapeutic effectiveness. This last rule is elucidated by the following case, which was reported to us via the hotline for MH emergencies:

In a 5-year-old boy, MH was diagnosed after a substantial delay caused by rather subtle signs of the oncoming catastrophe. It was 190 min after induction of anesthesia with succinylcholine and isoflurane when the anesthetist—using his own words—"suddenly" noticed a redness in the face of the child and a temperature of 42.7°C. Just as "suddenly," there was tachycardia of 200 beats/min. Now it took a further 10 min before the dantrolene reached the patient. After 10 min more the patient had received 3 mg/kg but his heartrate had not yet changed. After a short discussion how to proceed, dantrolene infusion was resumed, but at a slower rate. At 45 min after the diagnosis was made, base excess was −12 mEq/l and potassium 8 mEq/l; 65 min after diagnosis 10 mg/kg dantrolene had been infused, but potassium had peaked with 8.7 mEq/l. Circulatory collapse ensued, and, typical for continued hypermetabolism, immediate rigor mortis ensued.

This course of events illustrated why Ryan [5] of the Harvard Medical School stated that only aggressive therapy will save the life of a patient suffering from full-blown MH. In the case just reported, the diagnosis was late and very little time was left for treatment. Therefore, rapid administration of dantrolene must not be slowed down before convincing signs of therapeutic effectiveness occur: normalization of heartrate and end-tidal CO_2, lessening of muscular rigidity, cessation of blood gas abnormalities, normal concentration of potassium, and decrease in body temperature. This means that *during* the fast infusion therapeutic success must be confirmed by clinical observation and by laboratory findings. A significant and sustained decrease in heartrate, reduction of ventilation necessary to produce an end-tidal CO_2 of 5%, cessation of the acidifying metabolic process, reduction of hyperkalemia, and finally fall of the elevated body temperature—all these changes give evidence that therapy is on the right track. Now, and not earlier, transition to a slower rate of dantrolene infusion may be considered. If changes toward normal are not observed within 30 min with a cumulative dose of 10 mg/kg, diagnosis must be doubted. If definite signs of improvement are observed, however, a further increase of the dose of dantrolene must be considered. Although the stated upper dose is 10 mg/kg per 24, this is NOT a magic limit [6]!

After Emergency Measures, the Following Steps Should Be Taken

1. Administer 1.5 mEq/kg of bicarbonate and repeat according to blood gas analysis.
2. Use intravenous calcium if hyperkalemia is life threatening; it does not precipitate MH.
3. Change the breathing tubes and CO_2 absorber; changing the anesthesia machine is not necessary.
4. Treat hyperthermia with ice packs to the body surface.
5. Administer a diuretic (acetazolamide) if, in spite of mannitol (3 g of mannitol is contained in the 20-mg dantrolene bottle), diuresis of 1.5 ml/kg per hour is not attained [7].
6. Add intravenous lines: central venous catheter, intraarterial catheter.
7. To prevent recrudescence, continuation of dantrolene is prudent, even if the classical signs of MH have subsided following initial treatment. Give dantrolene every 6 h 1 mg/kg intravenously for 24–48 h [8]. Scientific data to support this recommendation quantitatively are lacking.
8. Transfer the patient to an ICU; observe the patient for at least 48 h.
9. "Smoldering" or relapse of the syndrome may be diagnosed by renewed tachycardia, increase of end-tidal CO_2, metabolic acidosis, hyperkalemia, and elevation of body temperature. Rapid infusion of dantrolene is again indicated.

Reasons for Persistence of a High Mortality Rate from Malignant Hyperthermia

It had been expected that the introduction of dantrolene into clinical practice would put an end to death from MH. Surprisingly enough, deaths from MH continue to be published, and while operating the hotline for MH emergencies in Germany we are aware that patients continue to succumb to MH. What are the reasons for this unsatisfactory state of affairs? An analysis of the ample personal information from fatal MH cases reported to our consultants revealed the following causes for the perplexing discrepancy between expectations and clinical reality [9]:

1. A lack of MH-sensitive monitoring (i.e., capnometry and blood gas analysis) has frequently been a problem. In addition, the widely held preoccupation with the name-giving symptom hyperthermia has been especially accountable for the delay of diagnosis: hyperthermia, which 30 years ago was thought to be the hallmark of MH, is not the cause but the result of biochemical derangements within the cells of striated muscle. Now, if the anesthetists should not wait for hyperthermia to develop, what should they look for? The first symptoms are those of an elevated metabolic rate. Almost simultaneously one can observe an unexplained tachycardia, soon followed by arrhythmia. Patients breathing spontaneously increase respiratory rate and tidal volume; patients fully relaxed and ventilated mechanically show a steep increase of end-tidal CO_2. There may be a rigor of the muscles. At this moment, the presumptive diagnosis of MH must be made and vigorous treatment started. Thereafter diagnosis must be confirmed by blood gas analysis, detection of hyperkalemia, and an increase of creatine kinase. It must be admitted that these symptoms are rather nonspecific,

but the magnitude of changes in the observed parameters is striking and may not be overlooked: the pulse rate in an adult jumps to 170 beats per minute, patients breathing spontaneously move minute volumes of 30 l, end-tidal CO_2 goes from 5% to 10% within minutes, and correspondingly arterial CO_2 tension reaches 100 mmHg as arterial pH plummets to 7.0.

2. The anesthetist often is concerned with treating symptoms only, instead of the cause of the disease. MH is not a failure of temperature regulation but a hypermetabolic syndrome. Measures of secondary importance include surface cooling or changing of the anesthesia machine. A host of useless drugs are often given *before* administration of dantrolene, including procaine amide, lidocaine, corticosteroids, aprotinin, muscle relaxants, fentanyl, beta-blockers, neuroleptics, antipyretics, and verapamil. In the presence of dantrolene, the latter results in dangerous hyperkalemia. Preoccupation with nonspecific facets of therapy may not only waste time, but delay the prime factor in therapy: intravenous administration of dantrolene.

3. The patient either did not receive dantrolene at all, or received too little or too late. The drug must reach the locus of its action as long as blood is circulating to the muscle. A full-blown MH episode approaches the point of no return very quickly; therefore, it is mandatory to have 36 vials of dantrolene available in every suite of operating rooms. Storing the drug in one central place for several hospitals means courting disaster. This arrangement has caused a deadly outcome for several cases; when the anesthetist made the diagnosis of MH, there was no dantrolene at hand.

4. At the beginning of MH, an oxygen deficiency will always develop: this episode must be kept as short as possible. Therefore minute ventilation must be increased without delay by a factor of 3. Although the patient's life will not be saved by this measure alone, this step is nevertheless important to prevent irreversible hypoxemic damage to vital organs before dantrolene takes effect [10].

5. Several cases of MH with extreme hyperthermia or even cardiac arrest as "first" signs of MH have led to the conclusion that earlier symptoms had been overlooked. Clinical observation and documentation are interrelated: the lack of both leads to late diagnosis and erratic therapeutic attempts.

Abandoning preoccupation with increase in body temperature, both as a means of recognizing the syndrome at an early stage and as a symptom to be treated with highest priority, could lead to a decrease in mortality rate. Patient safety would be enhanced if hospitals were to stock enough dantrolene wherever general anesthesia is administered. In light of the fact that all the fatalities analyzed revealed shortcomings in the diagnosis and treatment that are sufficient to explain each death, there is no need to have recourse to a hypothesis of types of MH not amenable to treatment with dantrolene. Therefore, as Beverley Britt from Toronto stated in 1985 [11]: Every patient must survive MH. It has to be added: under the condition that the diagnosis is made without delay and skillful treatment is instituted immediately.

References

1. Campling EA, Devlin HB, Hoile RW, Lunn JN (1992) The report of the national confidential enquiry into perioperative deaths 1990. National Confidential Enquiry into Perioperative Deaths, London, pp 120–121
2. Ellis FR (1992) Detecting susceptibility to malignant hyperthermia. Br Med J 304:791

3. Vitez TS (1989) Comment on: The incidence of venous air embolism during total arthroplasty. Surv Anesth 33:133
4. Kolb ME, Horne ML, Martz R (1982) Dantrolene in human malignant hyperther-mia. A multicenter study. Anesthesiology 56:254–262
5. Ryan JF (1993) Malignant hyperthermia. In: Coté CJ, Ryan JF, Todres ID, Goudsouzian NG (eds) A practice of anesthesia for infants and children, 2nd edn. Saunders, Philadelphia, pp 151–170
6. Rosenberg H, Fletcher J, Seitman D (1992) Pharmacogenetics. In: Barash PG, Cullen BF, Stoelting RK (eds) Clinical anesthesia, 2nd edn. Lippincott, Philadelphia, pp 589–613
7. Better OS, Stein JH (1990) Early management of shock and prophylaxis of acute renal failure in traumatic rhabdomyolysis. N Engl J Med 322:825–829
8. Malignant Hyperthermia Association of the United States (1993) Emergency therapy for malignant hyperthermia.
9. Schulte-Sasse U, Eberlein HJ (1991) Reasons for the persistence of a high mortality due to malignant hyperthermia and proposals for its reduction. Anaesthesiol Reanimat 16:202–207
10. Fukui T, Morio M, Fujji K, Kikuchi H, Ohtani M, Kishida T, Kitoh S, Suemor I (1982) Effectiveness of dantrolene on rigid type of malignant hyperthermia with chronic recurrent course. Hiroshima J Anaesth 18 (Suppl 1):41–48
11. Britt BA (1985) Malignant hyperthermia. Can Anaesth Soc J 32:666–678

Part 8. Free Discussion

"What Is Malignant Hyperthermia?"

VIIth International Workshop on Malignant Hyperthermia
10:00 a.m.–12:00 noon
Wednesday July 20, 1994

Contributors: Michio Morio, James R. Mickelson (Chairpersons), Kumar G. Belani, Beverley A. Britt, Michael Denborough, Thomas Deufel, Makoto Endo, Jeffrey E. Fletcher, Gerald A. Gronert, Hirosato Kikuchi, Genevieve Kozak-Ribbens, Renee Krivosic-Horber, Marilyn Green Larach, Tommie V. McCarthy, Kenjiro Mori, Henry Rosenberg, Vincenzo Tegazzin, Osafumi Yuge, Sang Ho Yuo et al.

Preface to Free Discussion

As the contents of this Free Discussion Session was to be transcribed from the tape recording and then become a part of the proceedings, the Chairperson requested at the outset that each speaker kindly give his name and affiliation. Unfortunately, not all gave their name as requested and therefore it was not possible to verify all the statements with the respective speakers. Thus, the organizers made editorial changes and identified speakers other than the chairpersons as A, B, C . . . and were not able to check the transcription with each of the speakers. However, utmost care has been exercised by the editors to ensure accuracy, but it is feared that the discussion made may not have necessarily been transcribed precisely.

The language used by the speakers at the Free Discussion Session was either Japanese or English. As all statements were simultaneously translated into the other language, the transcript made from the tapes is not necessarily polished, but we believe it faithfully conveys the essence.

The contents of the Free Discussion Session are extremely interesting, being a record of the present status (1994) of the clinical aspects and the research of malignant hyperthermia. The editors sincerely hope that this record will be useful in future studies of malignant hyperthermia.

Chairperson (Morio): Many authorities on malignant hyperthermia are gathered here at this workshop and so I thought that it would be important to record what they are thinking at the present moment. That is the reason why we conducted this questionnaire survey. I would like to thank everybody who responded to this questionnaire.

We distributed in advance a questionnaire to inquire the participants' view of malignant hyperthermia. I would like to present the results of this questionnaire survey.

Except for one participant, all the responders gave their names on the questionnaire sheet. Figure 1 shows that there were 49 responders, which accounted for 24% of the 204 participants. The response rate was slightly lower than we had expected. The nationalities of the responders are shown in this figure with 61% of the responders being Japanese.

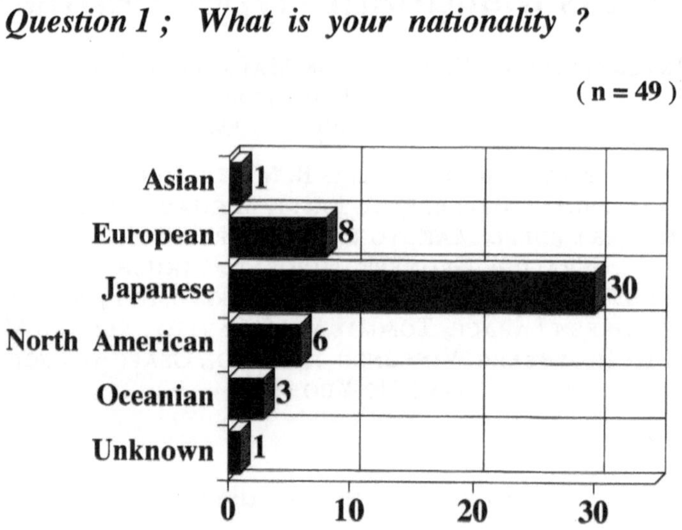

FIG. 1

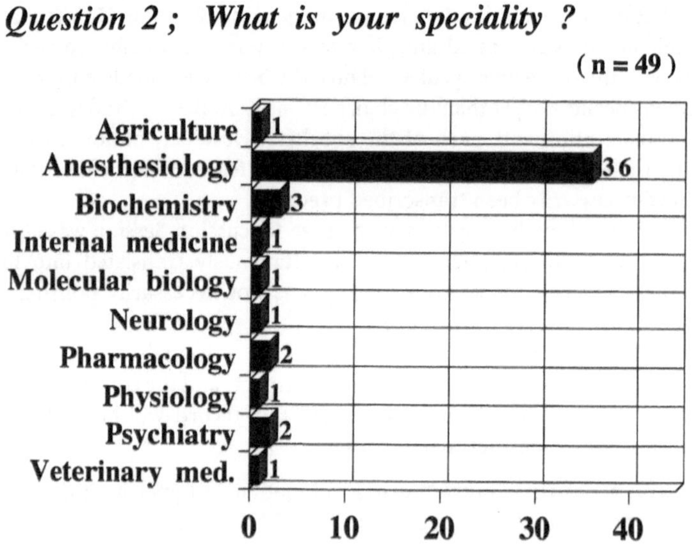

FIG. 2

Figure 2 shows the specialities of the responders. Anesthesiologists responding to the questionnaire accounted for 73% of replies. For the other specialities, we have one, two, or three people in each speciality, and for clinicians other than anesthesiologists, we had four.

MH Diagnosis

The question in Fig. 3. was: "What are the important symptoms and signs in the diagnosis of MH?" Each responder selected two answers. There were some comments that two symptoms were too few and so four people gave more than

Question 3 ; What is the most important symptoms in the diagnosis of MH ?

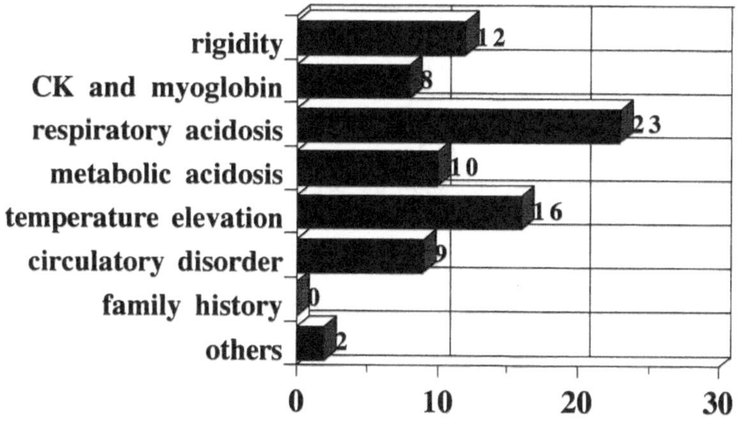

FIG. 3

two answers. Seven people gave only one answer, while three did not give any answer.

As for the purpose of the question, we know that MH signs and symptoms are all important, but we wanted to know what is most important to each of you in the diagnosis of MH. The question may have been too simple and so you may have found some difficulties in finding an answer. At any rate, here are the statistics. As you can see, respiratory acidosis including CO_2 output was selected by 23 responders, which accounted for 46.9% of all 49 responders, 16 (32%) selected temperature elevation, 20% selected metabolic acidosis, 18% circulatory disorder, and 16% CK and myoglobulinuria.

Rigidity appears in all the patients at the late stage of MH, but at the early stage, if succinylcoline is not used, the symptoms may be masked. Even without rigidity, MH cannot be refuted in some cases. We would like to start the discussion with these data on the clinical signs and symptoms. Any comments or suggestions from the participants?

Tachycardia

A: Why do we need to diagnose malignant hyperthermia in the operating room? It is because treatment must be commenced immediately. Proper diagnosis is essential for proper treatment. Since this is a disorder due to hypermetabolism, respiratory and metabolic acidosis is the background, and at a very early stage, tachycardia appears. In our department, we teach trainees that tachycardia without apparent cause is an early sign of malignant hyperthermia.

There are two possible causes of intraoperative tachycardia. The first is too light anesthesia, which is associated with tachycardia due to pain. The second is malignant hyperthermia. If the technique of anesthesia does not differ from the usual methods with ordinary doses of anesthetics and analgesics, malignant hyperthermia must be taken into consideration. I believe that tachycardia is the earliest clinical sign of malignant hyperthermia.

Chairperson (Morio): Is there any comment on this point?

B: I think when you ask what are the most important symptoms, the issue is which symptoms of MH come first; which symptoms are more likely to mean MH or not MH. It's a very complex thing. As far as tachycardia goes, there are many things that can cause tachycardia during anesthesia. There are enough for me to be able to teach a resident for more than an hour as to the possibilities, and unexplained tachycardia can be due to malignant hyperthermia.

I think the question of the most important symptom depends on the particular patient and what you are trying to find out. Are you trying to find out which is the earliest symptom so that you don't allow the patient to go on to a more fulminant episode, or which symptom or symptoms should be enough to trigger further investigation such as blood gases? There is a variable answer to this.

Clinical Grading Scale

B: I think that my opinion is reflected in the International Clinical Grading Scale. Because it is such a complex thing and because various symptoms mean different things, I get most upset about an unexplained respiratory acidosis. In terms of what will trigger my going to get a blood gas analysis, that's probably the first thing.

I guess the other symptom—or the other sign—is generalized muscular rigidity. If you see it,—you don't always see it—but if you do see it, there is very little else that gives you generalized muscular rigidity under anesthesia besides MH.

End-tidal CO_2

C: If we presume that one of the basic problems in malignant hyperthermia is hypermetabolism, and everybody agrees that this is a hypermetabolic syndrome, I think it has been adequately shown that the earliest change that you see would be an increase in carbon dioxide production, increase in oxygen consumption, which we generally measure by end-tidal CO_2, this can develop into respiratory acidosis if you don't change your ventilatory parameters. It is then presumed that tachycardia follows hypermetabolism.

The fever follows hypermetabolism, but the primary clinical sign would be hypermetabolism, which is manifested as the increase in end-tidal CO_2 level.

Recently, there was an article in *Anesthesia and Analgesia* by Steve Karren and Sierra Muldoon, which basically says that based on a number of reported cases—, and we have had these reported in other ways also—, that the anesthesiologist sees the increase in end-tidal CO_2, thinks that there is another problem happening, and increases the ventilation. So, to keep the end-tidal CO_2 within normal limits the anesthesiologist just keeps increasing the ventilation. In the particular case reported in the literature, they finally began to think about malignant hyperthermia when the ventilation was 20 liters per minute in a normal-sized patient.

So, increases in end-tidal CO_2 also have to be interpreted in context with ventilation, and we know there are other reasons that can cause an increase in end-tidal CO_2. So, there is nothing clear cut. It's something that should trigger your mind, and the first thing that may trigger your mind could be tachycardia or could be an increase in end-tidal CO_2. I think the more specific sign is an increase in end-tidal CO_2.

D: If we remind ourselves that the important point is to stop the malignant hyperthermia crisis early enough so that the patient will survive without any sequela, I think that the best sign is respiratory acidosis, which leads to the use of capnography.

I think that if all general anesthesia were done with capnography, there wouldn't be any possibility of death through malignant hyperthermia. Of course, we would have more false diagnosis of MH, but I am pretty sure that there wouldn't be any death. So, from that point of view, maybe, this sign is the best.

E: I think for an accurate diagnosis, anesthesiology expertise is very important because you have to look at the entire patient. As Dr. C said, if you see an increase in CO_2, you should try to find the reason for that increase and then you look for other signs. You put this picture together and then see if it fits into the diagnosis of malignant hyperthermia.

You can't just go by one sign, but you have to see if this progresses to something that fits into the picture of malignant hyperthermia. That's why I think you need to be very vigilant and observant. Just by looking at one sign you shouldn't make a quick diagnosis. I think you should look at the patient, look for other evidence of malignant hyperthermia, and then make the diagnosis.

B: I just wanted to say that we have had some deaths reported to the registry with end-tidal CO_2 monitoring, mostly because the patient was spontaneously breathing or on assisted ventilation and the increased $V_E$dt was significant and wasn't noticed, but they are very few. I always emphasize to the residents when I am teaching them that they must look at what the $V_E$dt is, and figure out what's appropriate, and get suspicious when it goes way up. But they kept the numbers fairly normal. They must look at the minute ventilation.

CHAIRPERSON (MORIO): As for the CO_2 output, if the accumulated CO_2 amount can be calculated, it is quite easy. But $Paco_2$ and/or F_ECO_2 depends on the ventilation. So, at an early stage, $Pvco_2$ may be increased, but $Paco_2$ is not increased so much. Therefore, if new devices are developed in the future, and not only the concentration but also the minute output of CO_2 could be monitored. We would appreciate it very much.

The next question is: If a CO_2 monitor is not used, as is the case in many countries, what is the next most important sign? What are the important indicators?

Oxygen Consumption

F: I just have a comment on measurement of CO_2 output, if I can go back to that for a moment. Most of us with good monitoring equipment can calculate oxygen consumption and CO_2 production under anesthesia, because if you have an expired–inspired gas readout, you know the difference between inspired and expired gases, and you can get a crude measure of oxygen consumption and CO_2 production by multiplying the difference by the expired minute volume.

Now, strictly speaking, the "expired" should be the mixed, expired volume. But also, if you have somebody who has normal lungs, and they have a flat plateau on the end-expired gases, then the very last reading will really be an alveolar gas measurement. If you then correct total minute volume for dead space, you will have a measure of oxygen consumption or CO_2 production that's fairly accurate. In fact, because of the case report that Dr. C was talking about, I wrote a letter to *Anesthesia and Analgesia* suggesting that we do this when we see problems that might be related to changes in minute volume, and I make my residents regularly calculate these things.

So, I don't think we need a device. I think the information is there if you are measuring expired minute volume and if you are measuring inspired and expired gases.

CHAIRPERSON (MORIO): Dr. F, do you think you can use that in routine clinical practise, because it's somewhat expensive?

F: Well, if you are already using capnography, and it's measuring both inspired and expired, you have the information. Now, for CO_2, it's easy, because inspired CO_2 is zero. So, with your expired CO_2 value, you can already get a rough calculation if you know the minute volume. It's nice to see the waveform, because then you know whether the number is relatively good for you, or if it's a lung pattern that expresses differences in distribution or emptying and things like that.

So, if you already have a capnograph that has a waveform, you are close to being able to calculate this. It really depends, of course, on the equipment you are using, but much of the equipment in the United States now has this capability, at least in most of the academic institutions. Many of the private hospitals just give you a numerical readout, and that's much more difficult to know the accuracy. But you can get a crude idea from the calculation, even with just a digital readout.

G: I think the question can be put in two parts: in a real clinical situation, what are the signs that lead to a diagnosis of MH and at what stage can we start detecting those signs?

There are many kinds of monitoring equipment that can provide data leading us to suspect an incidence of MH. That has a great effect on diagnosis. We have to consider this point carefully.

So, the question is, what kind of signature should we look for and when do we have to start the treatment? As Dr. E mentioned, careful observation, of course, has to be made, and when we do start the treatment, we begin by stopping the anesthesia and stopping the operation.

Tachycardia is the first sign that makes me suspect the onset of MH. If there is unexplained tachycardia, then I will suspect MH, and then I will come up with a strategy for coping with the situation. Stopping the operation and stopping the anesthesia when I start treatment that will bring the patient out of the respiratory acidosis.

A: I completely agree with Dr. G. I would like to mention that we should bear in mind that the capacity of our body for CO_2 is very large in comparison with that for O_2. It is not a rare experience that, during induction of anesthesia, we find hypercapnea after tracheal intubation, and hyperventilation takes time to normalize the end-tidal CO_2 level. It sometimes takes almost half an hour. With regard to abnormal levels of oxygen saturation, an immediate response can be obtained when appropriate ventilation is commenced. When I mean is that the change of CO_2 level in the case of malignant hyperthermia may take time. I personally have had no experience of observing a capnograph in the case of malignant hyperthermia. Tachycardia in malignant hyperthermia may be triggered by an increase in oxygen consumption. Thus, when unexplained tachycardia is noted, you should watch the level of end-tidal CO_2. The elevation of CO_2 level may be small in the very early phase of malignant hyperthermia.

F: Dr. A is correct. The CO_2 stores are much larger than the oxygen stores. They range up to 100 or 120 liters in the whole body, so you can have dramatic changes in CO_2, particularly during a situation of unsteady state, which we see right after induction. That is why it's really better to calculate oxygen consumption directly, if your gas monitoring device also measures oxygen concentration.

What you would really like to do is calculate the oxygen consumption directly from the difference between inspired and expired oxygen and the expired minute volume.

Some people have only CO_2, though, and then you have to take into account both an unsteady state and the large CO_2 stores that can restructure your numbers.

CHAIRPERSON (MORIO): We would like to stop the discussion with regard to tachycardia and CO_2 monitoring, but do you have any final comments.

A: I am affiliated to the World Federation of the Societies of Anesthesiologists, and we are aware of the differences in the operating theater facilities. Those who are here today are from nations who have sufficient sophisticated monitoring devices, such as the capnograph, pulse oximeter, anesthetic gas monitors, etc. Such operating theaters are rare cases, and those that have no monitoring devices are the most common in developing countries. In this sense, I believe tachycardia is the best sign of initiation of malignant hyperthermia. I reiterate that unexplainable tachycardia is the initial sign of malignant hyperthermia. We do not need any sophisticated monitoring for this diagnosis.

Muscle Rigidity

C: I think there are two parts to the question; what do you do when you have CO_2 monitoring, and what do you do when you don't have CO_2 monitoring?

First of all, when you do have CO_2 monitoring, the diagnosis of malignant hyperthermia is made on a sustained increase in CO_2. If you don't have CO_2 monitoring, then certainly tachycardia is a very early sign and should begin to make you think about your differential diagnosis of malignant hyperthermia.

The other thing I thank we should mention in terms of emphasis is muscle rigidity, particularly with the use of succinylcholine. Then, of course, we get into the great controversy about masseter rigidity, which we have avoided nicely over the past couple of days, thank goodness, but I think that muscle rigidity becomes a very important sign in terms of increasing one's awareness to this problem.

We have heard that succinylcholine is a very commonly used drug throughout the world, still is, and will undoubtedly continue to be.

In my opinion, if you are going to alert people to the signs of malignant hyperthermia, particularly in those situations where dantrolene is not available, I think that those people should be told that if you have muscle rigidity with succinylcholine, it's your obligation to diagnose malignant hyperthermia at that point, and either continue to observe the patient by blood gas analysis and other means to make the diagnosis, or seriously think about stopping the anesthetic.

In other words, rigidity is a very specific sign for malignant hyperthermia, particularly after succinylcholine, and when that is observed, it's the obligation of the anesthesiologist to look for the other signs of malignant hyperthermia and/or stop the anesthetic.

Respiratory Acidosis and Tachycardia which Comes First

B: For those countries in which anesthetics are given without availability of capnometry monitoring, there are still some other ways of looking at whether or not the patient has a respiratory acidosis. Those anesthesiologists need to be taught to carefully observe respiratory rate, respiratory depth, and the color of the absorber in terms of turning colors with the CO_2 absorbent. It's not as good, but they need to be taught the major stave here.

As Dr. C said, if you do not have dantrolene, you have to convince people that they need to abandon anesthetics early basically, without any question. Tachycardia is certainly something that should raise your suspicion. But it's not really specific.

I: I myself am not an anesthesiologist. I would like to ask the anesthesiologist. With regard to this tachycardia, of course, it is not a very specific sign. If there is tachycardia at a time that CO_2 monitoring is not available, I think even if it is nonspecific we have to place importance on the signs of tachycardia.

So, whether it is a hypermetabolic state, and whether tachycardia occurs or not, we have to take all this into consideration.

CHAIRPERSON (MORIO): Yes, that is an important point; CO_2 increase and tachycardia, which comes first.

B: Tachycardia is not specific; it is a very sensitive sign of malignant hyperthermia. There is a difference in terms of the differential diagnosis of tachycardia. In addition, it includes hypoxia, it includes hypercarbia for any reason, it includes hypovolemia, it includes pain, and it includes anesthetic drugs, and it includes hyperthyroidism, pheochromocytoma, osteogenesis imperfecta, and arthrogryposis.

I mean, the list goes on and on, and MH is at the bottom of that list in terms of the likelihood of it occurring. But it is a sensitive sign.

CHAIRPERSON (MORIO): It is nonspecific, yes, but for the MH patient, does tachycardia appear at the same time as the CO_2 increase?

As far as I am concerned, we have not examined this so closely. Together with tachycardia, sweating comes out very early. As far as Pa_{CO_2} is observed, it does not go up. Sweating and severe tachycardia were experienced by myself at first.

J: I can give experiences in my country and some problems that have come to my laboratory. For most of them, the monitoring is usually ECG; temperature measurement is not usual, only in small children.

Then, the first sign discovered in the patients with MH was tachycardia and rigidity. Why? Because, usually the anesthesiologists know the patients very well, and knows if the patients have any problems and which kind of drug they received, for example just the atropine before to start the operation.

Then, when tachycardia started, he asked himself what has happened. And sometimes, the surgeon discovers that something is wrong maybe in the blood, or maybe because the patient starts to become rigid. I remember five patients whom at this time, the first sign appeared in the ECG monitoring, then, rigidity and temperature.

CHAIRPERSON (MORIO): Thank you. We will be discussing pathogenesis or etiology, so I am sure we will be coming back to this point again later.

Does the pathogenesis consist in the muscle alone, or could it be general membrane disease, and could there be sympathetic excitation in addition? In that case, we could expect tachycardia at a fairly early stage. I think we still have a lot to study in this area.

The next issue has to do with the CO_2 output monitor, when this is not available how to manage MH? And as Dr. C has pointed out, rigidity is very important. Because if there is no rigidity, we do not suspect MH, but if it appears, we should for sure suspect MH.

Temperature Monitoring

As for temperature monitoring, I would like to ask you for your views. In Japan we are recommending temperature monitoring. For more than 10 years we have been trying hard to recommend this, but only about half of the anesthesiologists are doing this. There are cases of people being sued, and perhaps because we do not do temperature monitoring, we have higher rate of deaths due to MH.

My earnest question is: why doesn't everyone do temperature monitoring, because it's so simple, anyone can do it.

C: Well, I think you are asking the wrong group of people that question. I think we are all convinced about temperature monitoring and its importance in anesthesia, not only for hyperthermia but also for hypothermia.

It is a mystery to me why the American Society of Anesthesiologists does not endorse temperature monitoring as a routine vital sign in anesthesia. The devices are available in every operating room; they are inexpensive. I can't understand why people don't want to do this. They are sometimes afraid, I guess, that they will get wrong information from that monitor, and yet they are not afraid to get wrong information from every other monitor that we use. Certainly, the morbidity from temperature monitoring is practically nonexistent, and certainly, also, there are now several medico-legal cases that show that if a patient has a problem, and that's related to malignant hyperthermia or a change in body temperature, and temperature monitoring was not used, then the law courts have always ruled against the anesthesiologists.

I must say that I think we have to ask another group of anesthesiologists why temperature monitoring is not important. I think to all of us here, it is vital.

K: I am a private practitioner in San Francisco. I am not teaching; only private practice. I don't monitor temperature in every case. I will tell you what I am doing.

To me family history is very important, and also a competent anesthesiologist. You have to know what you are doing. So, I apply all the monitors, including EKG and blood pressures and a temperature monitor if the patient has a history of malignant hyperthermia. And also it depends on situation. I don't monitor every single case, because it is expensive. For D&C, I don't monitor temperature. And for a broken bone, I don't monitor temperature. But of course, I observe the patient but don't apply the temperature probe. For a competent anesthesiologist surveillance is very important.

L: I wonder if I could ask a question.

Is it still important to monitor temperature if you are monitoring end-tidal CO_2?

C: I think it's important to monitor temperature on every patient, because we are not only concerned with hyperthermia, but we are also concerned with hypothermia.

A few years ago, the American Society of Anesthesiologists and some state societies were asked by peer review organizations that pay bills to anesthesiologists their viewpoint on temperature monitoring, which basically says that if you think you should monitor temperature in a patient, then you should monitor it. And these insurance companies didn't understand that.

So, a panel was put together, including Dr. Gronert, Dr. Cesler, and some other people. Because of concerns about the ability to monitor temperature in patients undergoing mask anaesthesia or very brief anesthesia, I think we came out with a very reasonable approach, which said that the data from Dan Cesler's laboratory at the University of California, San Francisco, show that in the first 20 to 30 minutes of anesthesia, body temperature fluctuates. Therefore, temperature may not be the best monitor of what the physiologic events are during that period of time.

So, the statement came out that this panel recommended that temperature monitoring should be done on every general anesthetic lasting longer than 30 minutes.

There were some other statements in that committee report, which unfortunately raised some other issues. But basically, when that was published, there were a lot of negative comments from anesthesiologists. A second committee was put together, and came out with a very complicated story about recommendations, which was rejected by the American Society of Anesthesiologists; so no changes took place.

Temperature monitoring is not expensive; it is easily available; I think should be done whenever a general anesthetic is used, because of not only hyperthermia but also hypothermia. I can understand, if for a 10- or 15-minute procedure, the temperature isn't monitored. I think that's not unreasonable.

B: As far as monitoring those patients who have a family history of malignant hyperthermia, that's fine and great. Their problem is not going to be of much risk, because you are not going to give them triggering anesthetic agents, anyway. Most of our patients, when they first have an MH event, have no known family history of malignant hyperthermia. And especially in the United States, you probably can't get accurate medical history on 50% of your patients, because one side of the family isn't talking to the other and they hate each other.

The other thing you said: Well, I don't do it for some procedures, like fixing broken bones. Well, 25% of our almost certain or very likely MH events occurred during orthopedic surgery. There is no greater tragedy than losing a young healthy child, teenager, young adult, undergoing totally elective, low-risk surgery to malignant hyperthermia because you failed to put a temperature probe in.

Now different people have differently calibrated hands, but I think most anesthesiologists are not going to pick up how hot a patient is until they are really high, like 104°F(40°C), and by then, it's too late.

K: I would like to respond to that. Yes, I may not put in a probe, but I do monitor temperature. I am very sensitive; when 99°F(37.3°C) to 100°F(37.8°C), I think I can feel it.

E: I think if you are not monitoring CO_2, it's even more important to monitor temperature. Because if you see a temperature rise in the first half hour of an anesthetic, then it certainly limits your differential diagnosis a great deal, because in routine cases, you will see a core temperature drop during anesthesia induction in the first hour, and only then you start seeing a rise. But if you see a rise in the first hour, and you have other evidence pointing to MH, then you can certainly limit your differential diagnosis and make the diagnosis of MH.

C: In regard to temperature monitoring, first of all, it's very difficult to accurately estimate changes in temperature by feeling the skin. You can really be misled.

Second, the onset of MH is generally associated with cutaneous vasoconstriction, so that skin temperature may go down at a time when core or rectal temperature is going up. Your palpation of the skin may mislead you.

In regard to the A.S.A. (American Society of Anesthesiologists) mandating temperature measurements, one of the reasons they became very skittish about this is that when the first statement came out, the people who manufacture skin temperature devices and want to sell these all over the country threatened lawsuits to the American Society of Anesthesiologists, and the letters that were distributed us from these companies came from their lawyers, not from the companies themselves.

So, the A.S.A. was particularly worried that the statement they put out had to reflect the wishes of the membership, including people in practice in San Francisco and Oakland and all over the country, and at the same time, reflect a real concern about good care. So they made temperature measurement desirable, but they won't mandate it. That's where we sit right now.

CHAIRPERSON (MORIO): We really must move on. So, at this workshop we should say that unless there is a special reason, core temperature monitoring should be recommended to the anesthesiologist.

Another point: Tachycardia and increased CO_2 output, which comes first? As far as this is concerned, please monitor very carefully and come back to the next workshop to report your findings. If you have more information, I think this will be very helpful in considering pathogenesis.

Diagnosis of MH Susceptibility

Now let us move on to Fig. 4. Again, this question is not very well put. MH-susceptible; what is MH-susceptible? First, we need to define the terminology. There are such terminologies as MH syndrome, MH disease, MH-susceptible, MH-like syndrome, and then there is the rhabdomyolysis syndrome. I don't know if it's related or not, but anyway, it's a terminology issue. So, we asked a very simple question. People responded that clinical episodes, if reinforced by muscle biopsies, are even more important.

Do you have any further comment on this matter or particular question?

N: Again, this is another big, difficult issue. A proband is usually selected based on clinical episodes.

Another question is from amongst the family members, all those related. We cannot do the challenge test for all those suspected people. So, for now, biopsy is done, and based on that result, to suspect and measure;—I think that should be the general approach, commonsense approach.

CHAIRPERSON (MORIO): So, MH-susceptibility is based on clinical episodes and muscle biopsy. Although muscle biopsy is nonspecific, for now this should be used.

Do you all agree to this, for now?

Clinical Episodes

A: Clinical episodes; I think we should more clearly define what clinical episodes are.

CHAIRPERSON (MORIO): Is it a clinical grading scale, or what?

Question 4 ; How do you diagnose a patient as MH susceptible ?

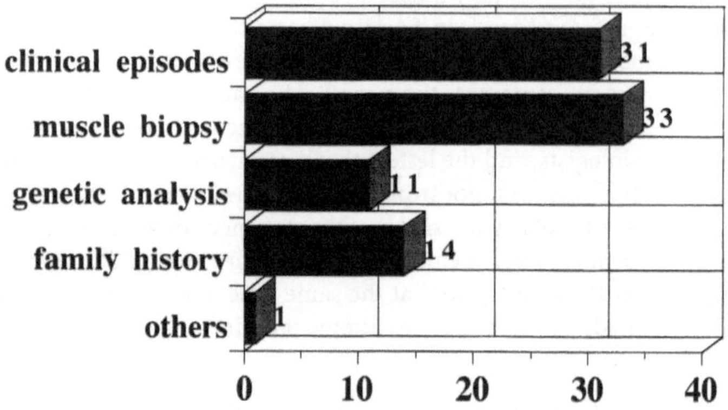

FIG. 4

A: I think we should have a simple (comprehensive, easy-to-understand) definition.

L: Sorry. My question wasn't related to the grading scale. I wondered if I can come back a bit later, if someone wants to talk about the grading scale now?

B: I agree that we should have a comprehensive, easy-to-understand clinical definition. When I came into this field, there was none that I could see. I think if people will try using the grading scale, they will find that it is really not difficult to use.

Contracture Test

L: My own view on the conversation on this topic so far is that the value of the contracture test is underestimated. I would say that it is really a specific diagnostic test. I think if it's carried out according to strict criteria, it is a pretty accurate test.

CHAIRPERSON (MORIO): Dr. L, which test do you think is most effective and helpful? There are a number of different tests; how do you evaluate each test?

L: Well, I have a bias in this respect, Mr. Chairman. But I tend toward the 3% halothane test and the graduated doses of caffeine, using the cutoff point of 2 m*M* of caffeine as diagnostic, and that has given us very reliable results since 1972.

C: The problem with making a simple, comprehensive statement about diagnosis of malignant hyperthermia, clinical malignant hyperthermia, is difficult, because I think it was Dr. Roses who referred to this disorder as "malignant hyperthermias." In fact, there are several different clinical presentations, some of which we feel confident to call malignant hyperthermia, some of which we have questions about as to their relationship to malignant hyperthermia.

How you make your definition will reflect your viewpoint. It really is difficult because in one situation, if you have muscle rigidity after succinylcholine, that's pretty specific. If you don't use succinylcholine, then you may or may not get muscle rigidity, depending on the length of exposure to the inhalation agent.

I think it's really difficult to make a comprehensive statement, and I agree with Dr. L that the muscle biopsy contracture test is as good as any clinical test that's used in the rest of the medical field.

Molecular Genetics

CHAIRPERSON (MICKELSON): I wonder if this is a chance to have another very exciting discussion on the relationship between molecular genetics and the clinicians.

It seems that both sides on this issue have very high expectations of the other side. I wonder if we would like to talk about that a little bit.

Q: I think that now that we seem to have mutations that may account for up to 20% of cases, that if these mutations are screened in the at risk people once they start appearing, and then when we try to correlate that, both the clinical and physiological criteria, that we may be able to come up with a clearer picture.

So, I advocate strong checking for these mutations in the different populations, especially for the ones that are present in more than one family. And I don't think that there's much use screening for the ones that are limited to one family yet.

R: Now, actually I might have to tone down some of the arguments which I think we have very productively had over the last few days.

I think, actually, what we all should bear in mind is that any diagnostic test has a dual mode of application, and this agrees also with what Dr. Rosenberg said yesterday. I think it's very important for the clinician or for the anesthesiologist to have a test with very high sensitivity that would actually give him absolute certainty that a patient is at risk, if he is at risk. I think this could be done by several different tests. It can be done according to a North American protocol; it can be done according to the European protocol, and it can be done applying mutation analysis, if you use the precautions Dr. Q uses. But that always means you've given an extra margin of safety, which is, you always produce false-positives.

Then, if you are interested in the pathobiology of MH, if you want to find out what actually happens if you are interested in MH as a model for disorders of excitation contracture coupling, I think then what we as molecular geneticists need is a very specific test. And then I think the differences might start.

In my hands, the European protocol has the advantage that it has this dual mode, that you can either get very high specificity by excluding the MHE from your diagnosis, or you can get a very high sensitivity by having this extra margin of safety and including the MHEs as MHSs.

These are the only certain terms of the clinical diagnosis. I think it's absolutely utilitarian to use either mode of this test. I think we should agree on that.

There is one last remark. I think we might actually be much better off if we started to think about anesthesia-related complications, rather than all those different types of MH. I think we should stick to the phenotype of MH, which among others I think should include hyperthermia, also triggering by halothane or the relational anesthetics. We then could talk about other diseases that also maybe have specific reactions that are related to anesthesia, like myotonias.

I am saying this because your reaction as an anesthesiologist to these events might be different from the classical MH. We would like to think about leaving off suxamethonium for the myotonics, but you can use halothane; for various and classical MH, you wouldn't. I think that's about it.

B: I agree with Dr. R, and I would point out that there are other cut points that can be used by molecular geneticists with the North American protocol that give a very specific, though less sensitive, test. By doing the studies that we have, we know very clearly what they are, and I have shared them with some of the geneticists.

CHAIRPERSON (MICKELSON): I think that's an important comment. I am more a protein chemist than anything else, I can see an incredible number of mutations in a particular protein; all lead into slightly different functional effects on that protein. And so, to have lots of different, although subtle, effects on that protein; to try to have them all grouped together into a single cutoff on a contracture test is asking a little bit too much.

C: Well, I am not sure whether it's asking too much of the contracture test or not, because we don't know what it's really measuring inside the functioning intact cell. That may or may not be asking too much.

I do believe, of course, that the molecular genetic analysis is very exciting, very important, and should obviously be continued in the research manner that we are doing it in terms of screening patients, screening the genome. I think it's absolutely critical that we continue to do that. I think it's still a research tool.

I think, however, there may be possibilities where there are families that are large enough—I am not a statistician, I am not sure what that large enough number is—where in that specific family where there has been a good correlation between the contracture test and the specific defect to generate a high enough LOD score to make some influence in that specific family for that specific defect based on the molecular genetics.

But we need to gather more information about it, which we are doing. At this point I regard it in probably 99% or more of cases as a research tool. Hopefully it will be extended to be available to other families once characterization is done.

CHAIRPERSON (MORIO): As to muscle biopsy we have the European method, North American method, and a Japanese method. So each method should be compared.

CHAIRPERSON (MICKELSON): Well, as I have shared with Dr. Endo, I think the Japanese method is the best test available to look for a specific defect that's in the ryanodine receptor gene. I mean, that's the most elegant and highest quality evidence when looking for a specific genetic defect.

But it's also possible that by using that test alone you are missing other possible genes that could be causing an MH symptom.

CHAIRMAN (MORIO): Yes, both high-specificity and high-sensitivity tests are needed; I think we need to continue to study in this direction.

Pathogenesis

I would like to have the next slide (Fig. 5). This is the question about pathogenesis. Forty-nine responded and 96% said that it is the SR (sarcoplasmic reticulum) membrane abnormalities. Two persons said that there is no need to check the SR but rather a cytoplasmic protein or contractile protein system may be affected. Multiple answers were permitted; therefore, at least 14 people, that is 28% of the responders, said that there may be other causes additional to SR membrane abnormalities. Fourteen said that the sarcolemma might also be affected. So, at least 14 people said that there are other reasons that the SR membrane abnormalities. Dr. Mickelson, do you have any comment on that?

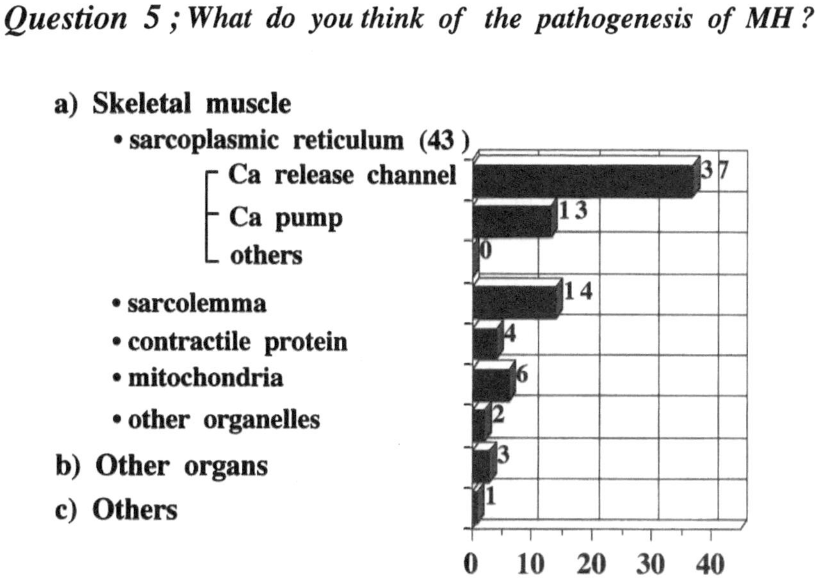

FIG. 5

CHAIRPERSON (MICKELSON): I think all of these possibilities have been shown by reasonable people over the years to have value. Certainly in one regard the sarcoplasmic reticulum has the weight of scientific evidence in support of it. And I think it's up to others who have other abnormalities in mind to support them at the same level of research and quality.

I mean, it's the SR evidence that has allowed the identification of a candidate gene that has been very useful for human MH research. I think we are still waiting for evidence from the other organelles possibly involved to provide a strong candidate gene. Dr. U, would you care to respond?

U: I am pretty much in agreement with what you have said. I think we have clearly demonstrated in our laboratory that with a normal excitation contraction system you can totally throw off the ability to regulate calcium by the addition of other modulators. I just present the example that we have worked with, and that's fatty acids.

In this case you get a total upset of calcium regulation. I know that people really don't understand what's been done in a lot of isolated tissue studies, because usually people look specifically at calcium release or specifically at calcium uptake. We have looked at what we feel is a more physiologically relevant system that incorporates both. We have shown that you can override the ability to regulate calcium in a normal SR system at physiological concentrations, or not physiological but anesthetic concentrations of halothane.

And so, again I agree with Dr. Mickelson. A lot of interesting information has come from pursuing the ryanodine receptor, but we can't rule out other systems, because there is no need to have to look at those proteins involved in excitation contraction coupling. I think the antioxidant system, the phospholipase C story, the free fatty acid story, these are all very viable hypotheses that include a huge number of proteins. Phospholipase C, there are at least nine isoforms that I am aware of. There is a lot yet to be done and there are a lot of potential answers.

Chairperson (Mickelson): Well, I think we need to be careful in evaluating all these studies; the extent to which in vitro work can be extrapolated to the in vivo situation is uncertain.

And related to that comment is what are the so-called primary alterations responsible for an MH episode, and what are secondary alterations that result from the primary? You know, we can see lots of abnormalities associated with in vitro preparations of muscle or organelles, but how do we know what is actually occurring in the intact muscle? That's going to be a very difficult question to get around.

B: I don't know what the right answer is, but whatever the answer is, it needs to go toward explaining why MH events occur more often in the young and in the male, and what's being affected where, and why that changes differently with time and with sex.

Q: I think if we look at the disorders where malignant hyperthermia or anesthetic-related events are a problem, if you look at it like a jigsaw and look at it from a clinical point of view, I think the idea of calcium dysregulation may fall into place as being the main event. Because if you will just list off the different diseases that fit, first of all, if we take the known defects in the ryanodine receptor, they do lead to altered calcium fluctuations. And those people, some of them do develop malignant hyperthermia.

The same type of mutations are present in central-core disease, although it seems that in there they lead to atrophy as well as the MH.

If you look at Brodie's disease, which is a pretty rare disorder where there is a defect in the calcium pump from fast-twitch muscle, those patients have what looks like a really strong contracture test, and as far as I can gather, the fast-twitch fibers are atrophied or somewhat degraded in those people, too.

Then, if you look at the familial cardiomyopathy, they have central cores in their skeletal muscle, which indicates the same type of mechanism whereby calcium leads to hypertrophy; if it goes an extra bit long, that leads to atrophy. And the same holds true, I think; for the pig, because there seems to be a link between the mutation in the pig and hypertrophy.

So, it would seem to me that it's possible that by exercising or by having abnormal calcium release channels, we can get hypertrophic situations. Then, if that goes a little bit past the threshold, that leads to atrophic situations. And I think that also could explain things like Duchenne muscular dystrophy, where one of the early signs as hypertrophied calf muscle, and then atrophy follows after the system breaks down.

And so, I think, looking at it from a clinical point of view, that just abnormal calcium with all the consequential side effects of that may explain a lot of the symptoms that we see.

C: Well, I think that almost everybody agrees that this is a problem of calcium regulation. I think that we also know that, like many other substances within the cell, there are elegant systems to regulate calcium, redundant systems, control systems.

I think that we should obviously still keep an open mind as to how those complicated interactive systems can lead to a syndrome, which in one case may be malignant hyperthermia, which in another case may be central-core disease and malignant hyperthermia, and another case central-core disease without malignant hyperthermia, Duchenne dystrophy. But it's a calcium dysregulation syndrome, and without being terribly specific all these cases relate to one defect.

R: My comment is actually that I think we shouldn't be too worried about one and the same mutation having such different phenotypes. Because I think that's the lesson that we are learning in all the inborne errors of metabolism, which seem very dull for molecular genetic in the beginning, because we know this was the enzyme defect, and that's why I should look at the permutation that's causing it.

But what you are learning there now is that you have absolutely no correlation between a given mutation and the phenotype of this patient. So we have, like in Gaucher's disease, or in other of these enzyme defects, there is one mutation that gives all the possible clinical phenotypes described.

So, I think we shouldn't be too worried about that, and of course, we should look also for other factors that influence the development of the phenotype in the patient.

CHAIRPERSON (MORIO): Any other question or comment?

Gene studies, gene research, and muscle tests and clinical episodes, and clinical signs and symptoms, the relationships between these things should be continuously studied.

Exercise-Induced Hypermetabolism, MH and Molecular Biology

This is whether the exercise-induced hypermetabolism has an effect or not. Most people responded that there was some relation. This question may not be very good, but in the porcine MH or PSS (porcine stress syndrome), this is true.

Why do we ask this question? Muscles or factors other than the skeletal muscles, is there any possible relation to other factors in the pathogenesis of an MH?

So, if instead of an MH, we say neuroleptic-induced malignant syndrome, we may have different answers. But exercise-induced hypermetabolism, most people say there is a correlation. Probably this is based on the result of the study on the porcine MH model. Dr. Mickelson, do you have any comment on this question?

CHAIRPERSON (MICKELSON): I wish I had an insightful comment, but we have always been puzzled why the syndrome in pigs can be called "porcine stress syndrome." Based just on an abnormality solely in skeletal muscle, we didn't know why that would

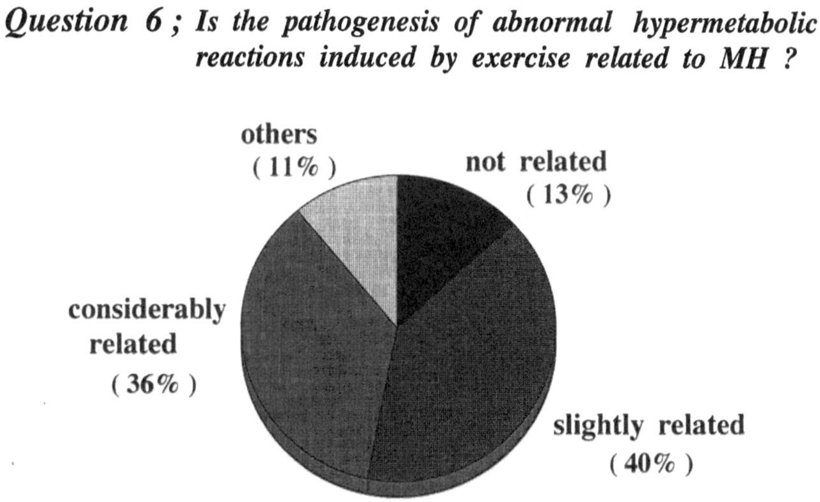

FIG. 6

cause a stress response. And so, that's what initiated our studies looking for abnormalities in other tissues, that could be also expressing the RYR1.

So, I am not enough of a neurobiologist yet to try to make a hypothesis as to how expression of the abnormal ryanodine receptor in tissue such as the brain could cause hypermetabolic reactions or stress-induced MH. But I think it's a possibility for further discussion or further experimentation. And as to how that could play a role in human MH, well, that requires a lot more investigation.

L: I think there is one very important practical reason why the association between MH and heat stroke should be recognized. And that is that dantrolene can be lifesaving in serious cases of heatstroke. So I think that is a very important reason for the association to be recognized.

Q: I have one question, actually. There are two known homozygote patients genetically with malignant hyperthermia or positive IVCT (in vitro contracture test). So, you might predict that they may have something related to the pig stress syndrome.

And maybe, you could comment. Dr. D, or I don't know if Dr. R. has any indication in his patients that fit that category. Because it will be nice to know what those patients look like.

D: We have studied one family of Italian origin, who seem to be the first to have both parents with the MH susceptibility proven by IVCT test, and they have five children. Two of them seem by the DNA analysis to be homozygous, and they seem clinically quite normal. They are now about 20 years old, and they have some sort of cores in their muscle, but clinically they are normal, and they have absolutely no problem with effort.

T: Please excuse me for my difficulties in speaking, but what I think is: A large gene could have many mutations. If I take the example of the R.E.T. gene, you can have, according to the mutation, four different clinical syndromes, heatstroke, renal, and so on, based on the mutation of one large gene.

I think that it is possible that mutation in the RYR1 gene could induce not only MH and CCD (central-core disease), but also according to the site of the mutation, heatstroke. That is my opinion.

In the opposite case, as in sarcomyelitus disease, many loci could give the same clinical picture that we have in the difficulty of linking genetic to clinical matters.

Chairperson (Mickeison): It seems to me that researchers who identify mutations in the human ryanodine receptor, and particularly the so-called porcine mutation, are asking too much of the humans that have this mutation that their clinical findings look like the pigs. There is a large number of other genes that are going to be influencing the expression of that RYR1 mutation.

One other minor point maybe we shouldn't be calling that the "pig mutation" in human medicine any more. For one, I wouldn't like to be a member of a family that is referred to as having a pig mutation.

C: I think this is a very important clinical question. I think it really calls for some very specific research to be done.

I don't know this aspect of the literature all that well, but I know that a number of years ago, both in Sweden and England, studies were done of MH-susceptible patients where there was sophisticated monitoring of oxygen consumption and lactate production during exercise. Those exercises were carried out in controlled environmental situations. They were, as far as I remember, carried out in heat stress-related environments.

I think whether we use the patients with the ryanodine receptor mutation to test this out, or just MH-susceptible patients in general, we need that kind of controlled clinical study. Maybe it has been done in Marseilles where I know there has been very strong and sophisticated research done into heat-related syndromes.

Until we do that kind of high-quality metabolic research with these patients, I think a lot of the answers we get will be colored by one's viewpoint of the clinical episode and the validity of the contracture test.

So, I think we really need more data here.

X: I don't think that will necessarily be valid, because it's like the anesthetic reaction; sometimes the patient has a reaction, and sometimes with the same triggering agents no reaction occurs.

I know from the exercise-induced reactions I have seen, the patient is the most surprised person, because he will say: "I have engaged in this exercise a thousand times before. Why did it happen this time?" Something that we don't understand has happened one time to trigger the reaction, and you can't call it up at will.

R: Actually I have one comment again back to the homozygote patients. We have seen one patient who is homozygote for this mutation, which should not be called the "pig mutation" any longer, but R6-14C. And in terms of IVCT clinical data she, was like her relatives studied, who only carried one copy of this mutation. So, in this patient we don't have any evidence that the homozygotes are different from the heterozygotes. This is with a defined mutation.

On the other hand, we had one patient, which ought to have been published, who actually does have a myopathy, and has both parents affected with MH. In this case we can't prove it homozygous because it's on two different genes. It's two different chromosomes. So, I think this is still open, somehow.

I think it actually may be too much to expect different mutations or to find different mutations to explain these different phenotypes. Because, again learning from the example of the metabolism in muscle, for example, you have one mutation, and it depends on the conditions under which your patients, at a given moment, is exercising whether or not he has a crisis.

So, I think that it is again the influencing factors we are to look at.

A: With regard to this issue, I would like to point out two things that are of clinical importance. The first is exertion-induced hyperthermia. Exertion induces an elevation of temperature to various degrees. This high temperature is followed by two outcomes. One is the common case where the patient falls down, collapses, and if he or she is brought to a cool place, the body temperature falls, and the patient recovers spontaneously. Because of perspiration, total circulating blood volume is decreased, cardiac output is decreased, and the cerebral circulation may be decreased. The other, rarer type is that the body temperature does not decrease even after the victim is taken to a cool place. The fever continues for several days, and sometimes a week. Consciousness is not recovered, and the patient dies. Rigidity of skeletal muscle to various degrees is concomitant. At Osaka University, a exertion-induced hyperthermia sufferer was administered dantrolene, and the fever disappeared instantaneously. The patient finally died several days later without recovering consciousness.

The second is anesthesia-induced and neuroleptic hyperthermia. The difference between these two entities is that in anesthesia-induced hyperthermia, the patient dies within several hours unless proper treatment is given, while in neuroleptic hyperthermia, the body temperature will stay around 40°–41°C for several days. We don't know whether anesthesia-induced hyperthermia occurs in patients who are susceptible to neuroleptic hyperthermia. However, the clinical features are very different from each other.

F: In regard to exercise-induced MH, there are some professional football players in the United States who are susceptible to malignant hyperthermia, and yet play the sport every week, every Sunday. It's a hot, sweaty sport, and they drive themselves hard. Perhaps the most famous case was MacLey Hill, who was a fullback for the Kansas City Chiefs. He died under anesthesia in the late 1960s, when he underwent knee surgery. He was about 5 feet 8 inches tall, and 200 pounds. I can't convert that quickly to centimeters and kilograms, but 90 kilograms or so, and probably 170 centimeters. A very muscular individual, who tolerated a very markedly stressful sport without ever having problems, and yet died under anesthesia with a classical case of MH at a time before it was easily diagnosed and certainly before dantrolene was available.

Furthermore, we had the uncle of a member of the Minnesota Vikings football team die with malignant hyperthermia at the Mayo Clinic during a surgical procedure. There is no doubt that his nephew, who played professional football, was also susceptible upon testing.

So that I think that you can have malignant hyperthermia susceptibility; you can have stressful exercise. It doesn't mean you will necessarily respond with an MH episode with exercise, but it is possible.

The only other point that I would make is that regardless of the cause of an uncontrollable fever, dantrolene will be effective in helping to control the body temperature if that's what you need to do. So, you don't need to worry about the MH. Simply, you can help to treat the fever and/or associated rigidity if you use dantrolene. It's particularly effective, as I said the other day, in someone who is either highly muscular or obese, because they are very difficult to cool by surface techniques, and we don't have many other easy ways to cool them in a clinical situation.

L: Since Dr. F has introduced football into the conversation, I would like to mention one other case.

A large American footballer developed heatstroke during a game of football. He recovered from this, and he was subsequently biopsied and shown to be susceptible to MH. I think this case was reported by Henry Rosenberg, and he may remember more details about it.

N: I think there are a number of such cases. In Japan we are looking at CICR (calcium-induced calcium release). The other day, Dr. Kawana reported on this as well—a CICR-enhanced group, I think there are two different groups.

Usually it's okay, but when it's high, perhaps it's risky. I say this because the calcium release channel becomes easier to open. A volatile anesthetic agent does this, so does calcium. With succinylcholine, depolarization takes place, and the calcium release channel is opened.

Another thing we should not forget is this: with the rise in temperature also, the channel is opened. So, when the speed or reaction of the calcium release channel is increased by exercise, symptoms can develop very similar to those of MH. I think that's not surprising to find.

R: I am relating to this last question. I must apologize because I forgot the name of the presenter, but he presented the CICR results and the grouping. There were two very red bars to the right of the graph. I was wondering if these were patients with a very high response. I was wondering: Has anybody done mutation research into these few patients, because they might, of course, be the homozygotes?

C: Well Dr. L. reads the out-of-the-way literature, obviously, on malignant hyperthermia, and we did report that in some sports journal.

I think that if you look at the literature about this issue of heatstroke or heat-related events in MH-susceptible patients, you will find some cases. I think in Dr. Muldoon's report, where she was looking at calcium ATPases, she had two patients who had heat-related syndromes who turned out to be positive. Dr. Kozak-Ribbens has reported patients with heat-related syndromes who turned out to be MH susceptible in biopsy testing. There were some very suggestive associations there.

I don't think we have the whole story about this by any means, because high body temperatures do funny things to calcium channels and other enzymes. I think the one thing we can state with certainty is that the heat-related syndromes in humans certainly do not occur with the frequency that they do in the pig.

Q: I have one comment, or one question, really. I guess if some of these patients who have familial cardiomyopathy have central cores and some muscle weakness, and central-core patients are susceptible to malignant hyperthermia, then, do the heatstroke or exertional problems have any cardiac problem that could explain some of its aspects of it? Has anybody looked at those patients?

T: Generally, central-core disease people do not move easily. So, I don't see any patients with central core disease with exertional heatstroke.

But I have heard something about research on NMR spectroscopy, because I have investigated patients with heatstroke and MH syndrome. And both groups have defective exercise metabolism. The differences between the two groups is that for heatstroke patients it is necessary to produce a relative ischemia or decrease in oxygen on blood output in the muscle, to see these defects. For the MH patients it is not necessary to see the same defect.

So, it is only a metabolic approach. But they also have hypermetabolic state in certain conditions.

Q: Dr. T, have you ever investigated any of your heat stroke patients for hypertrophic cardiomyopathy?

E: Talking about high temperature and triggering of MH, we are studying pigs that are MH susceptible and making them hyperthyroid. Hyperthyroidism alone has not triggered MH in these pigs, and then, under anesthesia with nontriggering agents, when we made them thyroid-toxic into the crisis stage where they have reached very high temperatures and extremely hyperdynamic, they still did not trigger, until they were challenged with halothane and succinylcholine. And then they triggered very quickly, much faster than without the hyperthyroid preparation.

We are still thinking about this. I don't know what we can say, but I thought I would share this initial preliminary information, since we are talking about hypermetabolism.

D: Does anybody know one or more families where the Mendelian transmission of this syndrome, abnormal hypermetabolic fraction induced by exercise, exists?

T: I haven't read about this in the literature and I don't know, because it is very difficult to investigate this type of family. For this family, it was a personal syndrome of the patient. They didn't agree with concept of a family-wide defect.

I followed up one family where one member suffered from exertional heatstroke, and one of his cousins had malignant hyperthermia. So, in this one family the two syndromes appeared. That I can say.

B: I guess one problem worth exploring further is that, as anesthesiologists we probably don't see most of the patients that come to the hospital for exertional heatstroke. At least, not in the United States. They remain in an emergency room; very rarely is an anesthesiologist called. So, we have no idea of the extent of the problem or a real opportunity to investigate it.

T: It is the same problem in France, because we receive many patients from the Army. But in civilian events such as the triathlon, marathon, and so on, incidents are not diagnosed as heatstroke. And nobody can say how many there are each year.

Dantrolene

CHAIRPERSON (MICKELSON): To my knowledge, dantrolene is specific to skeletal muscle in its relaxing effect. Is that true? Does it have any effects on cardiac muscle? Is that secondary to its effects on muscle, however?

B: It has definite effects on cardiac arrhythmia, and in fact, it was explored initially as an antiarrhythmic. Whether it has effects on contractility, I don't remember. The original reason why Norwich Eaton developed dantrolene was because they were looking for a new antiarrhythmic.

CHAIRPERSON (MICKELSON): I was going to run this hypothesis by the audience, that if dantrolene was specific for the RYR1 form of the ryanodine receptor and if the RYR1 is expressed in brain, maybe part of dantrolene's effects could be on brain metabolism.

E: I would just like to share some cases that I had. We had a patient with severe cardiomyopathy, who also had central-core disease, and came to us for cardiac transplantation because of severe cardiac failure. I monitored his hemodynamic function while I was giving him dantrolene. He did not show significant cardiac decompensation during dantrolene infusion.

So, I think if it has any effects at all, they were not manifested in this patient with severe cardiac failure needing cardiac transplantation.

L: I think Dr. B. has hit the nail on the head, that anesthesiologists are interested in patients with anesthetic problems.

But being a minority of one here, being a physician, the patients that are referred to me not only include anesthetic problems but include heat stroke patients, they include patients with rhabdomyolysis and renal failure, they include unusual myopathies, people with continuously elevated CKs (creatinc kinase). So I see a rather different pattern.

CHAIRPERSON (MORIO): Thank you very much, time is over. These issues should be further investigated and studied.

We would like to close this session. Thank you very much for this interesting discussion.

Closing Address and Summary

MORIO: First of all, I appreciate everybody's participation and cooperation in the symposium and workshop for these 5 days.

The highest environmental temperature, 38°C, reported since the establishment of the Hiroshima Meteorological Station has occurred during this conference. So I think everybody brought malignant hyperthermia from their country to Japan. Fortunately, this conference room itself was quite cool. However, this hot environment might affect the Japanese participants to some extent. It is quite clear that you are not susceptible to heatstroke, because you have not had heatstroke even in this hot situation.

We had a total of 204 participants in this symposium and workshop, and we had 60 visitors from overseas countries. Thank you very much for coming here all the way from your home countries.

In this workshop, in the first session, Session A, there were nine papers presented on MH-related syndrome. These MH-related syndromes should be differentiated from MH disease; however, these seem to be quite important syndromes for studying the pathogenesis of true MH disease. Many signs and syndromes were presented.

In Session B, five papers on masseter spasm were presented. In addition to the succinylcholine-induced masseter spasm, succinylcholine-induced rhabdomyolysis was reported. There were some comments that we need to warn against the use of succinylcholine, or if possible, the use should be banned, according to some participants. From this workshop we would like to hear support for such comments.

In Session C, on biochemical studies, there were ten papers. Dr. Nelson said that at low concentrations, dantrolene can accelerate RYR1, so we need to be careful in clinical practice. This was as very important presentation. We should all be careful when we use dantrolene in practice.

In addition to this comment, radical scavengers and other substances that may be effective in the treatment of MH, were presented.

In Session D, nine papers were presented on clinical classification and the incidence of MH. There were also presentations from Korea, China, and Kuwait, which were not represented in the previous workshops. Everybody said the incidence of MH is low in these three countries.

Dr. Larach and co-workers presented and proposed the clinical grading scale. This was compared with the Japanese criteria, and we found that the correlation between the two criteria is quite high. From France, another scale was proposed, and there were very interesting discussions. In the future, discussions on the scale or other criteria should be held.

For diagnosis of MH syndrome, in Japan we put emphasis on body temperature, but in the clinical grading scale from the United States, emphasis is not given to body temperature. If you would have one criterion for high body temperature, say, above 40°C, and if additional points are given for the high temperature, maybe your scale will become closer to our Japanese criteria. But anyway, the clinical grading scale is valuable as a scale, and it can be used to compare the criteria employed in various countries.

I think comparison of the different scales in necessary, and the clinical grading scale will give us the key for good comparison. However, before we clarify what MH disease is we need to modify some criteria.

We discussed why core temperature monitoring is not used during anesthesia. We need to propose some recommendations for the need of temperature monitoring during anesthesia.

The recent mortality rate for MH is 2% or 3% in Western countries, but in Japan, for these same 5 years, the mortality still exceeds 10%. This is a great difference. I am rather embarrassed by such figures. One of the reasons for this is related to the Japanese situation in the hospitals. More than 50% of mortality cases given general anesthesia by doctors other than anesthesiologists. Also, temperature monitoring is done only in about 50% of anesthesiology. However, the most important point is the difference in MH criteria. Our mortality rates are calculated based on cases of fulminant MH only.

If the core temperature were monitored, then there would be few deaths. In Hiroshima University, as I reported at an early stage, there were three deaths, but after the introduction of core temperature monitoring, there have been no deaths due to MH in our hospital.

In addition to the restriction on the use of succinylcholine, and also for promotion of the use of dantrolene, we need to review the regulations on anesthetics.

There are some countries where dantrolene is not yet available, and for such countries, we need to disseminate the results of the discussions in this workshop, and I hope you would be cooperative in such efforts.

In Session E on muscular testing, twelve papers were presented: there is the North American method, the European method, and also in Japan we do CICR measurement. We need to study the possible correlation between these three methods. A meeting was proposed to discuss this matter during the workshop.

In Session F, genetic studies, eleven papers were presented. New MHS genes were reported. We need to reinforce the cooperation with clinicians and to have further studies on genetics. This was confirmed during the session.

We have had some very active discussions. Thank you very much for your cooperation.

Next Workshop

M.C.: Thank you very much, Dr. Morio.

We would now like to announce the next workshop.

Morio: As you may know, last night, the participants of the farewell party proposed that the next venue should be Minnesota, in 1996 and Dr. Mickelson from Minnesota University accepted it. Does anyone object, or second?

I now propose that the next workshop will be held at Minnesota in 1996. Do you support this motion?

—(APPLAUSE)—

Mickelson: Thank you very much. I hope we can live up to the standards that you have set for us. I think in our discussions over the last several days, we have made it clear that what you want from the next workshop is a high-quality scientific meeting. We will do our best to provide that atmosphere, and stimulate the discussions.

M.C.: Thank you very much. This concludes the Seventh International Workshop on Malignant Hyperthermia. Thank you for coming to Hiroshima. We hope to see you in 1996 in Minnesota. Thank you.

Appendix

Program of the VIIth International Workshop on Malignant Hyperthermia

HIROSHIMA, JAPAN
July 17–20, 1994

Session A: MH Related Syndrome

(MODERATORS: F. LEHMANN-HORN and S. YAMAWAKI)

Oral Presentations

Association Between Malignant Hyperthermia and Severe, Chronic Muscle Pain
B.A. BRITT, C.A. MILDON, W. FRODIS, and E. SCOTT
(MH Investigation Unit, University of Toronto, Toronto, Canada)

Development of an Animal Model for Neuroleptic Malignant Syndrome
H. TANII, N. TANIGUCHI, M. TAKEDA, and T. NISHIMURA
(Department of Neuropsychiatry, Osaka University Medical School, Osaka, Japan)

Pediatric Perianesthetic Cardiopulmonary Resuscitation: Epidemiology and Prediction of Survival
M.G. LARACH[1], H. ROSENBERG[2], G.A. GRONERT[3], and G.C. ALLEN[1]
(Department of Anesthesiology, [1]Pennsylvania State University College of Medicine, Hershey, PA, USA, [2]Hahnemann University, Philadelphia, PA, USA, [3]University of California, Davis, CA, USA)

Hyperthyroid Hyperthermia—Metabolic Manifestations USA, in a Pig Model
K.G. BELANI, R.J. CARR, R. REARDON, D.S. BEEBE, V. KOMANDURI, M.F. SWEENEY, and P.A. IAIZZO
(University of Minnesota Medical School, Minneapolis, MN, USA)

No Association Between Neuroleptic Malignant Syndrome and Malignant Hyperthermia
P.J. ADNET, H.G. REYFORD, T. ETCHRIVI, G. HAUDECOEUR, F. SAULNIER, and R.M. KRIVOSIC-HORBER
(MH Investigation Unit, D.A.R.C. 1-Hôpital B, CHU Lille, France)

Poster Presentations

Exertional Heat Stroke: Clinical and Laboratory Findings Concerning 79 Patients. Relationship with MH

G. Kozak-Ribbens[1], L. Rodet[1], O. Deslangles[2], D. Figrella Branger[3], R. Petrognani[2], D. Bendahan[1], S.C. Gouny[1], C. Desnuelle[4], J.F. Pellissier[3], M. Aubert[2], and P.J. Cozzone[1]
([1] CRMBM-CNRS URA 1186, Faculté de Médecine, Marseille, France, [2] DAR, HIA A LAVERAN, BP 50, Marseille Armées, France, [3] Laboratory of Anatomy and Neuropathology, [4] Laboratory of Muscle Tissue Investigation, LA TIMONE University Hospital, Marseille, France)

Muscle Pathology in a Case of Fatal Neuroleptic Malignant Syndrome

S. Katsuragi, K. Teraoka, K. Ikegami, K. Ishizuka, H. Nakajima, S. Yasugawa, K. Yamashita, and T. Miyakawa
(Department of Neuropsychiatry, Kumamoto University Medical School, Kumamoto, Japan)

A Case Report of General Anesthesia for a Patient with "Minimal Change Myophaty"

Y. Takano, Y. Kuno, Y. Utsumi, K. Murata, and I. Sato
(Department of Anesthesiology, Koshigaya Hospital, Dokkyo University School of Medicine, Koshigaya, Japan)

Intraoperative Hyperthermia in a Patient with Oculo-Dento-Digital Syndrome

M. Tanaka, T. Sano, T. Yamamoto, S. Oshita, T. Sakabe, Y. Soejima, A. Tateishi, and T. Maekawa
(Departments of Anesthesiology-Resuscitology and Critical Care Medicine, Yamaguchi University Hospital, Japan)

Session B: Masseter Spasm

(Moderators: H. Rosenberg and K. Hanaoka)

Oral Presentations

The Effect of Dantrolene on Twitch Tension and Succinylcholine-Induced Jaw Muscle Contracture in the Rat

M.M. Keykhah, R.J. Storella, Y. Shi, and H. Rosenberg
(Department of Anesthesiology, Hahnemann University, Philadelphia, PA, USA)

The Effect of Very Low Dose Vecuronium on Succinylcholine-Induced Jaw Muscle Contracture in Rats

Y. Shi, M.M. Keykhah, R.J. Storella, and H. Rosenberg
(Department of Anesthesiology, Hahnemann University, Philadelphia, PA, USA)

In Vitro Human Masseter Muscle Hypersensitivity: A Possible Explanation for Increase in Masseter Tone

H. Reyford, P.J. Adnet, B. Tavernier, T. Etchrivi, G. Haudecoeur, and R. Krivosic-Horber
(Lille MH Unit, DAR I; Hopital B, CHR Lille, France)

Masseter Muscle Spasm as the Sole Presenting Sign
F.R. Ellis, P.J. Halsall, and P.M. Hopkins
(University of Leeds, Leeds, UK)

Deadly Rhabdomyolysis in Children with Occult Myopathy
U. Schulte-Sasse
(German Hotline for Malignant Hyperthermia Emergencies, Department of Anaesthesia and Critical Care Medicine, Heilbronn Community Hospital of the University of Heidelberg, Heilbronn, Germany)

Session C: Biochemical Study of MH

(Moderators: T. Nelson and K. Mori)

Oral Presentations

Free Radical Scavenging Enzyme Activity in Patients Undergoing Diagnostic Muscle Biopsy for Malignant Hyperthermia
G.E. DeBoer, C.E. Pippenger, H. Mitsumoto, and R. Solano
(Departments of Anesthesiology and Neurology, The Cleveland Clinic Foundation, Cleveland, OH, USA and FRESA Biomedical Laboratories, Inc., Redmond, WA, USA)

Treatment of Porcine Stress Syndrome Using N-Acetylcysteine
J. Peacock, S.J. Valentine, K.A. Williams, C.P. McPhee, and E. Papadopulos-Eleopulos
(Department of Anaesthesia, Royal Perth Hospital, Australia)

Can Dantrolene Sodium Cause Recrudescence of Malignant Hyperthermia?
T.E. Nelson and M. Lin
(Department of Anesthesia, The Bowman Gray School of Medicine of Wake Forest University, NC, USA)

Effect of Dantrolene and Hsp-29 on the Experimental Model of Malignant Hyperthermia
S. Cozzolino[1], A. Mancim[1], G. Bellezza[2], D. Scala[1], A. Venturelli[1], A. Borrelli[1], A. Ambrosio[2], A. Mancini[1], and A. Ruggiero[1]
([1] MH Investigation Unit, P.M.P. U.S.L., Naples, Italy, [2] N.C.I. Naples, Italy)

Expression of the Mutant Skeletal Muscle Ryanodine Receptor Allele in the Brain of Malignant Hyperthermia Susceptible Pigs
J.R. Mickelson, M.W. Ledbetter, and C.F. Louis
(Department of Veterinary PathoBiology, University of Minnesota, St. Paul, MN, USA)

Functional Effects of the Ryanodine Receptor Mutation on Single Channel Activity
N.H. Shomer, J.R. Mickelson, and C.F. Louis
(Department of Veterinary PathoBiology, University of Minnesota, St. Paul, MN, USA)

Effects of Induced Hyperthermia on Hepatic Circulation and Metabolism
H. Iwasaka, T. Kitano, H. Miyakawa, T. Noguchi, S. Oda, K. Taniguchi, and N. Honda
(Department of Anesthesiology, Oita Medical University, Oita, Japan)

Poster Presentations

Dantrolene Stabilizes Neuroblastoma Plasma Membrane

T. Hayashi, A. Kagaya, Y. Okamoto, N. Motohashi, and S. Yamawaki
(Department of Psychiatry and Neurosciences, Hiroshima University School of Medicine, Hiroshima, Japan)

Purification and Reconstitution of the Human Skeletal Muscle Ryanodine Receptor

A. Herrmann-Frank, M. Richter, and F. Lehmann-Horn
(Department of Applied Physiology, University of Ulm, Ulm, Germany)

Changes in Myoglobin and CK During Halothane Anesthesia—Different Effects of Thiobarbiturates on Age

I. Noguchi, G. Suzuki, and Y. Amemiya
(Department of Dental Anesthesiology, Tsurumi University, School of Dentistry, Japan, Department of Anesthesiology, Hyogo Children's Hospital, Hyogo, Japan)

Session D: Clinical Classification and Incidence of MH

(Moderators: G.A. Gronert and H. Kikuchi)

Oral Presentations

A Review 10 Cases of Malignant Hyperthermia Reported in Korea

H.J. Kim and S.J. Choi
(Department of Anesthesiology, Chungnam National University Hospital, Taejon, Korea)

Malignant Hyperthermia in Kuwait

A.R.M. AL-Qattan, Z.A. Omarah, and J.G. Jordanov
(Department of Anaesthesia and Intensive Care, Sabah Hospital, Kuwait)

Malignant Hyperthermia in China

J.X. Ni
(Department of Anesthesiology, Fourth Affiliated Hospital, Hebei Medical College, P.R. China)

Epidemiology of Malignant Hyperthermia Events in North America

M.G. Larach, L.J. Fuhrmann, and G.C. Allen
(North American MH Registry, Pennsylvania State University College of Medicine, Hershey, PA, USA)

Assessment of Clinical Grading Scale System for Malignant Hyperthermia in Japan

Y. Maehara[1], K. Mukaida[1], M. Kubota[1], M. Nakao[1], M. Kawamoto[1], O. Yuge[1], and M. Morio[2]
([1]Department of Anesthesiology and Critical Care Medicine, Hiroshima University School of Medicine, Hiroshima, Japan, [2]Chugoku Rousai General Hospital, Hiroshima, Japan)

Correlation Between Clinical Phenotype and Genotype in Malignant Hyperthermia (MH)

R. Krivosic-Horber, H. Reyford, A. Gerard, P. Adnet, and Y. Nivoche
(Lille MH Unit, DAR I, CHR Lille, Hôpital B, Lille, France)

Poster Presentations

Successful Management of Labor in a Patient with a History of Malignant Hyperthermia with Epidural Anesthesia

J. SAITOH, S. AKAZAWA, Y. NAKAIGAWA, and R. SHIMIZU
(Department of Anesthesiology, Jichi Medical School, Tochigi, Japan)

Malignant Hyperthermia in Second Cousins

G.E. DEBOER, M.M. GRIFFITHS, and H. MITSUMOTO
(Divisions of Anesthesiology and Neurology, The Cleveland Clinic Foundation, Cleveland, OH, USA)

Malignant Hyperthermia: A Diagnostic Approach

A. SCOGNAMIGLIO, A. BRACCO, D. MONTANARO, C. PLACIDA, and A. AMBROSIO
(MH Investigation Unit, P.M.P. U.S.L., Naples, Italy)

Application of a Clinical Grading Scale for Prediction of MH Susceptibility in Persons Suspected to Have Had an MH Episode

D. BENDIXEN, T.D. POULSEN, and H. ØRDING
(Danish MH Register, Herlev University Hospital, Herlev, Denmark)

Session E: Muscular Testing of MH

(MODERATORS: H. ØRDING and M. ENDO)

Poster Presentations

Malignant Hyperthermia Testing in Western Australia by the European Protocol

R.D. JOHNSEN[1], M.R. DAVIS[1], N.G. LAING[1], P.V. HURSE[2], D. PERLMAN[3], P.F. JACOBSEN[1], and B.A. KAKULAS[1]
(Departments of [1]Neuropathology, [2]Neurology, and [3]Anaesthetics, Royal Perth Hospital, Perth, Australia)

Studies of European and North American Malignant Hyperthermia Diagnostic Protocols and the Effects of Bay K 8644 on the Contracture Test

J.E. FLETCHER, T. DAWSO, and H. ROSENBERG
(Department of Anesthesiology, Hahnemann University, Philadelphia, PA, USA)

Analysis of Calcium Induced Calcium Release in Skinned Muscle Fibers from 164 Patients with Malignant Hyperthermia of Diseases Associated with Malignant Hyperthermia

K. MUKAIDA[1], Y. MAEHARA[1], M. KUBOTA[1], M. TANAKA[1], Y. FUJIOKA[1], M. NAKAO[1], O. YUGE[1], and M. MORIO[2]
([1]Department of Anesthesiology and Critical Care Medicine, Hiroshima University School of Medicine, Hiroshima, Japan, [2]Chugoku Rosai General Hospital, Hiroshima, Japan)

The Influence of Local Anesthetics on Ryanodine-Induced Contracture in Rat Skeletal Muscle

A. NISHIKAWA[1], S. OKU[2], and Y. AMAKATA[1]
([1]Department of Anesthesiology, Shiga University of Medical Science, Japan, [2]Surgical Center, Hospital of Shiga University of Medical Science, Shiga, Japan)

Screening for Dystrophin Defects in a Pediatric Population Referred for MH Testing

J.E. Fletcher and H. Rosenberg
(Department of Anesthesiology, Hahnemann University, Philadelphia, PA, USA)

Calcium-Induced Calcium Release May Be a More Sensitive Test for Malignant Hyperthermia Susceptibility than the Caffeine Contracture in the Skinned Muscle Fibers

S. Akazawa, J. Saitoh, Y. Nakaigawa, and R. Shimizu
(Department of Anesthesiology, Jichi Medical School, Tochigi, Japan)

Reproducibility of the Halothane and Caffeine Contracture Test for MH

J.E. Fletcher[1,2], L. Tripolitis[1], T. Dawso[1], H. Rosenberg[1], and J. Beech[3]
(Departments [1] Anesthesiology, [2] Biochemistry, Hahnemann University, Philadelphia, PA, USA, and [3] Department of Clinical Studies, University of Pennsylvania School of Veterinary Medicine, New Bolton, PA, USA)

Halothane-Induced Calcium Release from Sarco-Plasmic Reticulum in Malignant Hyperthermia Human Skinned Muscle Fibers

B. Tavernier, T. Etchrivi, P.J. Adnet, Y. Nivoche, and R. Krivosic-Horber
(MH Unit, DARC 1, CHRU Lille, Hôpital B, Lille, France)

Central Core Disease and Malignant Hyperthermia: A Clinical, In-Vitro Contracture Test and Histological Study of Six Familles

P.J. Halsall and F.R. Ellis
(Academic Unit of Anaesthesia, St. James's University Hospital, University of Leeds, Leeds, UK)

Oral Presentations

Comparative Results of Halothane/Caffeine Contracture Tests from Biceps and Vastus Muscle Biopsies for the Diagnosis of MH

G. Kozak-Ribbens[1], L. Rodet[1], A. Baeta[2], D.F. Branger[2], J. Figarella Pellissier[2], and P.J. Cozzone[1]
([1] CRMBM-CNRS URA 1186, Faculté de Médecine, Marseille, France, [2] Anatomy and Neuropathology Laboratory, University Hospital LA TIMONE, Marseille, France)

A Review of IVCT Results from Elderly Patients

F.R. Ellis, P.J. Halsall, and P.M. Hopkins
(University of Leeds, Leeds, UK)

Identification of Porcine Halothane Genotypes Using Biopsy Muscle Homogenates

K.S. Cheah[1], A.M. Cheah[1], and C.P. McPhee[2]
([1] Department of Agriculture, University of Melbourne, Victoria, Australia, [2] Queensland Department of Primary Industry, Animal Research Institute, Yeerongpilly, Australia)

Detecting Functional Alteration in Calcium-Induced Calcium Release Mechanism in Malignant Hyperthermia Syndrome
Y. KAWANA[1,2], M. IINO[2], H. OYAMADA[2], K. MATSUI[1], S. UCHIDA[1], K. SATA[1], H. KIKUCHI[1], and M. ENDO[2]
([1] The First Department of Anesthesiology, Toho University School of Medicine, Tokyo, Japan, [2] Department of Pharmacology, Faculty of Medicine, The University of Tokyo, Tokyo, Japan)

Session F: Genetic Study of MH

(MODERATORS: D.H. MACLENNAN and T.V. MCCARTHY)

Poster Presentations

Screening for RYR_1 Mutations in Malignant Hyperthermia Susceptible Individuals
A.M. ADEOKUN[1], P.J. HALSALL[2], P.M. HOPKINS[2], F.R. ELLIS[2], A.D. STEWART[2], J.L. HALL-CURRAN[2], and S.P. WEST[1]
([1] University of Newcastle upon Tyne, UK, [2] University of Leeds, UK)

Genetic Heterogeneity in Malignant Hyperthermia
J.L. CURRAN, P.J. HALSALL, P.M. HOPKINS, F.R. ELLIS, S.P. WEST, A.M. FOROUGHMAND, and A.D. STEWART
(University of Leeds and Northern Regional Genetic Services, Newcastle upon Tyne, UK)

RYR_1 Mutations in MHS Individuals from the United Kingdom
F.R. ELLIS, P.J. HALSALL, P.M. HOPKINS, A.D. STEWART, J.L. HALL-CURRAN, S.P. WEST, and A.M. ADEOKUN
(University of Leeds, Leeds, UK)

Contracture Phenotypes Versus Putative Mutant RYRI Subunit Distributions in MH Heterozygous Muscle Cells
T.E. NELSON
(Bowman Gray School of Medicine, Wake Forest University, Medical Center, Winston-Salem, NC, USA)

Exclusion of the Reported Arg^{614} and Gly^{341} Ryanodine Receptor Mutations as Being Responsible for MHS in Two Families That Map to Chromosome 19q13.1
A. OLCKERS[1,2], G.M. VITA[1], H. ROSENBERG[3], J. FLETCHER[3], H. ISAACS[4], and R.C. LEVITT[1]
([1] Johns Hopkins Medical Institutions, Baltimore, MD, USA, [2] University of Pretoria, South Africa, [3] Hahnemann University, Philadelphia, PA, USA, [4] University of the Witwatersrand, South Africa)

Oral Presentations

Evidence for a Genetic Cause of Porcine MH in Addition to the RYR_1 Mutation
IAN HUGHES
(Department of Farm Animal Medicine and Production, University of Queensland, Australia)

Modification by Breed and Sex of the Phenotype Attributable to the Causative Mutation for Porcine Malignant Hyperthermia
P.J. O'Brien[1], R.O. Ball[2], and D.H. MacLennan[3]
([1] Department of Pathology, [2] Department of Animal & Poultry Science, University of Guelph, Canada, [3] Banting & Best Department of Medical Research, University of Toronto, Canada)

The Ryanodine Receptor in Central Core Disease and Malignant Hyperthermia
K.A. Quane[1], J.M.S. Healy[1], K.E. Keating[1], B.M. Manning[1], F.J. Couch[1], L.M. Palmucci[2], C. Doriguzzi[2], T.H. Fagerlund[3], K. Berg[4], H. Ording[5], D. Bendixen[5], W. Mortier[6], U. Linz[7], C.R. Muller[7], and T.V. McCarthy[1]
([1] Department of Biochemistry, University College, Cork, Ireland, [2] Centro per le Malattie Neuromuscolari "Paolo Peirolo", Universita" Degli Studi di Torino, Italy, [3] Department of Anaesthesia and [4] Department of Medical Genetics, Ulleval Hospital, Oslo, Norway, [5] Herlev Hospital, University of Copenhagen, Denmark, [6] Kinderklinik, Kliniken der Stadt Wuppertal, Germany, [7] Department of Human Genetics, University of Wurzburg, Germany)

Screening for Proposed Mutations in the Ryanodine Receptor in a Large Population Referred for Testing by the North American MH Group Protocol
J.E. Fletcher, L. Tripolitis, and H. Rosenberg
(Department of Anesthesiology, Hahnemann University, Philadelphia, PA, USA)

Detection of a Novel Common Mutation in the Ryanodine Receptor Gene in Malignant Hyperthermia: Implications for Diagnosis and Heterogeneity Studies
K.A. Quane[1], K.E. Keating[1], B.M. Manning[1], J.M.S. Healy[1], K. Monsieurs[2], J.J.A. Heffron[1], M. Lehane[3], L. Heytens[2], R. Krivosic-Horber[4], P. Adnet[4], F.R. Ellis[5], N. Monnier[6], J. Lunardi[1], and T.V. McCarthy[1]
(Departments of [1] Biochemistry and [3] Anaesthetics, University College Cork, Ireland, [2] Department of Intensive Care, University Hospital of Antwerp, Belgium, [4] MH Unit, Hospital B, Lille, France, [5] MH Unit, Academic Unit of Anaesthesia, St. James's University Hospital, Leeds, UK, [6] Laboratory of Biochemistry, Grenoble Medical School, DBMS/Biochimie, Grenoble, France)

Sensitivity and Specificity of the Caffeine-Halothane Contracture Test: A New Approach
R. Krivosic-Horber[1], H. Reyford[1], P. Adnet[1], J. Lunardi[2], and T.V. McCarthy[3]
([1] Lille MH Unit, DAR I, CHR Lille, Hopital B, France, [2] Grenoble, France, [3] Cork, Ireland)

Genetic Analysis in Japanese Malignant Hyperthermia Susceptible (MHS) Families
Y. Maehara[1], K. Mukaida[1], E. Hiyama[2], O. Yuge[1], M. Kawamoto[1], M. Morio[3], D.H. MacLennan[4], R.G. Worton[5], and E.F. Gillard[5]
([1] Department of Anesthesiology and Critical Care Medicine and [2] Department of General Medicine, School of Medicine, Hiroshima University, Japan, [3] Chugoku Rousai General Hospital, Japan, [4] Banting and Best Department of Medical Research, University of Toronto, Charles H. Best Institute, Canada, [5] Department of Genetics, Hospital for Sick Children and Department of Molecular and Medical Genetics, Toronto, Canada)

Index